INDIRECT IMAGING

Measurement and processing for indirect imaging

Proceedings of an International Symposium
Sydney, Australia

30 August to 2 September 1983

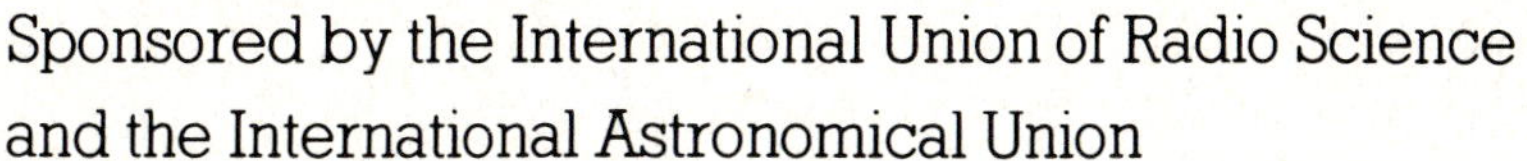

Sponsored by the International Union of Radio Science
and the International Astronomical Union

Edited by J.A. ROBERTS

CAMBRIDGE UNIVERSITY PRESS
Cambridge
London New York New Rochelle
Melbourne Sydney

Published by the Press Syndicate of the University of Cambridge
The Pitt Building, Trumpington Street, Cambridge CB2 1RP
32 East 57th Street, New York, NY 10022, USA
296 Beaconsfield Parade, Middle Park, Melbourne 3206, Australia

First published 1984

Printed in Great Britain at the University Press, Cambridge

Library of Congress catalogue card number: 83-26348

ISBN 0 521 26282 8

INDIRECT IMAGING

Contents

Keyword Index

Numbers refer to the section numbers in the Table of Contents.

Some of the delegates at the Indirect Imaging Symposium
Sydney, Australia, 30 August to 2 September, 1983.

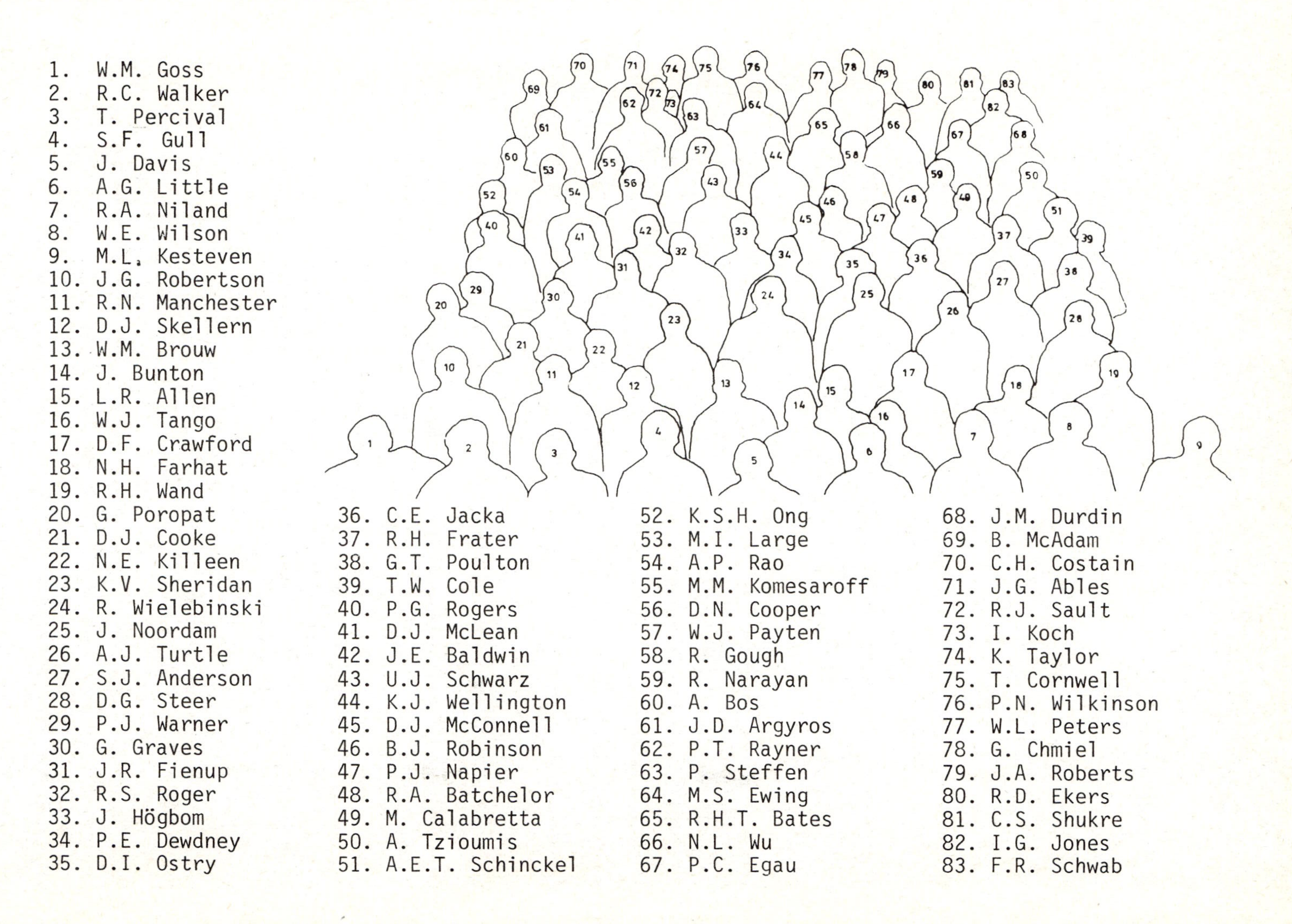

1. W.M. Goss
2. R.C. Walker
3. T. Percival
4. S.F. Gull
5. J. Davis
6. A.G. Little
7. R.A. Niland
8. W.E. Wilson
9. M.L. Kesteven
10. J.G. Robertson
11. R.N. Manchester
12. D.J. Skellern
13. W.M. Brouw
14. J. Bunton
15. L.R. Allen
16. W.J. Tango
17. D.F. Crawford
18. N.H. Farhat
19. R.H. Wand
20. G. Poropat
21. D.J. Cooke
22. N.E. Killeen
23. K.V. Sheridan
24. R. Wielebinski
25. J. Noordam
26. A.J. Turtle
27. S.J. Anderson
28. D.G. Steer
29. P.J. Warner
30. G. Graves
31. J.R. Fienup
32. R.S. Roger
33. J. Högbom
34. P.E. Dewdney
35. D.I. Ostry
36. C.E. Jacka
37. R.H. Frater
38. G.T. Poulton
39. T.W. Cole
40. P.G. Rogers
41. D.J. McLean
42. J.E. Baldwin
43. U.J. Schwarz
44. K.J. Wellington
45. D.J. McConnell
46. B.J. Robinson
47. P.J. Napier
48. R.A. Batchelor
49. M. Calabretta
50. A. Tzioumis
51. A.E.T. Schinckel
52. K.S.H. Ong
53. M.I. Large
54. A.P. Rao
55. M.M. Komesaroff
56. D.N. Cooper
57. W.J. Payten
58. R. Gough
59. R. Narayan
60. A. Bos
61. J.D. Argyros
62. P.T. Rayner
63. P. Steffen
64. M.S. Ewing
65. R.H.T. Bates
66. N.L. Wu
67. P.C. Egau
68. J.M. Durdin
69. B. McAdam
70. C.H. Costain
71. J.G. Ables
72. R.J. Sault
73. I. Koch
74. K. Taylor
75. T. Cornwell
76. P.N. Wilkinson
77. W.L. Peters
78. G. Chmiel
79. J.A. Roberts
80. R.D. Ekers
81. C.S. Shukre
82. I.G. Jones
83. F.R. Schwab

Contributors

ABLES, J.G.: Division of Radiophysics, CSIRO, P.O. Box 76, Epping, N.S.W. 2121, Australia

ATHERTON, P.D.: Kapteyn Laboratory, University of Groningen, Nettelbosje 2, Postbus 800, 9700 AV Groningen, Netherlands

ASUMA, K.: Department of Science, School of Education, Waseda University 1-6-1 Nishiwaseda, Shinjuku-ku, Tokyo 160, Japan

BALDWIN, J.E.: Mullard Radio Astronomy Observatory, Cavendish Laboratory, Madingley Road, Cambridge CB3 OHE, U.K.

BATES, R.H.T.: Electrical and Electronic Engineering Department, University of Canterbury, Christchurch 1, New Zealand

BOS, A.: Netherlands Foundation for Radio Astronomy, P.O. Box 2, 7990 AA Dwingeloo, Netherlands

BRACEWELL, R.N.: Electrical Engineering Department, Stanford University, Stanford CA 94305, U.S.A.

BROUW, W.N.: Netherlands Foundation for Radio Astronomy, P.O. Box 2, 7990 AA Dwingeloo, Netherlands

CARDEN, D.A.: Mount Stromlo and Siding Spring Observatories, Research School of Physical Sciences, Australian National University, Private Bag, Woden P.O., A.C.T. 2606, Australia

CHEN, C.F.: Ultrasonics Institute, 5 Hickson Road, Millers Point, N.S.W. 2000, Australia

CHIKADA, Y.: Nobeyama Radio Observatory, Tokyo Astronomical Observatory, University of Tokyo, Nobeyama, Minamimake, Minamisaku, Nagano-ken, 384-13, Japan

CHU, T.H.: University of Pennsylvania, The Moore School of Electrical Engineering, 200 S. 33rd St., Philadelphia, PA 19104, U.S.A.

COLE, T.W.: School of Electrical Engineering, University of Sydney, Sydney, N.S.W. 2006, Australia

COOMBS, J.A.G.: Royal Signals and Radar Establishment, Malvern, WR14 3PS, U.K.

CORNWELL, T.J.: National Radio Astronomy Observatory, Very Large Array, P.O. Box 0, Socorro, NM 87801, U.S.A.

CRAWFORD, D.F.: Astrophysics Department, University of Sydney, Sydney, N.S.W. 2006, Australia

DAISHIDO, T.: Department of Science, School of Education, Waseda University, 1-6-1 Nishiwaseda, Shinjuku-ku, Tokyo 160, Japan

DAVIS, J.: Chatterton Astronomy Department, University of Sydney, Sydney, N.S.W. 2006, Australia

DEWDNEY, P.E.: Dominion Radio Astrophysical Observatory, Herzberg Institute of Astrophysics, P.O. Box 248, Penticton BC V2A 6K3, Canada

DULK, G.A.: Astrogeophysics Department, University of Colorado, Boulder CO 80309, U.S.A.
DURDIN, J.M.: Astrophysics Department, University of Sydney, Sydney, N.S.W. 2006, Australia
EKERS, R.D.: National Radio Astronomy Observatory, Very Large Array, P.O. Box 0, Socorro, NM 87801, U.S.A.
EWING, M.S.: Owens Valley Radio Observatory, California Institute of Technology, Pasadena CA 91125, U.S.A.
FARHAT, N.H.: University of Pennsylvania, The Moore School of Electrical Engineering, 200 S. 33rd St., Philadelphia, PA 19104, U.S.A.
FIENUP, J.R.: Environmental Research Institute of Michigan, P.O. Box 8618, Ann Arbor MI 48107, U.S.A.
FRATER, R.H.: Division of Radiophysics, CSIRO, P.O. Box 76, Epping, N.S.W. 2121, Australia
FRIGHT, W.R.: Electrical and Electronic Engineering Department, University of Canterbury, Christchurch 1, New Zealand
GORHAM, R.A.: Mount Stromlo and Siding Spring Observatories, Research School of Physical Sciences, Australian National University, Private Bag, Woden P.O., A.C.T. 2606, Australia
GREENAWAY, A.H.: Royal Signals and Radar Establishment, Malvern, WR14 3PS, U.K.
GULL, S.F.: Mullard Radio Astronomy Observatory, Cavendish Laboratory, Madingley Road, Cambridge CB3 OHE, U.K.
HOBBS, T.I.: Mount Stromlo and Siding Spring Observatories, Research School of Physical Sciences, Australian National University, Private Bag, Woden, P.O., A.C.T. 2606, Australia
HÖGBOM, J.A.: Stockholm Observatory, S-13300 Saltsjöbaden, Sweden
HOOK, R.N.: Sussex University, Brighton, E. Sussex, BN1 9RF, U.K.
HIRABAYASHI, H.: Nobeyama Radio Observatory, Tokyo Astronomical Observatory, University of Tokyo, Nobeyama, Minamisaku, Nagano-ken, 384-13, Japan
INOUE, S.: Nobeyama Radio Observatory, Tokyo Astronomical Observatory, University of Tokyo, Nobeyama, Minamimake, Minamisaku, Nagano-ken, 384-13, Japan
ISHIGURO, M.: Nobeyama Radio Observatory, Tokyo Astronomical Observatory, University of Tokyo, Nobeyama, Minamimake, Minamisaku, Nagano-ken, 384-13, Japan
ITO, M.R.: Electrical Engineering Department, University of British Columbia, Vancouver, BC V6T 1W5, Canada
IWASHITA, H.: Nobeyama Radio Observatory, Tokyo Astronomical Observatory, University of Tokyo, Nobeyama, Minamimake, Minamisaku, Nagano-ken, 384-13, Japan
KANZAWA, T.: Nobeyama Radio Observatory, Tokyo Astronomical Observatory, University of Tokyo, Nobeyama, Minamimake, Minamisaku, Nagano-ken, 384-13, Japan
KESTEVEN, M.: Division of Radiophysics, CSIRO, P.O. Box 76, Epping, N.S.W. 2121, Australia
KIKUCHI, H.: Department of Science, School of Education, Waseda University, 1-6-1 Nishiwaseda, Shinjuku-ku, Tokyo 160, Japan
KOMATSU, S.: Department of Applied Physics, School of Science and Engineering, Waseda University, Nishiohkubo, Shinjuku-ku, Tokyo 160, Japan

KOMESAROFF, M.M.: Division of Radiophysics, CSIRO, P.O. Box 76, Epping, N.S.W. 2121, Australia
KOSSOFF, G.: Ultrasonics Institute, 5 Hickson Road, Millers Point, N.S.W. 2000, Australia
LABRUM, N.R.: Division of Radiophysics, CSIRO, P.O. Box 76, Epping, N.S.W. 2121, Australia
LARGE, M.I.: Astrophysics Department, University of Sydney, Sydney, N.S.W. 2006, Australia
LITTLE, A.G.: Astrophysics Department, University of Sydney, Sydney, N.S.W. 2006, Australia
McLEAN, D.J.: Division of Radiophysics, CSIRO, P.O. Box 76, Epping, N.S.W. 2121, Australia
MANCHESTER, R.N.: Division of Radiophysics, CSIRO, P.O. Box 76, Epping, N.S.W. 2121, Australia
MIYAZAWA, K.: Nobeyama Radio Observatory, Tokyo Astronomical Observatory, University of Tokyo, Nobeyama, Minamimake, Minamisaku, Nagano-ken, 384-13, Japan
MORIMOTO, M.: Nobeyama Radio Observatory, Tokyo Astronomical Observatory, University of Tokyo, Nobeyama, Minamimake, Minamisaku, Nagano-ken, 384-13, Japan
MORITA, K.-I.: Nobeyama Radio Observatory, Tokyo Astronomical Observatory, University of Tokyo, Nobeyama, Minamimake, Minamisaku, Nagano-ken, 384-13, Japan
MURATA, K.: Nobeyama Radio Observatory, Tokyo Astronomical Observatory, University of Tokyo, Nobeyama, Minamimake, Minamisaku, Nagano-ken, 384-13, Japan
NAGANE, K.: Nobeyama Radio Observatory, Tokyo Astronomical Observatory, University of Tokyo, Nobeyama, Minamimake, Minamisaku, Nagano-ken, 384-13, Japan
NARAYAN, R.: Theoretical Astrophysics Department, California Institute of Technology, Pasadena CA 91125, U.S.A.
NILAND, R.A.: Wills Plasma Physics Department, University of Sydney, Sydney, N.S.W. 2006, Australia
NITYANANDA, R.: Raman Research Institute, Bangalore 560 080, India
NOORDAM, J.E.: Royal Greenwich Observatory, Herstmonceux Castle, Hailsham, E. Sussex BN27 1RP, U.K.
NORTON, W.A.: Electrical and Electronic Engineering Department, University of Canterbury, Christchurch 1, New Zealand
OHKAWA, T.: Department of Science, School of Education, Waseda University, 1-6-1 Nishiwaseda, Shinjuku-ku, Tokyo 160, Japan
OSTRY, D.I.: Division of Radiophysics, CSIRO, P.O. Box 76, Epping, N.S.W. 2121, Australia
O'SULLIVAN, J.D.: Division of Radiophysics, CSIRO, P.O. Box 76, Epping, N.S.W. 2121, Australia
POULTON, G.T.: Division of Radiophysics, CSIRO, P.O. Box 76, Epping, N.S.W. 2121, Australia
RAO, A.P.: Radio Astronomy Centre, Tata Institute of Fundamental Research, P.O. Box 8, Ootacamund 643001 India
ROBERTS, J.A.: Division of Radiophysics, CSIRO, P.O. Box 76, Epping, N.S.W. 2121, Australia

ROGERS, A.W.: Mount Stromlo and Siding Spring Observatories, Research School of Physical Sciences, Australian National University, Private Bag, Woden P.O., A.C.T. 2606, Australia
ROGERS, P.G.: Division of Radiophysics, CSIRO, P.O. Box 76, Epping, N.S.W. 2121, Australia
ROSENFELD, D.: School of Electrical Engineering, University of Sydney, Sydney, N.S.W. 2006, Australia
SAULT, R.J.: School of Electrical Engineering, University of Sydney, Sydney, N.S.W. 2006, Australia
SCHWAB, F.R.: National Radio Astronomy Observatory, Edgemont Road, Charlottesville, VA 22901, U.S.A.
SCHWARZ, U.J.: Kapteyn Laboratory, University of Groningen, Nettelbosje 2, Postbus 800, 9700 AV Groningen, Netherlands
SKELLERN, D.J.: School of Electrical Engineering, University of Sydney, Sydney, N.S.W. 2006, Australia
SKILLING, J.: Department of Applied Mathematics and Theoretical Physics, Silver Street, Cambridge CB3 9EW, U.K.
STAPINSKI, T.E.: Mount Stromlo and Siding Spring Observatories, Research School of Physical Sciences, Australian National University, Private Bag, Woden P.O., A.C.T. 2606, Australia
STEENSTRUP, S.: Division of Chemical Physics, CSIRO, P.O. Box 160, Clayton, Vic. 3168, Australia
STEER, D.G.: Dominion Radio Astrophysical Observatory, Herzberg Institute of Astrophysics, P.O. Box 248, Penticton BC V2A 6K3, Canada
TAYLOR, K.: Anglo-Australian Observatory, P.O. Box 296, Epping, N.S.W. 2121, Australia
TOJO, A.: Nobeyama Radio Observatory, Tokyo Astronomical Observatory, University of Tokyo, Nobeyama, Minamimake, Minamisaku, Nagano-ken, 384-13, Japan
VAN GORKOM, J.H.: National Radio Astronomy Observatory, Very Large Array, P.O. Box 0, Socorro, NM 87801, U.S.A.
VARGHESE, J.N.: Division of Protein Chemistry, CSIRO, Royal Parade, Parkville, Vic. 3052, Australia
VELUSAMY, T.: Radio Astronomy Centre, Tata Institute of Fundamental Research, P.O. Box 8, Ootacamund 643001, India
WALKER, J.G.: Royal Signals and Radar Establishment, Malvern, WR14 3PS, U.K.
WALKER, R.C.: National Radio Astronomy Observatory, Edgemont Road, Charlottesville, VA 22901, U.S.A.
WIELEBINSKI, R.: Max-Planck-Institute für Radioastronomie, Auf dem Hügel 69, D-5300 Bonn 1, F.R.G.
WILD, J.P.: CSIRO, P.O. Box 225, Dickson, A.C.T. 2602, Australia
WILKINS, S.W.: Division of Chemical Physics, CSIRO, P.O. Box 160, Clayton, Vic. 3168, Australia
WILKINSON, P.N.: Nuffield Radio Astronomy Laboratories, Jodrell Bank, Macclesfield, Cheshire SK11 9DL, U.K.
WU, N.-L.: School of Electrical Engineering, University of Sydney, Sydney, N.S.W. 2006, Australia
YOKOYAMA, T.: Department of Science, School of Education, Waseda University, 1-6-1 Nishiwaseda, Shinjuku-ku, Tokyo 160, Japan

Foreword

It is now some two years since the URSI General Assembly at Washington in 1981 agreed to our request to convene a symposium on Measurement and Processing for Indirect Imaging in Australia in 1983. The symposium is now behind us and from all the comments received, it proved to be a most successful and rewarding gathering for our international band of indirect imagers.

We are grateful for the financial support and encouragement given to the symposium by URSI (the International Union of Radio Science) and IAU (the International Astronomical Union). Their joint sponsorship provided the impetus for attracting to Australia the substantial overseas contingent and to provide support for three young scientists from developing countries, Dr P.C. Egau from Kenya and Dr A.P. Rao and Dr R. Narayan from India. Thanks are due also to the CSIRO Division of Radiophysics and the University of Sydney, School of Electrical Engineering for hosting the technical sessions and workshops.

Symposiums are not complete without a published proceedings and we are most fortunate that Dr. Jim Roberts agreed to take on the role of editor. This volume bears testimony to the dedication, gentle coercion and thoroughness which Jim has exercised in his role.

On behalf of the Organizing Committee I take this opportunity to express my sincere thanks to Dr Andrew Pik who acted as symposium secretary, and to all those who contributed towards the success of the symposium - to the CSIRO and University staff for the smooth and efficient conduct of the meeting, to the speakers for their excellent presentations and to all the delegates for their spirited participation.

Bob Frater

Editor's Preface

An international symposium entitled Measurement and Processing for Indirect Imaging sponsored by the International Union of Radio Science (URSI) and the International Astronomical Union (IAU) was held in Sydney, Australia from 30 August to 2 September 1983. This volume contains the text of 45 of the papers presented, together with summaries of the remaining four papers. Reports of two of the four workshops which concluded the symposium are also included. Contributors to the discussion which followed the presentation of each paper were invited to record their remarks in writing; these records of discussion will be found immediately following the relevant paper. To assist readers in locating material the contents are grouped in sections; as a result the order of appearance of contributions in this volume does not follow precisely the order of presentation at the symposium. As a further assistance in locating material an index of 'keywords' is provided.

Indirect imaging has assumed importance in many fields, including geophysics, medicine, crystallography, acoustics and astronomy. Contributions from a number of these fields will be found in this volume, but papers related to aperture synthesis in radio astronomy dominate. Indeed it is clear that a number of the contributors regarded the symposium as a sequel to IAU Colloquium No. 49, on the formation of images from spatial coherence functions in astronomy, which was held at Groningen in 1978 (van Schooneveld 1979).

It is of interest to note some of the developments in the astronomical field since the Groningen conference. There have been considerable advances in the technique of constructing images from the amplitude of the Fourier transform without the phase (see papers 3.2, 3.3 and 3.4) and of improving measured phases and amplitudes using internal consistencies and simple assumptions about the character of the image (paper 6.1). Computer packages for constructing the 'maximum entropy' image consistent with a set of measurements have become more widely available, but it is clear from Section 8 that this subject is still full of contention. A version of the CLEAN algorithm which overcomes earlier difficulties with CLEANing extended images has also been developed (paper 7.2 and the discussion which follows). On the hardware side the completion of the Very Large Array and the developments in hardware Fast Fourier Transformers (paper 10.1) and digital correlators (papers 10.2, 10.3 and 10.6) should be mentioned.

This symposium demonstrated that indirect imaging is still developing rapidly in many fields. However, it also underlined the continuing poor communication between workers in different disciplines. Clearly we can look forward to further advances in indirect imaging, but, as John Baldwin reminded us in the concluding words of this symposium, the production of an image is but one step towards our real goal. Perhaps our aim should be to extract directly from our measurements information in that form which is most likely to advance our understanding of the particular physical process under study.

I wish to thank the referees who provided assessments of the papers, and to record my indebtedness to the staff of the Publications Section and General Office of the CSIRO Division of Radiophysics, without whose considerable editorial and typing assistance this volume would not have been published.

J.A. Roberts

October 1983

Reference

van Schooneveld, C. (1979). Image Formation from Coherence Functions in Astronomy. Dordrecht, Holland: Reidel.

1

Opening address

1. MEASUREMENT AND PROCESSING FOR INDIRECT IMAGING

J. P. Wild

Commonwealth Scientific and Industrial Research Organization
Limestone Avenue
Canberra ACT

It is to me a tremendous pleasure and honour to be asked to give the opening address at this symposium, jointly sponsored by the IAU and URSI, on Measurement and Processing for Indirect Imaging. The reason for this great pleasure has to do with nostalgia, because for one reason or another I have been out of the game for more than a decade, apart from fleeting glimpses of what was going on in the radio astronomy world during the IAU meetings in Sydney, Montreal and Patras. I also managed to visit the VLA and marvelled at its superb ingenuity, as I had earlier at Westerbork and before that at Cambridge.

I am not going to account for my doings over the last decade except to say that I spent the first half of it participating in an international quest to decide how best to scan a radio beam as part of an aid for landing aeroplanes. The decision reduced to the choice between two scanning systems one of which was the Fourier dual of the other. The two were therefore theoretically completely equivalent.

You might imagine that such a decision could be arrived at following a brief study and half an hour's discussion round the table. But things aren't like that in the aviation world. It took 5 years of intensive international activity - not to mention thousands of hours of flight-testing time - to make the decision. I personally attended meetings, each lasting between 2 and 3 weeks, at Melbourne, Washington, Paris, The Hague, London and two at Montreal before finally a United Nations vote decided upon which of the Fourier pair, frequency or time, should win the day. All manner of delegates, including chieftains from darkest Africa, were shipped in to participate in the vote - for each such vote counted the same as that of, say, the USA. The Chairman of the fateful meeting was from Iceland and I must say he did a superb job of chairmanship of a very difficult meeting. It is hard to convey to you the intensity of feelings, the national pride, the joy and the anguish that was engendered by this decision-making process. After announcing the final vote in favour of <u>time</u>, the President of the International Civil Aviation Organization said amid prolonged applause that this was the greatest moment in the history of the International Civil Aviation Organization.

The Fourier transform is indeed very central to the theme of this symposium. And I suppose that although the subject covers diverse applications the most pressing, sophisticated and historically significant one is aperture synthesis for radio astronomy.

Like one or two others in this room I lived through the whole drama of the emergence of this truly magical technique. I would like to recall the major steps - as I remember them. Those of you who attended a meeting on the history of this subject at Patras will recall that there were as many different versions as there were speakers. There were some priceless things said at that meeting. I shall never forget Hanbury Brown's description of how Cambridge used to respond to the latest Jodrell Bank discovery in the early 1950s : there were three standard answers to choose between:-

1. It's wrong.
2. We've already done it.
3. It's irrelevant.

As seen through my eyes the birth of aperture synthesis was a sentence in the pioneering 1947 paper of McCready, Pawsey and Payne-Scott from this laboratory. It was written underneath an equation. It read "... the second term is in the form of a fourier cosine transform. A little consideration indicates that this term varies sinusoidally with an amplitude given by a modulus of the component of the fourier transform ... It is possible in principle to determine the actual form of the distribution in a complex case by fourier synthesis using information derived from a large number of components."

Tracing the history from here on I shall treat aperture synthesis in the strict sense that it requires the summation of many observations to form the image - so I don't include things like the Mills Cross or the Culgoora radioheliograph. This will put me on the sidelines and help to keep me honest.

The first practical demonstration of one-dimensional aperture synthesis was done on the Sun by Stanier at Cambridge in 1950. The first two dimensional aperture was by O'Brien in 1953, also at Cambridge, and it is interesting that O'Brien gives credit to McCready, Pawsey and Payne-Scott for suggesting the method. O'Brien made partial use of the rotation of the earth in achieving his two dimensional synthesis of the solar image. A year or so later Christiansen and Warburton made the first sophisticated picture of the sun at 21cm wavelength using earth rotation synthesis. It also appears that at about this time Jan Högbom used earth rotation synthesis on radio sources as part of his PhD. thesis. The next step was made by Ryle and Neville in 1962 who performed a complete earth rotation synthesis on the north polar region of the celestial sphere. This spectacular map was the prototype of modern aperture synthesis mapping and it provided the justification for the Cambridge

1-mile telescope. This proved a great success and was followed by the Cambridge 5-kilometre telescope in August 1972. Most appropriately Sir Martin Ryle received the Nobel Prize for his outstanding contribution to aperture synthesis. I should remark that the total development of aperture synthesis took place hand in hand with the growing art of electronic computing and it was greatly to Ryle's credit that he recognized the potential of the computer in very early days.

Meanwhile the Westerbork synthesis radio telescope had been inaugurated in June 1970 and was a brilliant success with attention being given to system detail as had never happened before. Then in 1980 the VLA was inaugurated and this of course has become one of the wonders of the world. This makes it difficult for us in Australia to bring something out of the hat to celebrate Australia's 200th anniversary in 1988 but I am sure that Bob Frater and his colleagues will do their best to do so.

In giving this brief history of aperture synthesis mention should also be made of some more easily forgotten landmarks. In particular I should mention the theoretical paper by Bracewell and Roberts in 1954 who clarified to us all the classical fourier picture of indirect imagery and we should remember the paper of Roger Jennison in 1958 which founded the principle of closing the phase loop which was later to become a vital part of aperture synthesis especially to VLBI. One might mention also the important contribution of Jan Högbom in 1974 for inventing the image-improvement process known as "clean", and more recently we have the introduction of maximum entropy into radio astronomy of which I know my colleague Jon Ables is a constant evangelist. But I think I should leave the subject there because I am already out of my depth.

Whilst aperture synthesis or indirect imaging has reached a remarkable degree of sophistication in radio astronomy, it has many other applications and I want to touch on some of these. Historically the first application, preceding radio astronomy by several decades, was x-ray diffraction for crystallography - indeed by the time radio astronomy came along the Cavendish lab had been steeped in x-ray diffraction for a long time and that helped them to get off to a flying start in radio astronomy. My undergraduate professor was Sir Lawrence Bragg. I remember he had little time for fancy ways of doing a fourier transform. He would say "lock yourself up for two weeks with a slide-rule, and you'll save yourself a lot of time".

With the introduction of high-powered computers modern x-ray diffraction can do marvels. A recent example is the work done at the CSIRO Division of Protein Chemistry in Melbourne in collaboration with the John Curtin School of the ANU, Canberra. This work resulted in the determination of the three-dimensional structure of the influenza virus neuraminidase. It must be fairly hot stuff because Nature devoted the brightly coloured cover of a recent issue to illustrate the result.

The technique of x-ray diffraction extends also to neutron and electron diffraction - the data from each is of complementary value to the others.

I would now like to turn to the field of medical imaging. Here we are concerned with the reconstruction of, for example, sections of the patient from a series of measurements. Three ways of doing this are by x-rays, ultrasound and nuclear magnetic resonance. I believe Dr Kossoff will be speaking on imaging by ultrasound, and I will say nothing about it except to note its widespread use and its suitability for the examination of soft tissue, for example the foetus in a pregnant woman. I will mention, instead, the rather more novel NMR method. The key point here is that individual nuclei (mostly protons, but also ^{13}C or ^{31}P) precess about an applied magnetic field at a frequency determined by the particular nucleus and the magnitude of the field. Then, if the field is not uniform but has a spatial gradient, the frequency of precession depends on position and this makes tomographic mapping possible from the NMR frequency data. Using digital processing the necessary fourier transform can be performed and the image recovered. By this technique images of superb quality and resolution can be obtained of direct sagittal cross-sections of the brain. Note that here we have a technique that allows us to see through the skull and with minute detail.

Finally, I would like to mention some work that is closer to home in two senses: it is being done in this Division (in collaboration with Macquarie University's Geophysics Department) and it has strong similarities with the problems that confront radio astronomers. I refer to a type of seismic imaging, applied to determine the condition of a coal seam. By measuring the acoustic energy transmitted from source to receiver one can map the variations in propagation velocity or in acoustic attenuation in the coal seam. The problem is a tomographic one, to reconstruct an image from projections, but the data is very expensive to acquire and so, unlike the medical examples I have just mentioned, the data is sparse. Thus the imaging problem is comparable with VLBI mapping.

I believe the subject you are dealing with in this symposium is of incalculable importance to modern technology and analytical methods. It enables us to extend what we used to do only with light to the use of x-rays, electrons, nuclei, radio waves, acoustic waves, ultra sonic waves, and seismic waves. The possibilities and applications are limitless; there is a revolution in progress, and you are in the middle of it here and now.

As long as the Chief of the Division of Radiophysics does not over-interpret the words of the Chairman of CSIRO, may I say I believe, that this field of work, the world over, deserves strong support and a very high priority for the scarce resources which are nowadays available for scientific endeavour. For it is a master key that can unlock many doors into the future.

Mr Chairman thank you for inviting me to give the opening address at this symposium, and may I wish you all every success in your discussions inside and outside the meeting room.

2

Radio imaging systems and design

2.1 THE RADIO TELESCOPE POPULATION

J.E. Baldwin
Cavendish Laboratory, Cambridge, England.

INTRODUCTION

The central topic of this meeting is how to obtain the best images from data which emerges from telescopes and other similar imaging instruments. The quality of these images depends partly on the cleverness of the analysis but most importantly on the properties of the telescopes themselves, which are not the main subject of our discussions. The aim of this brief review of radio telescopes is therefore to provide a summary of their properties as a background to the discussions. It may then be clearer how to incorporate any new ideas into the designs of the next generation of telescopes. I shall not discuss the relative merits of individual telescopes but rather the properties of the whole population of telescopes - those that existed in the past, those in active operation now and those under construction or planned for some indefinite future. It is of particular interest to see if there are any theoretical limits to the performance of telescopes and whether we are yet close to those limits.

There are many quantifiable parameters of a telescope which are vital to its success, and even to its existence, such as its cost, the chief mechanical engineer, the time taken to build it, the designer of the electronics, the weather at the site, who wrote the software, the operational budget and the observers who use it. But the only characteristics which affect our discussions here directly are those defining the outputs of the telescope. They are few in number:

Frequency of operation
Angular resolution
Flux density of the faintest detectable source
Field of view
Dynamic range of the images
Observing time necessary to complete an image
Frequency resolution for radio spectroscopy
Polarisation capability

There is no satisfactory way of displaying the whole of this many-dimensional data and I will concentrate on just two or three projections of it.

OPERATING FREQUENCY AND ANGULAR RESOLUTION

The most dramatic developments in telescope design in recent years have been in improved angular resolution and so the relation between angular resolution and frequency of operation is the most familiar plot. Fig. 1 illustrates this for telescopes past, present and future. It does not pretend to include all telescopes, all frequencies on one telescope nor even every telescope which has done important work; many famous early telescopes are omitted. Similarly in later figures I do not expect to have attributed the current performance of telescopes precisely; they change faster than one can read about them. But I think that the overall pattern of telescope performance is not misrepresented.

Fig. 1 shows the very wide range of instruments existing or proposed and the natural limits to the extension of that range. For ground-based telescopes the lower limit for practical high-resolution work is 10 MHz, although of course work has been done below 10 MHz, notably in Tasmania. For space-based telescopes the ultimate lower limit will be set by free-free absorption in the interstellar medium, probably at frequencies somewhat below 1 MHz. Even if we lie in a favoured position in this respect, the scattering in the interplanetary and interstellar medium sets severe limits on the usable angular resolution.

Three factors set limits to the useful resolution achievable at higher frequencies:

(1) The diameter of the Earth (for ground-based interferometers)

(2) Interstellar scintillation. Angular diameters due to scattering reach about 0.15 arcsec at 81.5 MHz at the galactic poles increasing to as much as 5 arcsec at lower latitudes. These scattered sizes vary as ν^{-2}.

(3) Synchrotron self-absorption. For sources of a given flux density there will be a lower limit to their angular size at any frequency (disregarding any relativistic beaming effects) which varies with frequency as $\nu^{-5/4}$. The limit for a 1 Jy source is shown.

The remaining boundary of the distribution is a man-made one separating us from the infra-red. Although undefined it must be close to 700 GHz. These various boundaries have not yet been reached in many places but they are not far away after the recent rapid developments which have filled in so large a part of Fig. 1. The main new thrusts since the Groningen meeting on image formation in 1978 have been, at centimetre wavelengths, the introduction of the VLA, Merlin, MOST and many-station VLBI, at low frequencies the Clark Lake and Cambridge 151 MHz telescopes, whilst at very high frequencies a large number of millimetric telescopes are in operation or under construction. Notable among these are the Hat Creek and Cal Tech interferometers, the IRAM 30m dish and

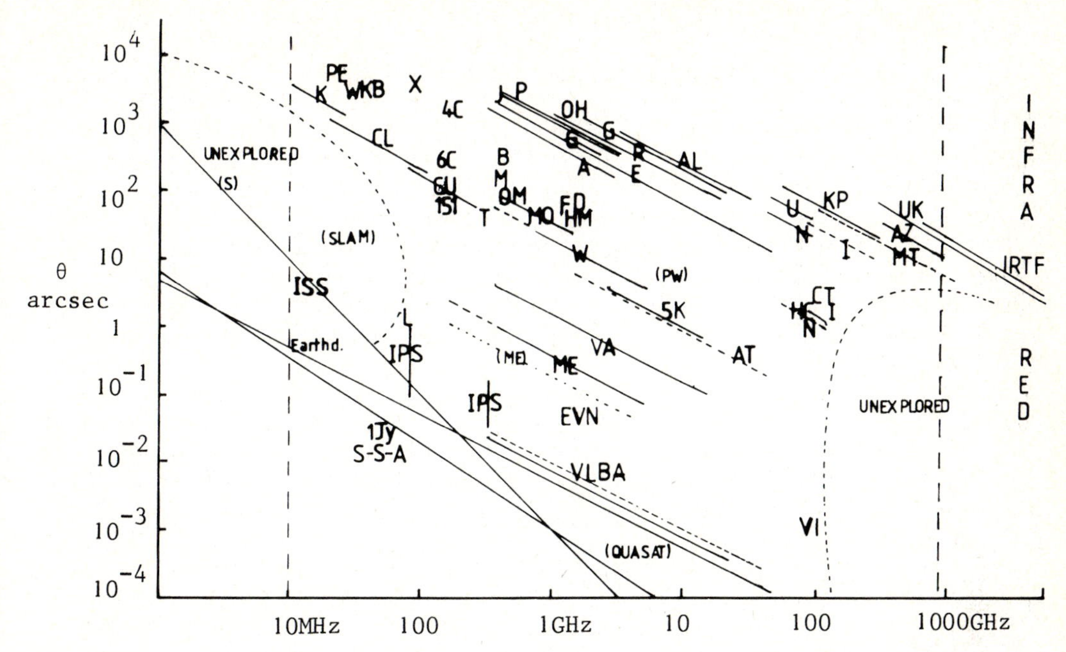

Fig. 1. Resolution of radio telescopes as a function of frequency of operation. For elliptical beams, $\theta = \sqrt{\theta_1\theta_2}$. Projects under development shown dotted; proposals in brackets. Key to symbols, which applies also to Figs 2 and 4.

A	Arecibo
AL	Algonquin
AT	Australia Telescope
AZ	MPI/Univ Arizona sub mm
B	Bologna
CL	Clark Lake
CT	Cal Tech, 10m & interferometer
CU	Culgoora
D	DRAO Penticton
E	Effelsberg
EVN	European VLBI network
F	Fleurs
G	Greenbank 140ft and 300ft
HC	Hat Creek
HM	Half Mile, Cambridge
I	IRAM 30m & interferometer
IRTF	Hawaii
J	Jodrell Bank 250ft
K	Kharkov
KP	Kitt Peak 12m
L	81.5 MHZ LBI Cambridge
ME	Merlin
M	Molonglo
MO	MOST
MT	UK/NL mm telescope
N	Nobeyama
OH	Ohio
OM	One Mile, Cambridge
OO	Ooty occultation & interferomete
PE	Penticton 10 & 22 MHz T.
P	Parkes
PW	Warner proposal
R	Ratan 600
S	Saunders proposal
SLAM	Swarup proposal
T	Texas
UK	UK Infra red telescope
UM	Univ. of Massachusetts
U	Uppsala
VA	VLA A
VB	VLA B
VLBA	VLB Array
VR	Vermillion River Observatory
W	Westerbork
WKB	38 MHz Cambridge
4C	Cambridge
5K	5 km
6C	Cambridge

interferometer, the Nobeyana 45m dish, the resurfaced Kitt Peak dish and the 15m UK/Netherlands telescope. The first VLBI measurements at 3 mm wavelength are an important extension of high resolution to very short wavelengths.

Three major unexplored regions remain:

(1) High resolution (1-100 arcsec) instruments at frequencies below 80 MHz. One proposal by Swarup in this region is for a few arcsec array at $\approx$ 50 MHz in India. At frequencies below 10 MHz resolutions of $\sim$ 1 arcmin are achievable at say 3 MHz using a space interferometer as proposed by Saunders.

(2) Resolutions better than 10^{-3} arcsec for ν > 1 GHz. The proposal for space VLBI, Quasat, fills this region.

(3) Resolution significantly better than 1 arcsec at frequencies above 100 GHz. This region is for technical reasons completely unexplored up to the present. We should be aware that this is not just an unknown island but the tip of an unexplored continent crossed by an arbitrary boundary to the radio astronomy domain similar to those which divide Antarctica.

It might seem that, when these gaps are filled, there are no new frontiers. But this is just one view of telescope parameters and there is plenty of space for development in other coordinates.

LIMITING FLUX DENSITY AND ANGULAR RESOLUTION

Virtually all extended radio sources have steep ($\alpha \sim 0.7$) radio spectra. The radio source population at weak flux densities is also dominated by steep spectrum sources, even at high frequencies. The very compact, flat spectrum ($\alpha = 0$) sources are only numerically important at high flux densities. Therefore in assessing the dependence of the limiting flux density achieved by a telescope on its angular resolution it is sensible to convert that flux density to a standard frequency. I have adopted $\nu = 1$ GHz, $\alpha = 0.7$ as a reasonable basis for Fig. 2. Thus $S(\nu)_{min}(\nu/\mathrm{GHZ})^{0.7}$ is the corrected flux density of the weakest reliable source observable by that telescope.

The very narrow distribution of points in the figure is no accident; the intensive efforts of designers and observers have made it so. The solid line is the locus of points at which there is, on average, one source of flux density $S(\nu)$ in 50 beam areas, θ^2, a value traditionally adopted for the source density at which confusion sets in. This provides a natural limit to the useful sensitivity achievable, not only as a measure of the reliability of source surveys but equally for observations designed to measure the detailed structure of a source. The sensitivity limitations of high resolution telescopes have, up to now, made this unimportant but the present and proposed sensitivities of some

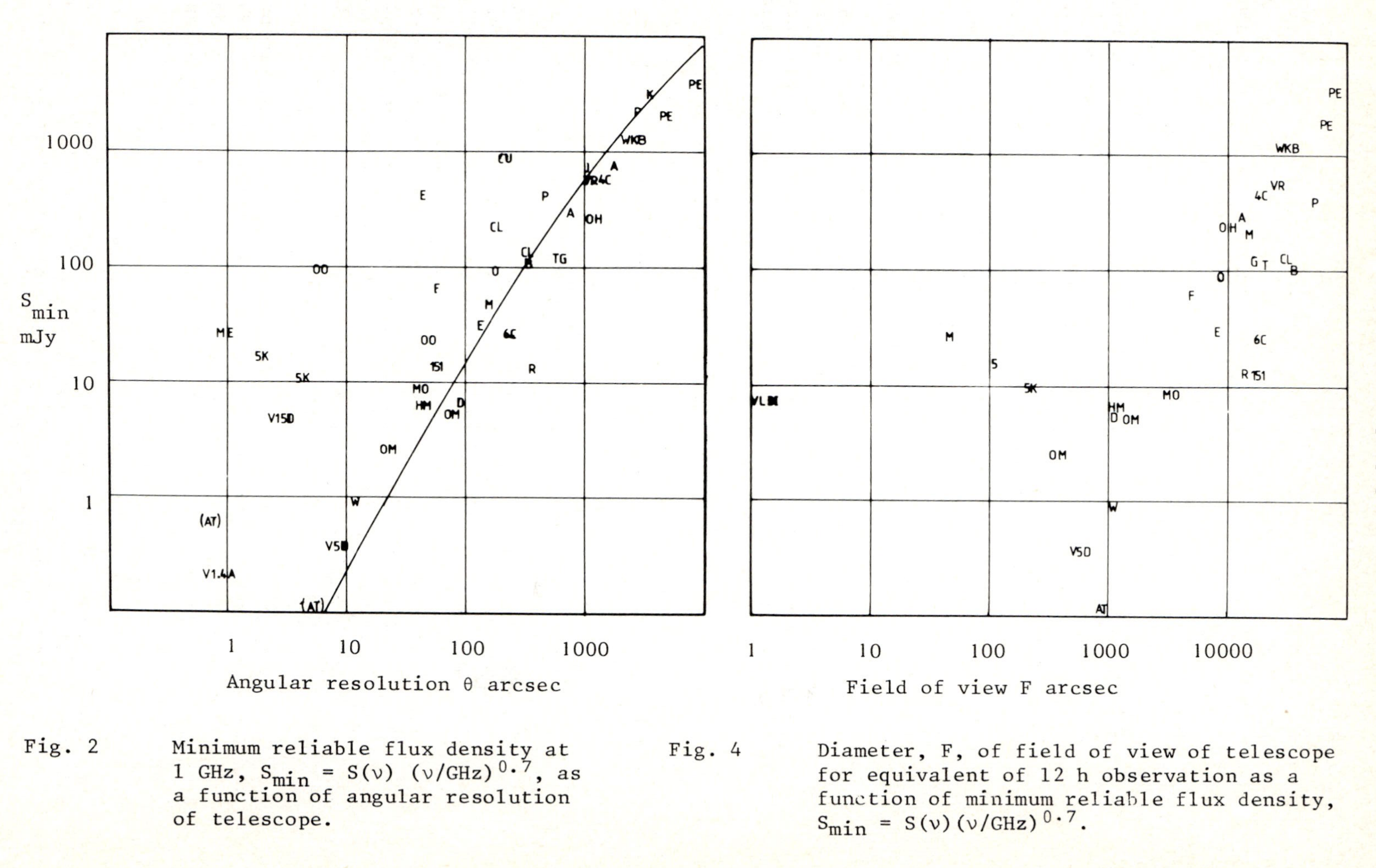

Fig. 2 Minimum reliable flux density at 1 GHz, $S_{min} = S(\nu)\ (\nu/GHz)^{0.7}$, as a function of angular resolution of telescope.

Fig. 4 Diameter, F, of field of view of telescope for equivalent of 12 h observation as a function of minimum reliable flux density, $S_{min} = S(\nu)(\nu/GHz)^{0.7}$.

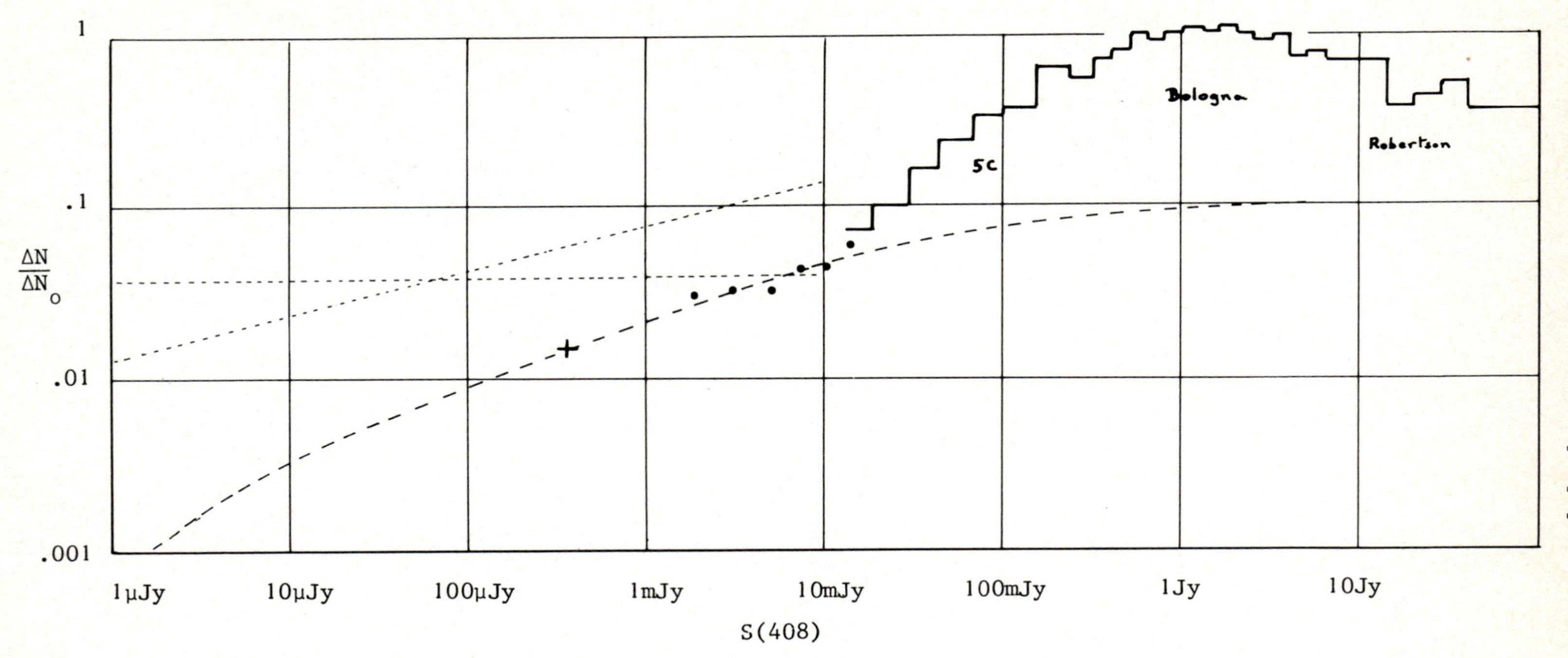

Fig. 3 Estimates of the differential source counts down to 1 μJy at 408 MHz relative to a static Euclidean model.

• Observed values at 1415 MHz from Westerbork surveys converted to 408 MHz with $\alpha = 0.7$.

+ Observed 5 GHz count from VLA similarly converted to 408 MHz.

.... Limits to counts set by sky background brightness. Best estimate is a factor of 5 below this line.

- - - - Extrapolated curve for a model in which the sources with $P(408) < 10^{25}$ W/Hz sr do not evolve.

telescopes (e.g. the VLA D configuration and the AT) show that it must be taken seriously as a limit on useful observational sensitivity in future.

The curve drawn evidently requires knowledge of the radio source counts to much weaker flux densities than is currently known. The best estimate I believe we can make is summarised in Fig. 3. It is based on the 408 MHz differential source counts relative to a static, homogeneous, Euclidean model which have been known for some years at flux densities greater than 10 mJy. Below that limit the Westerbork counts at 1415 MHz (Windhorst et al. 1981) and the VLA 5 GHz counts to 60 μJy of Kellermann (1982) and his coworkers have been added after conversion to 408 MHz using a mean spectral index of 0.7. The flattening of the slope of the counts below S(408) = 10 mJy is to be expected. If the broad outline of the deduced cosmological evolution of radio sources is correct, namely that powerful sources show strong evolution whilst weak ones show little or none, then the weak flux density part of the curve can be deduced directly from the local radio luminosity function. That curve is shown as the dotted line in Fig. 3 extending down to S(408) = 1 μJy. Two facts suggest that it is a reasonably good estimate:

(1) The VLA counts lie close to this line.

(2) The curve lies safely below the absolute upper limit set by the minimum sky brightness at 408 MHz. Two examples of source counts which would just equal that total brightness are shown. The best estimate made several years ago (Bridle, 1967) from spectral arguments is that 30 K of the minimum galactic background at 178 MHz is extragalactic in origin with $\alpha \sim 0.7$. If that is correct, the sky brightness it provides for sources weaker than S(408) = 10 mJy is consistent with the projected counts within a factor of 2.

This proposed $\Delta N/\Delta N_0$ - S curve has a very simple property; its mean slope between 1 μJy and 1 Jy is close to + 0.50. Thus in this region the integral source counts $N(> S) \propto S^{-1}$, explaining the almost constant slope of 2 of the line in Fig. 2 except at high flux densities. For this particular slope the rms error on sources whose number density is 1 per 50 beam areas is $1/\sqrt{50}$, i.e. 14%.

This accident of nature, if we dare call it that, has an interesting consequence. The limiting surface brightness S/θ^2 set by confusion is independent of the resolution, θ. For the curve drawn it is 6 mJy/arc-min^2 at 1 GHz. So, if one seeks sources of very low surface brightness, the only way to improve on this natural limit is to subtract compact sources from high resolution maps and then to smooth the resulting maps. This technique has been used very little so far, perhaps mainly because it requires initial maps with very large numbers of pixels to make significant advances.

The constancy of the limiting surface brightness set by confusion also has an implication for the design of telescopes. Given a telescope whose sensitivity is adequate to reach the confusion level, then a telescope of higher resolution using the same receiver technology and bandwidth will also reach the confusion level provided that the area filling factor of the aperture plane is identical in the two cases. This is at variance with current practice where the filling factor in one dimension of a synthesis array is at best uniform, i.e. falls off as r^{-1} in two dimensions, and the filling factors of high resolution arrays are very small compared to low resolution arrays. There are good practical reasons for these decisions but if ultimate levels of sensitivity are to be reached the penalty of building much denser arrays must be faced.

FIELD OF VIEW AND THE SPEED OF TELESCOPES

The field of view and the speed of a telescope are not well defined quantities. We all know from experience that telescopes having higher angular resolution have smaller fields of view and we have the feeling that telescopes which detect faint objects over wide fields are in some sense 'fast'. One definition of speed which quantifies this in a well-behaved way is

$$\text{Speed} = (\text{Field diameter, } F)/S_{min}$$

measured for some fixed total time T of observation, say 12 hours. It is a convenient definition since mapping it leads to a specific speed for a telescope regardless of how it is used. For example, with a single dish where the integration time t on a point can be varied by changing the speed of scanning, the total solid angle Ω (= F^2) mapped is proportional to T/t (i.e. $F \propto t^{-\frac{1}{2}}$) whereas the minimum detectable flux density set by noise is also $\propto t^{-\frac{1}{2}}$. The same would be true of a synthesis telescope where the time spent on one field could be traded against the number of fields studied. As before it is appropriate to take $S_{min} = S(1\ \text{GHz})_{min} = S(\nu)\ (\nu/\text{GHz})^{0.7}$. Fig. 4 presents the plot of F against S_{min} for some of the telescopes plotted in previous figures. Lines of slope + 1 are lines of constant speed. The fastest telescopes lie obviously in the right hand lower corner. I had genuine difficulties in deciding what the value of F should be for several telescopes. Indeed, in practice many instruments have used several scans of an area when in principle one was sufficient to reach the tabulated S_{min}. The results cannot be precise but that is unimportant. The message to be taken from Fig. 4 is not to see whose telescope if fastest, although I was careful to check that our hosts would be happy with the position of the AT, but to note that a large fraction of all telescopes lie within a narrow range ($\sim$ 10:1) in speed. Only the very high resolution instruments such as Merlin and VLBI are significantly slower and even for these the field of view F and hence the speed could be increased dramatically, if there was good astronomical reason to do so, although at the expense of a heavy computing load.

DYNAMIC RANGE OF MAPS

This quantity has become the focus for a large amount of work in the last three or four years. It used to be called the sidelobe level in a map and there are only two references to the phrase in the Groningen symposium. It is still a quantity which is loosely defined and therefore difficult to compare from telescope to telescope. For instance, the non-repeating sidelobe levels close to bright sources show a strong dependence on angular distance from the source so that the dynamic range may be only 100:1 close to the source but 10^4:1 if the comparison with the peak of the source is made at a radius of 100 beamwidths. This is not yet taken into account in the definitions.

Several techniques have been shown to provide dramatic improvements in the quality of images including phase closure, the use of redundant baselines, self-calibration, Clean and Maximum Entropy, some involving changes in the operation of the telescope whilst others are concerned only with improved analysis of the data output. An assessment of the relative merits of these and other techniques is one of the most important aspects of this symposium and one which we should use in turn to answer the question 'What is the correct design for a telescope to give the best possible dynamic range in its images?'. There seems no obvious limit to the dynamic range which is achievable in principle.

CONCLUSIONS AND FUTURE PROSPECTS

This rapid retrospect on the last few years of telescope development makes it evident that the greatest advances have been made in angular resolution. Relative to ten years ago the factors of improvement in resolution of high quality maps vary between 10 and 1000 over the whole frequency range from 30 MHz to 300 GHz. The improvement in sensitivity measured in Jy of nearly 1000 is also impressive. However, if the typical spectrum of a source is taken into account, as for many purposes it should be, that factor is reduced to only 30 over the last 15-20 years. Similarly, there has been progress in making faster telescopes but the factors of improvement have been quite small by the measure of speed that I have suggested.

Some prospects for the future are clear. They include:

(1) Exploration of wholly new fields at very low and very high frequencies and of very high resolution at centimetric wavelengths.

(2) Improvements in surface brightness sensitivity by the removal of confusing sources.

(3) Increase in the speed of telescopes. For single dishes this might be accomplished by the much greater use of multiple feeds whilst for synthesis telescopes extremely large numbers of elements are needed to provide large fields of

view without sacrificing the sensitivity. It seems difficult to envisage much larger bandwidths of operation than are already planned for the Australia Telescope.

Even with these improvements we shall have to work very hard to reach what I put before you as a modest goal, namely, a telescope having 1 arcsec resolution over a 6° diameter field with an rms sensitivity of 1 μJy in 1 hour's observation. It is not a wild dream; it exists in another branch of our subject and is known as a Schmidt telescope.

It is pleasant to speculate on a natural extension of even that goal, a broad-band array of tiny independent elements whose field of view is virtually the whole sky. It would be perhaps the ideal telescope. It has no moving parts, can accommodate all observing proposals simultaneously and solves the problem of data storage by keeping the data where it should be kept - in the sky - until you want to look at it. We should, of course, need two of them.

REFERENCES

Bridle, A.H. 1967. Mon. Not. R. astr. Soc., 136, 219.
Kellermann, K.I. 1982. In IAU Symposium 104, Early Evolution of the Universe and its Present Structure.
Windhorst, R.A., Kron, R.G., Koo, D.C. & Katgert, P. 1981. In IAU Symposium 97, Extragalactic Radio Sources p427.

DISCUSSION

R.D. EKERS

Are the normal galaxies included in your extrapolation of the source counts to low flux levels.

J.E. BALDWIN

Yes, the local radio luminosity function I used included normal galaxies down to spirals about as weak as M31. Indeed, at 1 mJy most of the sources are normal galaxies at $z \sim 0.5$ to 1. They are sufficiently numerous ($\sim 10^9$/sr) that they almost touch.

A.J. TURTLE

At what level do stars contribute to confusion?

J.E. BALDWIN

I don't know, and I'm not sure if anyone does. They are probably negligible at low frequencies to well below 1 mJy. At high frequencies (~10 GHz) they might become comparable with the extragalactic sources at 1 mJy where there are about 10^9/sr.

2.2 SPECTRAL LINE IMAGING WITH APERTURE SYNTHESIS RADIO TELESCOPES

R. D. Ekers
National Radio Astronomy Observatory, Socorro, New Mexico
USA

J. H. van Gorkom
National Radio Astronomy Observatory, Socorro, New Mexico
USA

Abstract Problems specific to broadband or multiple frequency channel aperture synthesis observations are discussed. These include the chromatic aberration resulting from the use of a finite bandpass, effects of continuum removal, self-calibration, and image deconvolution. Algorithms for optimizing the signal to noise ratio in spectral line processing are reviewed. Some aspects of the storage and display of spectral line data cubes are discussed

Chromatic aberration

Aperture synthesis radio telescopes have an inherent chromatic aberration which results from the formation of an image by adjusting the phase, $\Delta\phi$, of the correlated signals for each point in the image plane (the Fourier transform relation), instead of the arrival time of the wavefront, $\Delta t = \Delta\phi/2\pi f$, at each point. This aberration causes a radial smearing which increases linearly away from the point in the image for which the time delays have been equalized by the delay tracking system. It is commonly known as the delay beam smearing.

In order to overcome this defect it is necessary to subdivide the band into channels which are sufficiently narrow to have negligible aberration, and to use the correct frequencies when combining this information. In most spectral line observations the channel bandwidth criterion is automatically satisfied by the spectral resolution requirements, but for broadband continuum observations the number of channels required will be determined by this effect. This is especially true at the lower frequencies where the relative bandwidth is larger. It is interesting to note an analogy with the use of single dishes as mapping instruments. The much greater power of the monochromatic synthesis arrays for mapping is a result of using all the information available, rather than rejecting all except that coming from one position in the sky. However, to accept the narrow field of view imposed by the bandwidth smearing is an equivalent limitation and one which is often imposed by components of the system which are less costly than the rest of the array.

Even if there is negligible delay beam smearing in the individual spectral line channels, the sidelobe structure due to the array geometry will still change with frequency and this will have to be taken into account in the analysis of the spectral line data cube. The effects of frequency dependent sidelobes could be removed by use of the deconvolution procedures as discussed in a later section, but it is preferable to proceed as far as possible without introducing the additional deconvolution uncertainties. Consider two extreme cases. If the spectral line channels are computed using the correct frequencies for each channel the structure in the image will all be in the correct place but the sidelobe pattern will be changing from channel to channel by a radial scaling factor. If the same frequency is used for each channel the sidelobe patterns will all be the same but the structure in the image will now be distorted by the same radial factor. (If all these channels were added this would give the delay beam smearing.) The former situation is preferable if the spectral line images contain a lot of weak frequency dependent structure spread out over a large field. However if the dominant source is near the field center, and if the line and continuum emission do not coincide spatially, the latter situation will be better since the sidelobes will be correctly handled in linear map combinations.

High spectral dynamic range

In some spectral line observations very high spectral dynamic range is required. For example, since in a recombination line observation the line to continuum ratio may only be a few per cent and the line must generally be measured to at least 10% accuracy the spectral dynamic range would need to be better than 1000:1 at any point in the image. Fortunately the phase errors caused by the atmosphere, one of the main sources of error in a continuum observation, are independent of frequency and can be removed by subtracting a continuum image formed from frequency channels outside the line emission. However it is important to remember that any remaining small gain and phase errors will appear as much larger errors in this line-continuum image. For example a small frequency dependent phase error will be amplified by the continuum to line ratio in the line-continuum image.

Bandpass corrections. Astronomical calibration of the complex antenna based bandpass function is needed to correct for the remaining errors. Fortunately the standing wave modulation of the bandpass, which is one of the largest and most uncertain errors in single dish observations, is not a problem in an interferometer because the receiver noise is not correlated between elements.

Bandpass normalization by the autocorrelation spectrum for each antenna can be used to correct for the amplitude variations across the band but this does not correct for the phase variations.

Continuum subtraction. Before subtracting the continuum image from the line channels the possible effects of the chromatic aberration discussed in the previous section must be considered. If the effect of the frequency dependence of the sidelobe structure is important it may be necessary to subtract the Fourier transform of a model of the continuum from the measurements in the visibility plane using the exact frequencies and geometry to calculate the visibility coordinates. This model might consist of the positions and amplitudes of discrete continuum sources, or it could be derived using the deconvolution procedures discussed later.

Self-calibration

When applying the self-calibration (or closure) relations (Readhead & Wilkinson 1978; Schwab 1980) to correct for the atmospheric phase errors in spectral line observations the same set of phase corrections can be used at each frequency. The optimum procedure to determine these corrections depends on the nature of the image. If there is strong continuum the average of all the continuum channels can be used to obtain the self-calibration solution. However if the line emission is stronger than the continuum the channel with the highest signal to noise ratio in the line could be used (e.g. a channel containing a maser line). For cases in which all the channels have comparable intensity and spatial structure changing with frequency it may be necessary to develop a three-dimensional model to be used for a global self-calibration solution.

The quality of a self-calibratied spectral line image will ultimately depend on those frequency dependent errors which cannot be attributed to complex multiplicative antenna gains (Thompson and D'Addario 1982).

Deconvolution effects

The deconvolution of multichannel aperture synthesis images (e.g. spectral line, polarization, temporal sequences, etc) presents a number of special problems. The information in adjacent channels may be needed to optimize the deconvolution, the required channel combinations may involve negative quantities (e.g. HI absorption, Stokes parameters and Faraday rotation), and it may be important that the deconvolved channels have identical transfer functions. Consider the case of an absorption line experiment involving strong and complex continuum emission. If individual channels are deconvolved independently a small fractional error in the reconstruction can result in a substantial increase in the errors when these channels are subtracted to form the line-continuum image. This is especially noticeable with the CLEAN (Högbom 1974) algorithm (van Gorkom 1982). If the channels are subtracted before deconvolution this problem is avoided because the strong continuum signal and its sidelobes are removed, however the resulting image can now contain negative features so algorithms which rely on positivity to suppress sidelobes (e.g. MEM, Gull & Daniell 1978) cannot be used. In this case algorithms such as CLEAN which use spatial constraints can be used. Finally, it is necessary to determine the ratio of the deconvolved line-continuum to a deconvolved continuum image in order to compute the distribution of optical depth. For this purpose we have an additional constraint on the deconvolution algorithm; it must produce estimates of the real sky as seen through identical transfer functions (i.e. with the same beam). The MEM algorithm allows the spatial resolution and amplitude linearity to change with both the signal to noise ratio and the offset from zero so the result would be incorrect.

Ideally, the astronomer should specify exactly what he is trying to estimate and use a deconvolution algorithm designed to estimate this quantity using all available information. Two practical considerations make this approach untenable in general although it may be useful in some cases. 1) the method to do the deconvolution may not be known, and even if it is known the implementation may be difficult. 2) it is often not possible to specify the problem adequately in advance. For example, most deconvolution algorithms require information about the observational errors but the instrumental effect determining the errors may only become apparent after inspection of the deconvolved channels at an intermediate stage in the analysis.

Bandwidth synthesis. In a multiple channel broadband continuum observation it is important to take advantage of both the sensitivity and spatial coverage provided by the bandwidth when doing the deconvolution. This will be especially important for a VLBI array since even a 10% bandwidth corresponds to many hundreds of kilometers of spatial coverage. To avoid the radial smearing the observing band will be subdivided into narrow channels as described in the first section of this paper. Each of these channels will be at a different frequency so it will have a different beam shape. To optimize the signal to noise ratio and the use of information we need to combine all the channels before deconvolution (or self-calibration). If all the features in the image had the same spectrum this combined image could be correctly deconvolved by using a beam formed by appropriately weighting the beams for each channel. However if different features in the image have different spectra then each feature will have a different beam and this procedure will fail. Fortunately the extra information gained by the range of frequencies can greatly exceed that needed to determine the spectral index and the intensity of features so it is possible to devise deconvolution procedures which use all the data to produce both an intensity and a spectral index image.

Digital cross correlator effects

In almost all aperture synthesis telescopes the spectral line capability is provided by a digital cross correlation spectrometer and this introduces some additional effects which must be considered.

Quantization. The effects of the quantization in a digital cross correlator can be removed by applying the appropriate quantization correction (Van Vleck correction in the case of one bit quantization) to each point in the cross correlation measurement prior to the Fourier transform into the frequency domain. If the source is weak compared to the system noise the correction will be the same factor at all delays so that the spectrum is only wrong by a scale factor which will be removed by the normal calibration. If the cross correlation coefficient exceeds about 20% the spectrum will be distorted unless this quantization correction is made (D'Addario 1982).

Gibbs phenomenon. The Gibbs phenomenon is the ringing at the edges of the frequency spectrum which occurs because of the truncation of the temporal cross correlation measurements. The consequences of the Gibbs effect are more serious in spectral line aperture synthesis than in an autocorrelation spectrometer. For the autocorrelation spectrometer the correlation function is assumed symmetric (no negative lags are measured) to yield a power spectrum which is real. However the cross correlation function is not symmetric and yields a spectrum of complex source visibilities. Since the effect of the Gibbs phenomenon now depends on the visibility phase (figure 1) the frequency ripple will change with position in the image. The only place where the effect will be correctly calibrated by the bandpass calibration is at the position of the bandpass calibrator (normally the field center).

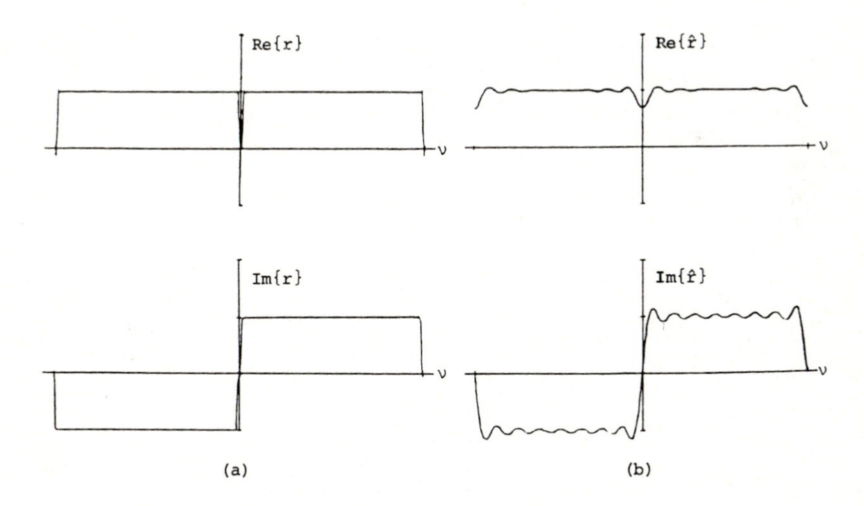

Figure 1. (a) Cross power spectrum resulting from a continuum source of unit flux in reference direction: "true complex gain". Note non-zero phase. (b) Computed cross power spectrum with 16 delays (from D'Addario 1982).

There are a number of possibilities to reduce the Gibbs effect. The amplitude of this ringing is inversely proportional to the number of frequency channels and it decreases away from the band edge, so it can be reduced by using more channels and more bandwidth. It can also be attenuated by suitable tapering of the cross correlation measurements prior to the Fourier transform (e.g. Hanning smoothing). Again this requires more frequency channels to obtain the same spectral resolution. Finally, since the effect can be calculated for the observed structure it could in principal be included in a three-dimensional deconvolution procedure! These effects are discussed in more detail by Bos (this symposium).

Spectral line processing

The optimization of the signal to noise ratio in many spectral line problems is equivalent to a matched filter problem where the filter characteristics are determined by the data to be filtered. Consider the problem of using a set of spectral line images to determine the integrated properties of the line emission at each point in a rotating object. Such properties would be the integrated emission, its velocity and velocity dispersion (the zero, first and second moments of the velocity profile). These are to be determined at each point in the image. In each channel the emission will appear at a different position because of the Doppler shift. One obvious way to get the total emission is to take the sum over all the line channels. However at any given point in the image the line emission is only present in a few channels, the rest containing noise, so this procedure will increase the noise level significantly over that in the individual channels. As an example the hydrogen observed in one channel of the spiral galaxy NGC 1097 is shown in figure 2(a) and the total hydrogen determined by summing all the channels is shown in figure 2(b). The increase in noise is clear. A number of methods have been developed to avoid this signal to noise degradation by including only the channels with line emission.

The CUTOFF method. This method was first described by Rogstad & Shostak (1971). It applies an acceptance gate in surface brightness to exclude points with no line emission. Figure 2(c) shows the result of using a cutoff of 2*rms on the same data used to produce figure 2(b). The image is considerably improved but the method has a serious disadvantage. The calculated moments are subject to systematic effects depending on the cutoff value used. The integrated emission will be too small because features below the cutoff will be omitted, and the velocity will be biased towards the center of the range because, on average, noise peaks above the cutoff will be closer to the center of the band than the emission peak (figure 3).

The WINDOW method. This was developed to overcome the biases introduced by the CUTOFF method (Bosma 1981). An acceptance window covering the velocity range of the emission in each profile is chosen by iterating on the window's position and width until the mean intensity outside the

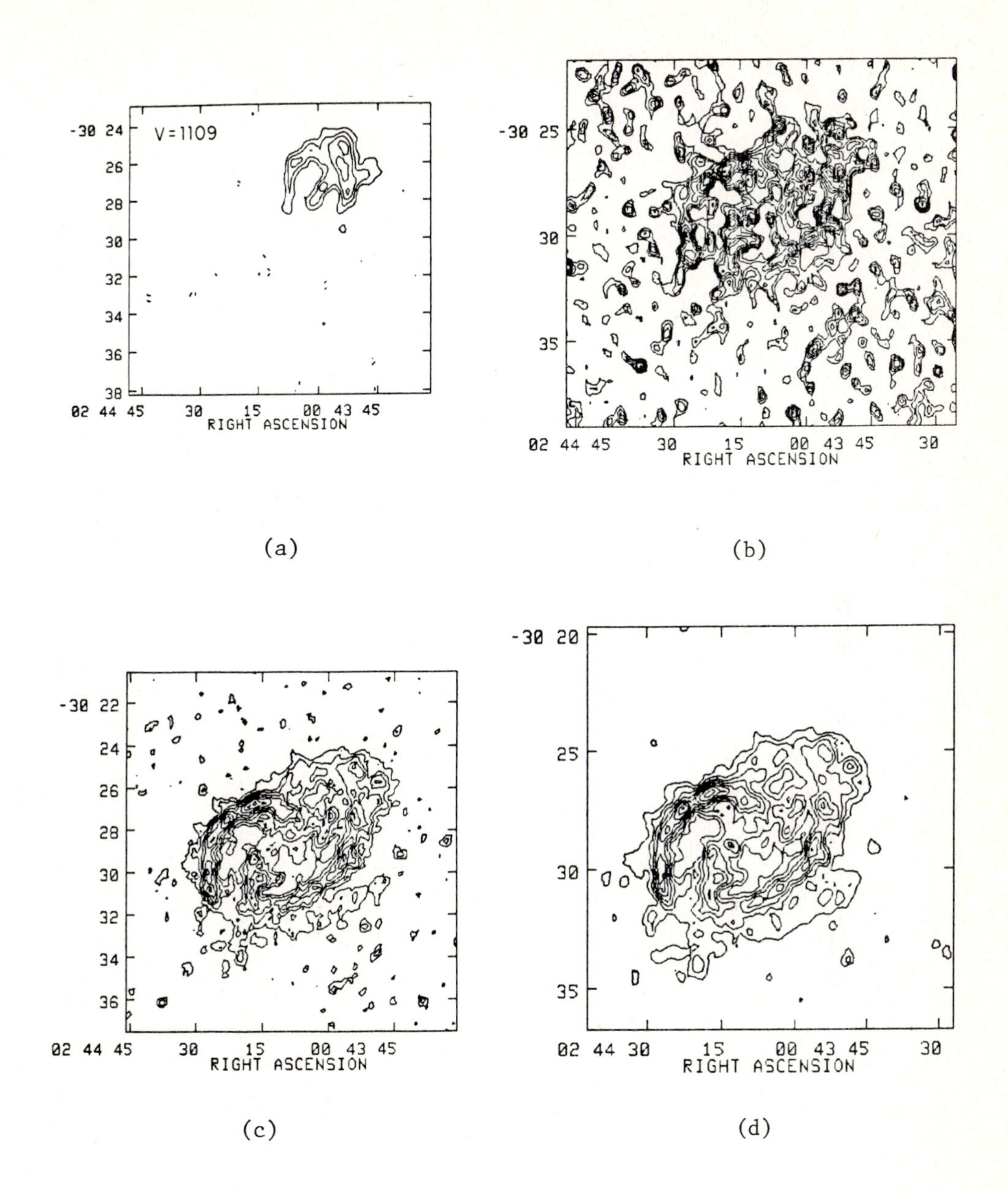

Figure 2. (a) An example of one channel of the neutral hydrogen emission from the spiral galaxy NGC 1097. The total hydrogen emission obtained from: (b) the sum of all channels, (c) the sum of all channels using a cutoff 2* rms, and (d) the sum of all channels using a spatial smoothing of twice the resolution to determine the acceptance window.

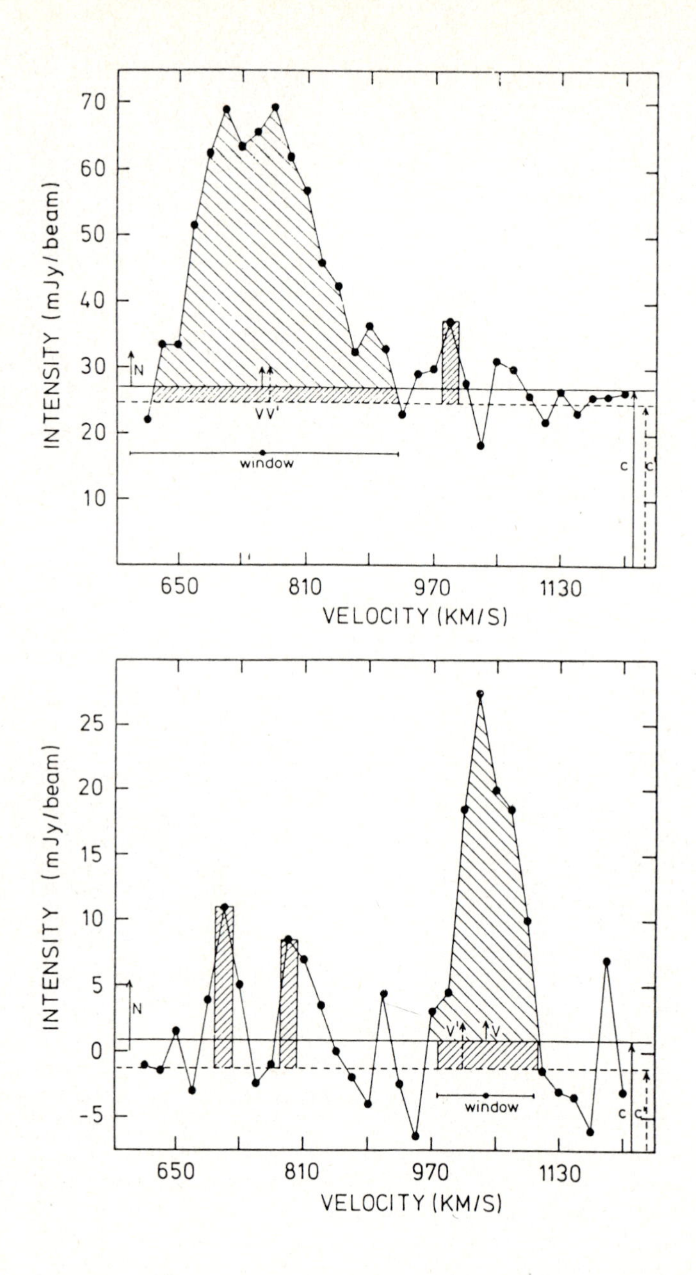

Figure 3. Some examples of WSRT profiles observed in the galaxy NGC 5033. N denotes the noise level (1σ), C and V are the results for continuum and velocity from the window method, while C' and V' are the results from the cutoff method. The profile integrals are shown with course shading. The fine shading indicates the additional contribution to the integral from the cutoff method. The cutoff level used is 1.5σ (Bosma, 1978).

window is constant. Although less biased estimates of the integrated emission and the velocity are obtained this method still gives a biased estimate of the profile width.

Interactive window. A variation on the WINDOW method in which each profile is inspected on a display and the windows set interactively using a suitable graphics input device. Although often superior this does introduce the possibility of personal bias, and it is very time consuming for a large database.

Profile fitting. In this case a preconceived function, usually a Gaussian function (or a Voigt profile in the case of recombination lines) is fitted to the profile at each point in the image. In this case no bias will be introduced unless the funtional form assumed is incorrect. Comparisons of profile fitting and WINDOW methods are discussed by van der Kruit & Shostak (1982).

An improvement to both the CUTOFF and the WINDOW methods is obtained by smoothing the data, either spatially or in velocity, before determining the points to be included, but then going back to the full resolution data for the actual moment calculations. Figure 2(d) is an example of the use of spatial smoothing to determine the acceptance window.

Display of the spectral line data cube

Typical three-dimensional images which are obtained from the spectral line aperture synthesis telescopes will contain from 10^6 to 10^7 resolution elements. For such large amounts of information careful attention has to be given to the image display techniques in order to obtain good impressions of what is in the data and to be able to recognize errors.

In the simplest display technique a set of two-dimensional slices through the three-dimensional cube are displayed either as a mosaic, or in sequence utilizing time as the third dimension. The cinematographic display is more powerful if the frame rate and direction can be controlled interactively. (Blinking a pair of two-dimensional images is a special case of this.) Normal computer graphics systems have insufficient bandwidth to display a sequence of large images at a sufficiently fast rate, but special purpose devices can do so. At the University of Groningen a computer controlled analog video disk is used for this purpose, and the National Radio Astronomy Observatory is currently developing a digital storage device capable of storing 128 512^2 images and displaying them at 6 frames per second.

Another possibility is to use a false-colour display in which one variable, e.g. velocity, is mapped into colour while the intensity is still determined by the intensity of the image. This works quite well when the brightness at any point in the image is a simple function of velocity. Image displays which can operate directly on intensity, hue and saturation (intstead of the traditional red, green and blue) have considerable advantage for the interactive manipulation of false-colour displays (Buchanan 1979). In order to interactively change the transfer function controlling the colour in a false colour display with an RGB system it must be capable of performing nonlinear image combinations at video rate. Examples of the use of false colour displays are given by Allen (1979).

The third possibility is to use a system which can generate a perspective display of the three-dimensional surface, If the display hardware has the further capability of smoothly rotating the image being projected the three-dimensional perception of the data structures is greatly enhanced. True three-dimensional effects can also be generated from stereo pairs (figure 4) but this requires use of special glasses and is less satisfactory for noisy data. Finally (through use of mirrors!) we now also have the possibility of generating space filling three-dimensional images with interactive graphics capabilities (Stover 1981).

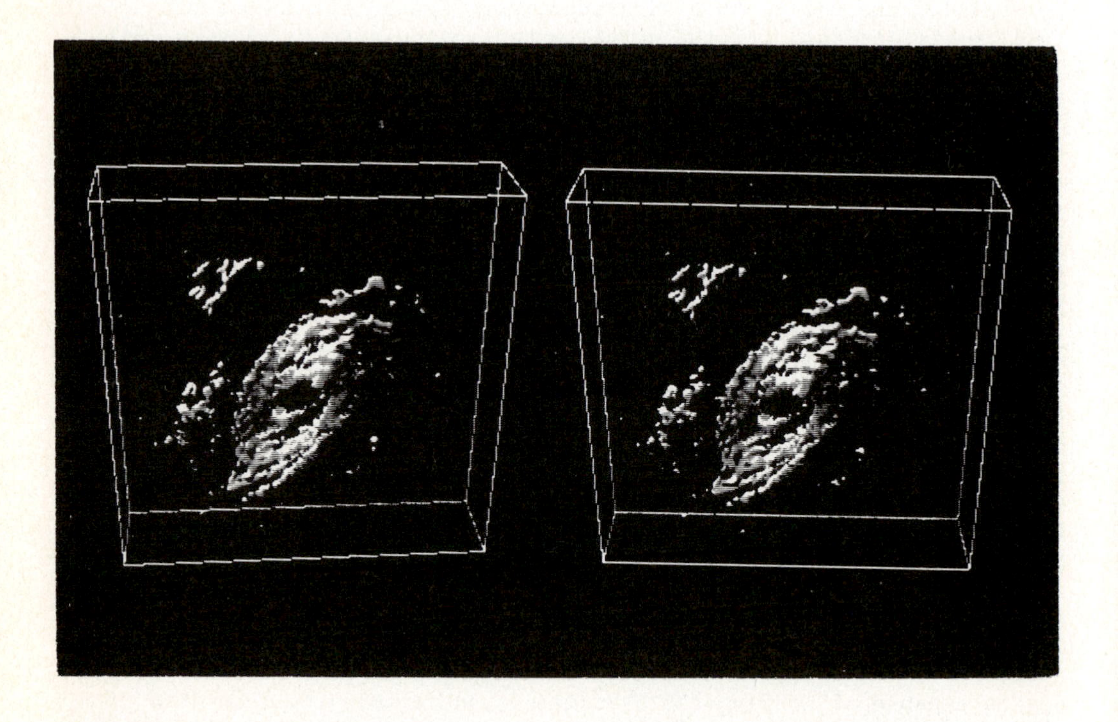

Figure 4. Stereo image pair for the HI distribution (x,y) and velocity (z) in the spiral galaxy M81. Data is from a VLA observation by B. Hine and the stereo images have been provided by A. Rots.

References

Allen, R.J. (1979) Exploring methods of data display in an interactive astronomical data-processing environment. In Image Formation from Coherence Functions in Astronomy, ed. C. van Schooneveld, pp. 143-155.

Bosma, A. (1981). 21cm line studies of spiral galaxies, I. Observations of the galaxies NGC5033, 3198, 5055, 2861 & 7331. Astron. J., 86, 1825.

Buchanan, M. (1979). Effective utilization of colour in multidimensional data presentations. Proc. S.P.I.E., 199, 9-18.

D'Addario, L.R. (1982). Digital cross correlators and cross correlation spectroscopy. NRAO, Proc. VLA Workshop on Synthesis Mapping, pp. 7.1-7.22.

van Gorkom, J.H. (1982). Advanced spectral line problems. NRAO, Proc. VLA Workshop on Synthesis Mapping, pp. 15.1-15.22.

Gull, S.F. & Daniell, E. (1978). Image reconstruction from noisy and incomplete data. Nature, 272, 686-690.

Högbom, J.A. (1974). Aperture synthesis with a non-regular distribution of interferometer baselines. Astrophys. J. Suppl., 15, 417-426.

van der Kruit, P.C. & Shostak, G.S. (1982) Studies of nearly face on spiral galaxies. I. The velocity dispersion of the HI gas in NGC3938. Astron. Astrophys., 105, 351.

Readhead, A.C.S. & Wilkinson, P.N. (1978). The mapping of compact radio sources from VLBI data. Astrophys. J., 223, 25-36.

Rogstad, D.H. & Shostak, G.S. (1971). Aperture synthesis study of neutral hydrogen in the galaxy M101. Astron. Astrophys. 13, 99.

Schwab, F.R. (1980). Adaptive calibration of radio interferometer data. Proc. S.P.I.E., 231, 18-24.

Stover, H. (1981). Graphics system displays true 3D images. Mini-micro Systems Dec. 121-123.

Thompson, A.R. & D'Addario, L.R. (1982). Frequency response of a synthesis array: Performance limitations and design tolerances. Radio Science 17, 357-369

DISCUSSION

R.H.T. BATES

It is wasteful to use windows to prevent glitches with truncated data. It is better to use smooth edge-extension, or joining. This has been shown to work with picture processing and in computer-assisted tomography.

R.D. EKERS

Two practical problems occur (1) an image 512×512 in size has 250,000 spectra to correct and (2) the transform from lag to frequency is done before the transform from the aperture to the image plane so the corrections have to be made to complex data.

2.3 PROJECTION IMAGING OF 3-D MICROWAVE SCATTERERS WITH NEAR OPTICAL RESOLUTION

N.H. Farhat
University of Pennsylvania, The Moore School of Electrical Engineering, 200 S. 33rd St., Philadelphia, Pa. 19104, USA

T.H. Chu
University of Pennsylvania, The Moore School of Electrical Engineering, 200 S. 33rd St., Philadelphia, Pa. 19104, USA

Abstract. Microwave projection imaging of complex shaped conducting objects from broadband far field data gathered in an anechoic chamber facility over the (6-17) GHz range is described. A novel synthetic *target derived reference* technique, which would greatly simplify data acquisition in practice by making the scatterer furnish the reference signal needed for coherent detection, is employed. Image enhancement through the use of polarization information and a priori knowledge of object symmetry is demonstrated. Finally the implications of the results for radar imaging of distant objects with near optical resolution and their extension to 3-D correlation imaging of broadband emitting objects are briefly discussed.

INTRODUCTION

Target shape estimation in the context of inverse scattering from far field data is a longstanding problem with considerable present day interest (Bojarski 1967, 1974; Lewis 1969; Raz 1967; Das & Boerner 1978; Farhat & Chan 1978; Chan & Farhat 1981; Boerner, Ho & Foo 1981). It can be shown from inverse scattering theory assuming physical optics and Born approximations hold, that monostatic or bistatic measurement of the far field scattered by an object as a function of illuminating frequency and object aspect can be used to access the Fourier space $\Gamma(\underline{p})$ of the object scattering function $\gamma(\underline{r})$. Here $\underline{p}$ and $\underline{r}$ are 3-D position vectors in Fourier space and object space respectively. In the context of this work, the object scattering function γ is taken to represent the 3-D geometrical distribution and strength of those scattering centers of the object that contribute to the measured field. The Fourier space data manifold $\Gamma_m(\underline{p})$ measured in practice is necessarily of finite extent which depends on the values of $\underline{p}$ realized in the measurement. These depend in turn on geometry and on the angular and spectral windows utilized. It is possible then to retrieve a diffraction and noise limited version γ_d of the object scattering functions by 3-D Fourier inversion of Γ_m. In particular tomographic or projective reconstruction of γ_d based on the *projection-slice theorem* or the *Radon transform* have been demonstrated computationally (Farhat & Chan 1978; Chan & Farhat 1981; Das & Boerner 1978; Boerner, Ho & Foo 1981).

In this paper the inverse scattering principles referred to above are used to demonstrate the first centimeter resolution projection imaging of microwave scatterers utilizing angular, spectral, and polarization information. A description of the measurement system used is given and a procedure for correcting the data for clutter, measuring system response, and for an unknown range-phase term in order to be able to access the Fourier space is demonstrated. Range-phase removal is achieved with a novel method for generating the equivalent of a phase reference point on the scatterer. This target derived reference method has practical advantages that were listed elsewhere (Farhat & Chan 1978; Farhat, Chan & Chu 1980). Examples showing the use of polarization information and symmetry in image enhancement are presented for two test objects. The implications of the results for radar imaging networks and their extension to tomographic and projective imaging of incoherently emitting objects are then briefly discussed.

DATA ACQUISITION AND IMAGE RETRIEVAL

An experimental Microwave Imaging Facility (Farhat & Werner 1981) was employed to access the 3-D Fourier space of the test object as described in detail elsewhere (Farhat, Chu & Werner 1983). Two test objects are used, a metalized 100;1 scale model of a B-52 aircraft with 79 cm wing span and 68 cm long fuselage, and a metalized 72:1 scale model of the space shuttle of 33 cm wing span and 50 cm body length. The test object is mounted on a computer controlled elevation-over-azimuth positioner situated in an anechoic chamber as shown in the simplified diagram of Fig. 1. Automated measurement of the scattered field over any band in the (6-18) GHz frequency range is provided by a coherent microwave measurement system consisting of a microwave sweeper and a broadband coherent receiver.

Fig. 1. Simplified diagram of measurement system.

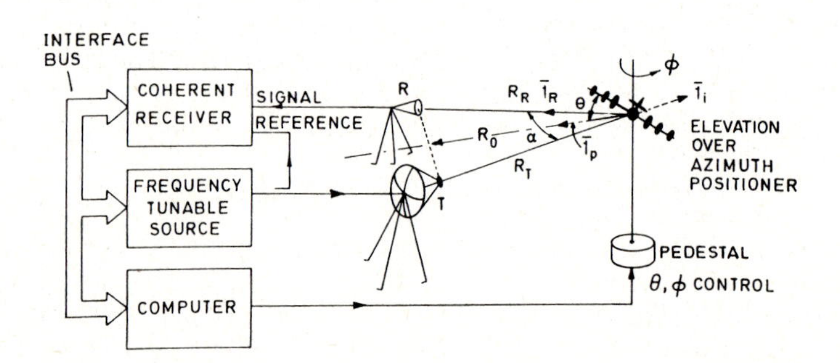

Single Fourier space slices of finite extent were obtained for each object for a fixed elevation angle of $\Theta = 30^\circ$ and azimuth angle ϕ altered between 0° to 90° in steps of $.7^\circ$ for a total of 128 looks corresponding to aspect angles extending from headon to broadside orientation of the models relative to the antennas. For the value of $\alpha = 20^\circ$ used in the recording geometry, the vector $\bar{p} = k\,(\bar{1}_i - \bar{1}_R) = 2k \cos\frac{\alpha}{2}\,\bar{1}p$ (see for example Chen & Farhat 1981) extends from $p_1 = 2k_1\cos\frac{\alpha}{2} = 503.25$

Rad/m to $p_2 = 2k_2\cos\frac{\alpha}{2} = 1402.5$ Rad/m in the direction of $\vec{1}_p$ in Fig. 1. In this fashion the $\bar{p}$-space is accessed in a polar format consisting of 128 radial lines covering an angular span of 90° each corresponding to one value of ϕ and with each line containing 128 data points for a total of 16,384 complex data points. This can be done for any desired state of polarization of the transmitter or the receiver.

In practice the system of Fig. 1 cannot measure the field components directly. Instead the measurement is corrupted by undesirable effects of chamber clutter and measurement system response. Error free determination of $\bar{\Gamma}(\bar{p})$ requires, therefore, not only the range-phase normalization referred to earlier, but also requires correction of the measured scattered field for clutter and system response.

Accordingly the inclusion of the effects of clutter and measuring system response requires modifying the expression for the measured field (Chan & Farhat 1981) as applied to the configuration of Fig. 1 to the form,

$$\psi_m(\bar{p}) = kA\, e^{-jk(R_T+R_R)}\, \Gamma_m(\bar{p})\, H(p) + C(p) \tag{1}$$

where A is a complex constant, k is the wavenumber, and H(p) and C(p) represent the system frequency response and clutter respectively. Note that H and C are functions of frequency alone though $p = 2k\cos\alpha/2$ and when polarization effects are taken into account that C(p) is also polarization dependent. Furthermore C(p) in eq. (1) contains implicitly the effect of system response H on clutter measurement. To determine Γ_m in eq. (1) from the measured field ψ_m it is necessary to determine the unknown terms H, C and (R_T+R_R) first. The clutter term C(p) is readily determined by recording the scattered field in the absence of the object. The system response is determined next by using a reference target such as a conducting sphere or cylinder of known radius a and known physical optics frequency response proportional to exp $(jk2a\cos\alpha/2)$ (Ruck 1970). With the reference target in place, the measured field will be,

$$\psi_r(p) = kB\, e^{-jk(R_T+R_R)}\, e^{jk2a\cos\alpha/2}\, H(p) + C(p) \tag{2}$$

and is independent of aspect angle ϕ (and Θ when a sphere is used). Here B is a proportionality constant. It follows from eq. (2) that,

$$\exp-j\left(\frac{R_T+R_R}{2\cos\alpha/2} - a\right)p\; H(p) = \frac{1}{kB}\left[\psi_r(p) - C(p)\right] = G_r(p) \tag{3}$$

All quantities on the right-hand side of eq. (3) are now known. Therefore its Fourier inversion yields,

$$g_r(\xi) = \frac{1}{2\pi}\int G_r(r)\, e^{jp\xi}\, dp = h(\xi-\xi_c) \tag{4}$$

where $h(\xi)$ is the "spatial" impulse response of the measurement system and,

$$\xi_c = (R_T+R_R)/2\cos\alpha/2-a \tag{5}$$

By taking ξ_c to equal the centroid of g_r defined as,

$$\xi_c = \int \xi g_r(\xi)\, d\xi / \int g_r(\xi)\, d\xi \tag{6}$$

and equating eqs. (5) and (6) the range term (R_T+R_R) can be estimated with an accuracy of $\delta R = c/2\Delta f = 1.37$ cm where c is the velocity of light and Δf (= 11 GHz) is the spectral bandwidth utilized in the measurement. Equation (3) can now be used to determine H(p). Having determined H(p) and C(p) in eq. (1) the target response $\Gamma_m(\bar{p})$ can be obtained, from,

$$\Gamma_m(\bar{p}) = \frac{1}{kB} e^{jk(R_T+R_R)} [\psi_m(p) - C(p)]/H(p) \tag{7}$$

in which all quantities on the right side are now known except the combined range (R_T+R_R) between the *phase center* of the target and the phase centers of the transmitting and receiving antennas. In the following, the range to the *scattering centroid* of the target is taken as a first estimate of the range to the phase center. However since the scattering centroid of an arbitrarily shaped object is aspect dependent, the term (R_T+R_R) must be estimated for every object aspect for which ψ_m is measured. This is achieved as in determining the range-phase term of the reference target namely by Fourier inversion of

$$G(p) = \frac{1}{kB} [\Gamma_m(\bar{p}) - C(p) /H(p) = e^{j(\frac{R_T+R_R}{2\cos\beta/2})p} \Gamma_m(\bar{p}) \tag{8}$$

to first obtain the space "spatial" impulse response $g(\xi-\xi_c)$ of the object for each aspect angle from which an estimate of the range (R_T+R_R) to the scattering centroid for that aspect can be obtained using eq. (6).

Because of the migration of scattering centroid with aspect angle, the above estimate of (R_T+R_R) requires "fine tuning". This is done by noting that the target echos for adjacent angular "looks" are caused essentially by the same scattering centers and therefore are highly correlated. A slight migration in the scattering centroid from look to look can thus be traced by cross-correlating the impulse responses of successive adjacent looks and using the displacement of the correlation peaks to correct the above range estimate. The results shown below were obtained using this *range tweeking* or *phase tweeking* method. With the range information determined, eq. (8) can now be used to extract the required Fourier space data $\Gamma_m(\bar{p})$. It can be shown that the above procedure for removing the range-phase term is equivalent to placing a fictitious coherent reference source in the target and using the signal at the receiver as a local oscillator for coherent detection of the scattered field.

The single slice of the Fourier space $\Gamma_m(\bar{p})$ of each of the two test objects was obtained using the procedures described above. Circularly polarized plane wave illumination was used and both the co-polarized and cross-polarized components of the scattered field were measured. The polar format in which the data is acquired was converted to a rectangular format suited for digital Fourier inversion through the use of a closest-four-neighbors interpolation scheme (Mersereau & Oppenheim 1978). The results obtained by applying the Fast Fourier transform algorithm to the interpolated data for both polarizations are shown in Fig. 2(a) and (b). It is evident that complementary information is contained in the two polarization images. Therefore image enhancement can be expected by adding the "orthogonally" polarized images as illustrated by the images in (c) of Fig. 2. Because man-made objects of interest in imaging radars are invariably symmetrical and their plane or planes of symmetry can be inferred from their heading, symmetrization of the polarization enhanced images about the vertical line of symmetry running through the fuselages was performed digitally leading to the polarization and symmetry enhanced images shown in (d). The projection images shown were actually magnified in the vertical direction by a factor $1/\cos\Theta = 1.155$ in order to obtain a properly scaled projection image of the scattering centers as they would be seen in a bottom view of the test objects. It is seen that features of the test objects used are delineated in the correct geometrical relation and relative size to enable recognition and classification of the scatterer. The image resolution achieved is of the order of 2 cm. The retrieved images are also seen to be nearly free of the speckle noise that plagues conventional coherent imaging systems.

CONCLUSIONS

The results presented here have important practical advantages in the imaging of remote scatterers with arrays of broadband monostatic coherent transmitter/receiver elements, as envisioned in future broadband radar imaging networks, since they indicate that maintenance of phase coherence between the sources at the various stations is not necessary during data acquisition. Phase coherence can be realized later in the data processing in the form of a synthetic target derived reference. This eliminates the need for reference signal distribution networks which are known to be costly and often impractical.

Finally it is worth mentioning that 3-D extension of the well known van Cittert-Zernike theorem can be used to show that tomographic or projective imaging of incoherent 3-D objects is feasible. Indeed projection imaging of a 3-D distribution of broad-band noise sources from cross-spectral power density measurements has recently been demonstrated (Farhat 1983).

Fig. 2. Examples of projection images.

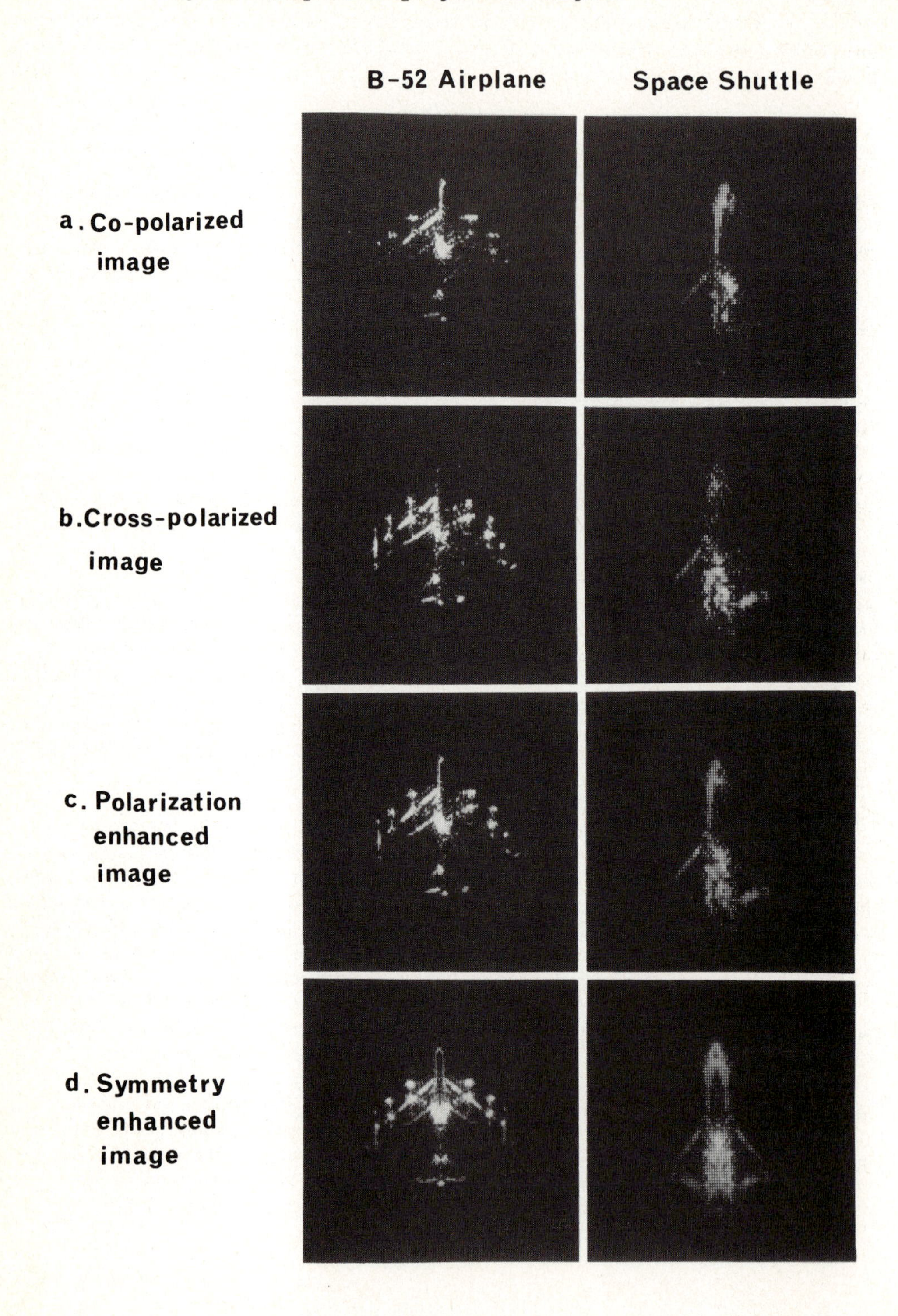

ACKNOWLEDGEMENT

This research was supported by the Air Force Office of Scientific Research, Air Force Systems Command under Grant No. AFOSR-81-0240B and by the Army Research Office under contract DAAG29-80-K-0024P02.

REFERENCES

Bojarski,N., (1967). "Three-Dimensional Short Pulse Inverse Scattering", Syracuse University Research Report, Syracuse, N.Y.

Bojarski,N., (1974). "Inverse Scattering", Naval Air Command Final Report N000 19-73-C0312F.

Boerner,W.M., Ho,C-M and Foo,B.Y., (1981). "Use of Radon's Projection Theory in Electromagnetic Inverse Scattering", IEEE Trans. on Ant. & Prop., AP-24, 360.

Chan,C.K. and Farhat, N.H., (1981). "Frequency Swept Tomographic Imaging of Three-Dimensional Perfectly Conducting Objects", IEEE Trans. on Ant. & Prop., AP-29, 312.

Dainty,J.C., (Ed.), (1975). *Laser Speckle*, Springer-Verlag Berlin.

Das,Y. and Boerner,W.M., (1978). "On Radar Shape Estimation Using Algorithms for Reconstruction From Projections", IEEE Trans. on Ant. and Prop., AP-26, 274.

Farhat,N.H. and Chan,C.K., (1978). "Three-Dimensional Imaging by Wave-Vector Diversity", 8th Int. Symp. on Acoust. Imaging, published in *Acoustical Imaging*, 8, A. Metherell (Ed.), Plenum Press, New York 1980, 499.

Farhat,N.H., Chan,C.K. and Chu,T.H., (1980). "A Target Derived Reference Technique for Frequency Diversity Imaging", Poster paper presented at the North American Radio Science Meeting, Quebec.

Farhat,N.H. and Werner,C.L., (1981). "An Automated Microwave Measurement Facility for Three Dimensional Tomographic Imaging by Wavelength Diversity", presented at the 1981 Int. IEEE AP-S Symp./National Radio Science Meeting, L.A.

Farhat, N.H., (1983). "High Resolution 3-D Tomographic Imaging by Wavelength and Polarization Diversity", Univ. of Pennsylvania report No. EO/MO 6 prepared under grant AFOSR-81-0240B., Phila., Pa.

Farhat,N.H., Chu,T.H. and Werner,C.L., (1983). "Tomographic and Projective Reconstruction of 3-D Image Detail in Inverse Scattering", Proc. 10th Int. Optical Computing Conference, Boston.

Lewis,R.M., (1969). "Physical Optics Inverse Diffraction", IEEE Trans. on Ant. and Prop., AP-17, 308.

Mersereau,R.M. and Oppenheim,A.V., (1978). "Digital Reconstruction of Multidimensional Siganls From Their Projections", Proc. IEEE, 62, 1319.

Raz,S.R., (1967). "On Scatterer Reconstruction from Far Field Data", IEEE Trans. on Ant. & Prop., AP-24, 66.

Ruck,G.T., (1970). *Radar Cross Section Handbook*, Plenum Press, New York, 50-54.

DISCUSSION

T.W. COLE

When these scanning radars go into use, will they include filters to avoid emissions in the radio astronomy bands?

N.H. FARHAT

Wide-band imaging radar networks based on the principles we describe will certainly have to avoid radio-astronomy bands and other prohibited bands. In fact broad-band data acquisition will also have to be confined to those bands where broad-band gear is available. The Fourier space is accessed then in nonconguent segments but we will still be able to form a projection or tomographic image using the techniques described.

2.4 VLBI: ONCE AND FUTURE SYSTEMS

Martin S. Ewing
Owens Valley Radio Observatory, California Institute of Technology, Pasadena, California, USA 91125

Abstract. Recording and correlating systems for VLBI have been improved dramatically since their beginnings in the late 1960s. Video cassette recorders now have the best operating economy, but two development efforts are aimed at transferring the high recording density of cassettes to large instrumentation recorders. The U.S. VLBA project is actively examining prospects for recording and correlating VLBI data from 10–19 telescopes.

1. *THE VLBI HARDWARE PROBLEM*

Systems for Very Long Baseline Interferometry must meet a number of special requirements beyond the needs of other radioastronomical interferometers. In this paper, the impact of the peculiar demands of VLBI observations on hardware systems are reviewed from the perspective of recording and correlation.

The critical technical problem in VLBI, distinguishing it from conventional connected-baseline radio interferometers, is the recording of large amounts of information at remote antennas. The recorded information must be conveyed to a central facility to be cross-correlated and filtered, with minimum information loss, into a data set of manageable volume. A signal detection process follows, in which the interferometer fringes are extracted. The fringe amplitude, frequency, and phase contain the essential data from the astronomical source.

Correlation and subsequent data processing for VLBI experiments are similar to the same functions for connected interferometers, but there are enough peculiar requirements to merit examination. In the main, however, this paper will analyze the hardware steps that lead to the production of visibility data, including software that is required for correlator control and fringe fitting.

Of course, VLBI comprises many technical disciplines in addition to data capture and correlation. Specific problems for VLBI include the need for providing independently stabilized coherent local oscillators, for precise time standards, for atmospheric calibrations, and for phase calibration across multiple observing bands. Other technical challenges lie in the "standard" radioastronomical areas of antenna and receiver design, radiometry, etc.

2. *HISTORY OF VLBI DATA SYSTEMS*

The history of VLBI—16 years—has been paced largely by developments in recording systems and correlator technology. Data recording capacity at each antenna and the number of baselines that can simultaneously be correlated at the data processing facilities have been the areas of greatest improvement.

In recent years, the later stages of data processing have also been recognized as being critical for the production of optimal visibility data. Consequently, new techniques more fully exploit the internal constraints of the correlator output—among delay variables, fringe rate and phase, for various baselines. These techniques, now gathered under the banner of "global fringe fitting," are iterative in nature; they are therefore subject to possible poor convergence, and they make heavy computing demands.

Nevertheless, hardware design is still very important to the success of VLBI systems. The achievement of the highest recorded bandwidth and the greatest frequency resolution on as many baselines as possible at the lowest contruction and operating cost are primary concerns for new VLBI arrays.

2.1 Recording Systems

There are two principal specifications for VLBI data recording systems: data recording rate and operating economy. Substantial improvements in both have occurred in the history of VLBI observations.

Mark I. Mark I was the first widely-used VLBI system. It employed 7-track computer tapes that recorded a 360 kHz band for only about three minutes per tape.

Mark II. Mark II is a second-generation recording system based on industrial and consumer video tape recording technology. The earliest form of Mark II utilized a 5 cm (2 in) helical-scan video recorder—the Ampex VR-660-C. For details of the early Mark I and Mark II systems, see Moran (1976).

The second generation recorder for Mark II was a 2.5 cm (1 in) reel-to-reel system, the IVC-825A; it was introduced in 1976–77. Tape alignment has been much better with this system, and reliability has also improved. The IVC system is still in use at Caltech's Jet Propulsion Laboratory (JPL). Mark II VLBI facilities have largely converted to the much more economical VHS cassette recorders which allow the 4 Mb/s data to be recorded for 4 hours on a cassette that costs only $10–14 (US).

Mark III. A significant improvement in sensitivity for VLBI has been achieved with the introduction (1979) of the Mark III system developed by Haystack Observatory in conjunction with the Goddard Space Flight Center of the U.S. National Aeronautics and Space Administration (NASA GSFC). Mark III employs Honeywell Model 96 instrumentation recorders with up to 28 longitudinal tracks recorded in parallel. At 340 cm/s (135 in/s) tape speed, 28 streams of 4 Mb/s data may be recorded at once on tape of 2.5 cm (1 in) width, giving an increase of $\sqrt{28}$ times in sensitivity compared with Mk II. Unfortunately, a single tape, costing about $200 (US), lasts only about 13 min at this speed.

The Mark III recorder is capable of operation at 8 Mb/s (4 MHz IF passband) per track when the unit is run at 680 cm/s. This capability is desirable in cases in which highest possible sensitivity must be obtained in a short integration time, such as at millimeter wavelengths. The Mark III IF systems are currently being upgraded to support this recording mode.

The Mark III record head system is being developed with the objective of realizing the tape packing densities of video cassette recorders on the instrumentation recorder. The proposed modified system has been termed "Mark III–A." (See discussion in §3.1.)

Other Systems. Several other systems have been used for VLBI observations. A series of *analog* recording systems was operated from 1967 to 1982 at the Canadian National Research Council. The IF signal (up to 4 MHz nominal bandwidth) was formatted by the insertion of standard television synchronizing pulses for recording on nearly unmodified video recorders (Moran, 1976). Both IVC (2.5 cm) and VHS recorders were used. Analog recording schemes, although they offer high bandwidth, have not been popular for VLBI because of the difficulty of adequate gain and bandpass calibration.

A new Canadian VLBI system—C2—is under development by the Earth Physics Branch of the Department of Energy and Mines. C2 is based on a 12 Mb/s digital recording system (see §3.1), with two video cassette recorders available at each telescope. A 16× burst mode allows a wide instantaneous bandwidth (up to 96 MHz per recorder) to be recorded. Intelligent controllers at each telescope will perform the delay and local oscillator offsets, simplifying the correlator requirements.

In Japan, the Kashima Branch, Radio Research Laboratories, Ministry of Posts and Telecommunications is developing a Mark III–compatible VLBI system—the K–3. The system is composed of both a recording and correlation system; it is described by Kawaguchi, *et al.* (1982).

Table 1 summarizes the parameters of the major recording systems discussed above along with the proposed U.S. VLBA system.

2.2 Correlators

Correlators for VLBI have evolved along with the recorder systems. The first Mark I VLBI experiments were correlated in general-purpose computers. At the NRAO, an IBM 360/50 computer required about 15 minutes to correlate a minute of Mark I data on a single baseline. It was immediately clear that dedicated hardware correlators would be required to keep up with the demand for VLBI observing.

Mark I Correlator. Haystack Observatory installed the first digital correlator for VLBI in 1973. It operated as a peripheral processor attached to a CDC 3300 general-purpose computer. The Mark I and early Mark II correlators are discussed further by Moran (1976).

Mark II Correlators. At the same time, the NRAO was developing the Mark II correlator. Over the years it has evolved into a system giving 256 lag channels on each of 3 baselines with special support for spectral line observations. A copy of the NRAO Mark II correlator has been installed at the Max-Planck-Institut für Radioastronomie (MPIfR) at Bonn.

In 1976, Caltech's Owens Valley Radio Observatory (OVRO) and JPL jointly began operating a Mark II correlator. The capabilities of this system were similar to the NRAO Mark II, except that the design permitted expansion to 5 recorders. The CIT-JPL Mark II system was expanded in stages, with 5-station operation achieved in 1979. This correlator, which is sometimes known as "Block 0," has specialized in source structure mapping and in geodetic baseline measurement, but has not supported spectral line observations (Thomas, 1981). Both IVC and VHS recorders continue to be supported at Caltech.

Mark III Correlators. Along with the Mark III VLBI recording system, Haystack Observatory and NASA GSFC developed a Mark III correlator system. This system, like the Mark II correlators, operates at a maximum clock rate of 4 Mb/s; however, it processes up to 28 data streams in parallel for each baseline. Its most distinguishing feature is its modular design in which all functions for a single baseline and IF data track are integrated in a single plug-in assembly. With a particular number of correlator modules available, it is then possible to trade off bandwidth (number of frequency channels) against baselines in a convenient and economical manner (Rogers, *et al.*, 1983). The Haystack system currently handles up to 4 telescopes in a single tape pass. A 3–station correlator based on the Haystack design has been built for MPIfR, Bonn.

Both the Haystack Mark III and the Mark II correlators were constructed with Medium Scale Integration Integrated Circuits (MSI ICs). With this technology, about 9 ICs were required to implement a single complex correlator channel—delay, multiplication, and accumulation.

Table 1. Summary of VLBI Recording Systems

VLBI System	Tape Min.	Bandwidth MHz	Economy Mbit/$
Mark I	3	0.3	9
Mark II IVC	60	2.0	360
Mark II VHS	240	2.0	4,800
Mark III Standard	13	56.0	500
Mark III Double	7	112.0	500
Mark III-A	84	112.0	6,000
C2	240	6.0	14,400
VLBA Standard	-	64.0	6,000
VLBA Double	-	128.0	6,000
VLBA Peak	-	256.0	6,000

As discussed above, the Kashima group in Japan is developing a Mark III–compatible correlator for the K–3 system. It is similar, but not identical, to the Haystack system.

JPL and OVRO are now developing a "Block II" VLBI correlator (Ewing, 1983). Based on Programmed Array Logic (PAL) devices and 80-bit counter ICs (AMD 9513), this is the first correlator to rely heavily on Large Scale Integration (LSI)—up to a few thousand gates per IC. With the LSI/PAL technique, a single complex correlator channel is reduced to about one equivalent IC.

The primary input for Block II is Mark III tape; the system is fully compatible with Mark III data recording. However, a secondary mode supports Mark II (4 Mb/s) input from VHS cassette or IVC recorders. In Mark III mode, up to 4 stations can be processed at once. In Mark II mode, because of the large number of correlator channels available, up to 11 stations can be correlated. Thus, Block II will be capable of very rapid processing of even the largest Mark II experiments.

The C2 Correlator. A 3–station, 2–channel correlator for the Canadian C2 VLBI system is under construction. Its task is simplified because the delay and fringe rate corrections are largely performed before the telescope data is recorded. First operation of this system, intended mainly for geophysical investigations, is expected in 1983.

VLBI correlators that are currently operating and those that are projected for the near future are summarized in Table 2. Where the maximum number of tapes is indicated by a range, the larger value may be obtained with a reduced number of lags per baseline. The total number of lags indicates the number of complex multiplier–accumulators available in the system. The number of tracks usually indicates the number of independent recording heads in the system, but in the case of the VLBA system, it indicates the (maximum) number of recordable IF frequency bands. "Quantization level" refers to the number of distinguishable IF states that are sampled and correlated. One-bit sampling corresponds to 2 levels, up to 4 levels per sample can be encoded into 2-bit samples.

3. CURRENT DEVELOPMENTS

VLBI recording and correlating technology is being developed in several promising directions. Higher recording data rates, especially with the instrumentation tape approach, require higher recording densities. In the United States, a group consisting of the NRAO along with universities including Caltech and MIT–Haystack is actively pursuing a dedicated array—the Very Long Baseline Array (VLBA)—which will comprise 10 new or upgraded telescopes plus as many other telescopes as possible. Canadian astronomers are proposing an 8–station Canadian Long Baseline Array—the CLBA. At the same time, the "traditional" *ad hoc* arrays in North America and Europe plus the Australia Telescope call for a new generation of large correlator systems.

3.1 Recording Technology

Commercial development of magnetic tape recording for consumer television has led to very high recording densities. The current Mark II system makes good use of the VHS cassette system, achieving about 4.3 Mb/cm^2.

However, there is some uncertainty about how best to apply the technology to wideband recording systems. Arrays of consumer-type recorders are one possibility, but upgraded instrumentation recorders may offer convenience and higher reliability. This section presents some of the developments which are underway for VLBI data recording.

Mark III-A System. The Mark III VLBI system as outlined above has successfully demonstrated the feasibility of VLBI data recording at 112 Mb/s and higher. Its greatest shortcoming has been its tape consumption rate. In the 14– to 20–antenna arrays now being discussed, including elements of the U.S. VLBA and Very Large Array (VLA), the Canadian CLBA, and European and Japanese stations, a tape inventory costing $9 million would be needed.

Haystack Observatory (H. Hinteregger and J. Webber) and Caltech-JPL (B. Rayhrer) are pursuing two related, but somewhat different approaches to the Mark III density upgrade problem. Both are based on the very high accuracy ($\pm 0.2\ \mu$m) with which the Honeywell Model 96 transport has been shown to control lateral tape motion.

The standard Mark III writes 28 tracks, each 640 μm wide. The objective at Haystack and JPL is to build "head stacks" composed of 28 much narrower heads. After a complete tape pass, the entire head stack is shifted by a precise offset and the tape is restarted in the reverse direction. With the 40 μm track width proposed by Haystack, a 2800 m (9200 ft) tape can be made to last for nearly 3 hours at 56 MHz bandwidth—a factor of 12 increase in recording density.

Table 2. Current and Projected VLBI Correlators

Correlator System	Tapes	Total Complex Lags	Tracks	Track Rate Mb/s	Quant. Level
(operational)					
Mark II—NRAO*	3	768	1	4	2
Mark II—Caltech	5	160	1	4	2
Mark III—Haystack*	3–4	672	28	4	2
(in construction)					
K–3—Kashima	2	256	32†	8	2
C2—Canada	3	96	2	12	2
Block II (Mark III tape)	4	1344	28	8	2
Block II (Mark II tape)	11	1344	1	4	2
(planned)					
VLBA	10–19	47104	32	16	4

**Duplicated at MPIfR, Bonn* *†28 Channels for Mark III Data*

This track width is similar to that in common use with the VHS and Beta VCRs. The performance requirement for good VLBI use is a fractional error rate of no more than 10^{-4}; this corresponds to a "slot" signal–to–noise ratio of about 53 dB or better. Preliminary head samples have encouraged the belief that this performance is attainable.

The fabrication approach favored by the Haystack group is to begin with production-type VHS "gapped bars," which are laminated ferrite–glass–ferrite bars about 23 mm in length. The width of the glass gap is about 0.3 μm. With precision grinding methods the bars will be processed into accurately aligned stacks of 28 heads.

An independent effort at JPL has followed a somewhat different approach to the Mark III density upgrade. This project seeks a factor of 20 in tape density. Unassembled VHS (or Beta) ferrite heads are to be obtained from the VCR production lines. These very inexpensive heads will be wound and mounted in a precisely dimensioned fixture.

Video Cassette Systems. As shown in Table 1, the VCRs now in use in Mark II VLBI already offer very high recording economy. Their principal drawback for wideband VLBI work is that they carry only a single 4 Mb/s channel of information. However, the VCRs are reliable, widely available and inexpensive—as low as $400 in the U.S. market. Considerable engineering effort is being invested, both in industry and in VLBI laboratories to develop even higher performance that may benefit VLBI systems.

P. Newby and J. L. Yen have evaluated advanced techniques for utilizing existing VCRs for VLBI recording. They have pursued NRZI recording with partial-response decoding and adaptive equalization to achieve good error performance—in the usual VLBI sense: $\leq 10^{-3}$ or 10^{-4} (Newby and Yen, 1982). The equivalent linear bit spacing is about 0.51 μm/bit (50,000 bits/in), considerably better than the current Mark III practice—about 0.77 μm/bit (33,000 bits/in). A single VCR can record 12 Mb/s with this technique; rates of up to 16 Mb/s or beyond appear feasible.

Industry is very actively pursuing new markets for VCRs and similar products. VHS and Beta systems have already been marketed for "digital stereo" recording; they are also being adapted for digital television. One firm (Hitachi) was recently reported to have built a prototype VCR, based on a VHS mechanism, that records 8-bit samples at a 10.7 MHz rate. With a metal tape, the recorded bit spacing is reduced to 0.36 μm/bit.

3.2 The U.S. VLBA Recording System

The most ambitious current proposal for a terrestrial VLBI system is the VLBA project in the United States. The VLBA project plan calls for 10 new 25-m antennas to be built on U.S. territory from Hawaii to Massachusetts to Puerto Rico. Each telescope will be instrumented for observations in 10 bands from 325 MHz to 43 GHz. Instantaneous observing bandwidth is about 64 MHz in "normal" mode, with operation possible at 128 MHz or higher bandwidths in high-speed modes. The remote antennas are to be staffed at the lowest feasible levels, requiring unattended data recording for up to 24 hr.

To accomodate bandwidth synthesis and spectral line observations, VLBA recording may be divided into as many as 32 separate channels, produced by 16 IF converters, each providing an upper- and lower-sideband output. Polarization analysis is possible if channels are paired with both right and left polarizations recorded for each frequency band. For the most easily calibrated continuum experiments, a few, higher bandwidth channels are preferred.

Eleven IF channel bandwidths are proposed, ranging from 125 kHz to 32 MHz. Two-, three-, or four-level sampling is provided up to 24 MHz, while only two levels are available at 32 MHz. Oversampling by a factor of 2 or 4 is permitted, as long as the bit rate per channel does not exceed 64 Mb/s. In principle, the IF can present $\sim$ 1GHz to the recording system. The high bit rate (2 Gb/s) required to record such a wide band is achievable with the proposed VLBA technology. Tape consumption would be very high, operators would have to be continuously present, and the correlator would be very expensive. For such practical reasons, the average VLBA recording rate is to be limited to about 128 Mb/s. Double-rate, and possibly quadruple-rate recording will be available under special circumstances.

The actual digital interface is 32 physical "channels," each operating at a 16 MHz clock rate. It appears possible to develop a single instrumentation recorder that will record up to 512 Mb/s over 32 tracks. A single 3,400 m (11,000 ft) tape might last for 200–300 min, depending on the bit density actually achieved.

In its initial operation, the VLBA is expected to operate at a much lower average recording rate, 100–128 Mb/s, because of economic limitations. The potential remains, however, for increasing sensitivity by a factor of two for special observations.

Both instrumentation tape and VCR approaches are being explored for the VLBA. A VCR system appears to be less expensive and requires less engineering development. It may depend, however, on the availability of an automated tape-changing system. One drawback of the VCR is that it operates at only one head-to-tape speed, so that the aggregate recorded data rate can be controlled only by changing the number of active recorders. The instrumentation recorder, by contrast, can record at any rate up to its maximum.

3.3 The VLBA Correlator System

Although 10 new elements are planned for the VLBA, it is clear that the new antennas will need to be linked with additional antennas in in the U.S. and other countries. The VLBA correlator is to provide full performance for 10 antennas, but significant capability to handle arrays as large as 19.

A 19-antenna array contains about half the number of baselines of NRAO's 27-element VLA. However, the VLBA must be operated with shorter integration times to give reasonable fields of view. The Array will also be used for wavelengths as short as 7 mm, as against the VLA limit of 1.2 cm. In all, the VLBA correlator data rates are comparable to the VLA.

VLBI correlation has several characteristics not encountered in most connected interferometers. Fringe and delay rates are high; data are usually Nyquist-sampled with only 1- or 2-bit resolution; and multiple passbands (especially in geodetic experiments) must be combined coherently in the "bandwidth synthesis" process. Furthermore, the IF data are recorded, so there is an opportunity to re-correlate data to gain more frequency resolution or a wider field of view.

Data Flow. One useful way to understand the VLBA correlator problem is through the quantity of data at each processing step. Burns and Ewing (1983) give a step-by-step analysis, but some of the highlights will be mentioned here. The tape transports, 10 to 19 in number, each produce 128 Mb/s. As a concrete example, 10 recorders will be assumed to be playing back a typical continuum source observation.

For each baseline, in the proposed design, there are 512 complex correlation channels (lags), operating at a clock rate of 16 MHz. The overall data rate at the multiplication stage is enormous—nearly 6 Tb/s—but is immediately reduced through accumulation. Each accumulator is read out at a 40 Hz rate, corresponding to a total data rate of 90 Mb/s.

Two successive stages of processing are planned for this data. First, a "firmware-driven" microprocessor will further accumulate and filter the correlations. This processor, whose operations are largely independent of the details of an experiment's geometry, can correct for local oscillator offsets (for spectral line data), transform from lag to frequency domain, and correct for delay errors (the "fractional bit shift"). This processor can accumulate data for a time limited only by the rate of change in fringe rate, typically dumping at a 5 Hz rate (total data rate of 6 Mb/s).

The second stage of data handling occurs in the "fringe processor," which is a general-purpose microprocessor with a large memory space and an FFT peripheral. The fringe processor will combine separate IF channels coherently, transform back to a "synthesized" delay function, and convert sequences of correlation samples into fringe rate spectra, delay channel by delay channel. (The net calculation is a two-dimensional FFT.)

Only data selected by "windows" in delay and fringe rate are output for further processing. Window parameters are determined through *a priori* experiment information and also through global fringe fit information derived from calibration observations. The overall correlator output can be reduced to about 30 kb/s, with data transferred at a 0.1 Hz rate. Such a data flow is acceptable to the mid-size computer that might be used to control the correlator and store results in a disk or tape data base.

Integrated Circuit Technology. The VLBA project is investigating several approaches to VLSI technology for a large correlator. One would simply be to use the Block II technique: to use Programmed Array Logic and high-density counters, which can effectively achieve about one complex correlator channel per IC.

A 3–5 μm n-MOS IC design technology is now becoming widely available to digital designers. J. Peterson and others at JPL are designing a full-custom 16-channel correlator with 32 20-bit accumulators (Ewing, 1983). This chip is expected to operate at a 16 MHz clock rate; it could also form the basis of a VLBA correlator. There are uncertainties associated with the production of such a chip through a "silicon foundry," since the yield, and therefore the end price, can not be accurately foretold. JPL and Caltech hope to make a trial production run to gain further experience.

Other IC technologies are available with intermediate densities and design costs. A "standard cell" approach is available from several sources in which the designer chooses standard circuits (gates, registers, etc.) from a library and combines them on silicon. "Gate arrays" are ICs with standard arrays of simple logic functions, needing only user-specified metalization layers to interconnect them.

Estimates for the cost of an entire correlator development project do not appear to depend critically on the particular digital technology that is chosen. All the approaches listed above seem to lie within about 30% of one another in terms of overall cost for the VLBA system. The full-custom approach may be favored, however, because of its potential for later expansion at low cost. Only about 0.1 IC is required to implement a complex correlator channel with the full-custom design.

4. *CONCLUSIONS AND PROSPECTS*

Digital recording and correlation are undergoing rapid development for current and future generations of VLBI systems. In the near future VLBI systems with bandwidths equal to or greater than those in use now at the most powerful connected interferometers will be practical.

Currently, the video cassette technology is the most economical means of storing bulk VLBI data. Over the next 5 years, as this technology is fully implemented on multi-track transports, enormous data rates—over 1 Gb/s—will be possible for VLBI. "Vertical" recording (H field normal to the tape) and metal tapes will increase the recording density by perhaps a factor of two.

Optical recording technologies are also being developed. They may become available soon, with read/write capability, for computer use. Optical recording has the potential of much higher density than magnetic. It is not likely, however, that optical systems will soon reach the favorable cost/performance ratio of current VCRs for VLBI applications.

The correlator for the VLBA will need about as many channels as the VLA correlator. New advances in IC and microprocessor technology allow the very large data flow to be handled economically with custom correlators and distributed VLSI microprocessors.

There are no apparent technological barriers to increasing VLBI recording bandwidths to 500 MHz or beyond. With the availability of low-cost VLSI correlator ICs, an economical correlator should be able to process the increased bandwidths.

Acknowledgements

Helpful discussions with Alan Rogers, John Benson, and the VLBA Correlator Design Group are gratefully acknowledged. This work was partially supported by U.S. National Science Foundation grant 82-10259.

References

Burns, W. R., and Ewing, M. S. (1983). Radio Astronomical Synthesis Arrays—Real Time Processing Needs. *In* International Society for Optical Engineering (SPIE) 27th International Technical Symposium, Proceedings. (In press.)

Ewing, M. S. (1983). The JPL–Caltech Block II Very Long Baseline Interferometry Processor. *In* Techniques d'Interférométrie à trés grande Base, Proceedings, pp. 293–301. Toulouse: Cepadues – Éditions.

Kawaguchi, N., Sugimoto, Y., Kuroiwa, H., Kondo, T., Hama, S., Amagai, J., Morikawa, T., and Imae, M. (1982). The K-3 Hardware System Being Developed in Japan and its Capability. *In* Proceedings of IAG Symposium No. 5, Tokyo.

Moran, J. M. (1976). Very Long Baseline Interferometer Systems. *In* Methods of Experimental Physics, v. 12, part C, pp 174–197. New York: Academic Press.

Newby, P., and Yen, J. L. (1983). IEEE Trans. Magnetics, in press.

Rogers, A. E. E., Capallo, R. J., Hinteregger, H. F., Levine, J. I., Nesman, E. F., Webber, J. C., Whitney, A. R., Clark, T. A., Ma, C., Ryan, J., Corey, B. E., Counselman, C. C., Herring, T. A., Shapiro, I. I., Knight, C. A., Shaffer, D. B., Vandenberg, N. R., Lacasse, R., Mauzy, R., Rayhrer, B., Schupler, B. R., and Pigg, J. C., (1983). Science, *219*, p 51–54.

Thomas, J. B. (1981). An Analysis of Radio Interferometry with the Block 0 System. JPL Publication 81–49, December 15. California Institute of Technology Jet Propulsion Laboratory: Pasadena.

DISCUSSION

P.T. RAYNER

What language is being used to program the fringe processor?

M.S. EWING

We are currently developing software in PASCAL and C for the 68000 using cross-compilers running on a VAX.

2.5 VLBI ARRAY DESIGN

R. C. Walker
National Radio Astronomy Observatory*
Charlottesville, Virginia 22901 U.S.A.

INTRODUCTION

Current Very Long Baseline Interferometer (VLBI) experiments use arrays of radio telescopes whose locations were chosen for reasons totally unrelated to their potential use as interferometer elements. The telescopes are distributed sufficiently well that maps of compact radio sources can be made, but the coverage of the sampling (u-v) plane is far from optimal. Dedicated array projects that have been proposed in the United States (VLBA) and Canada (CLBA) provide the first real opportunity to optimize the distribution of VLBI antennas for performance. The configuration searches for those projects are now nearly complete. The discussion below is based on experience gained during those searches, especially the search for the VLBA configuration.

The process of finding the configuration for an array has several basic steps. First the constraints under which the search will be made must be specified. Then methods of testing and comparing possible configurations must be selected. Finally, many possible configurations must be examined. The process can be simplified if any of the systematic antenna distributions that have been studied can be used, but this may not be allowed by the geography or the desired coverage.

CONSTRAINTS

The most important step in selecting the configuration for an interferometric array is to specify the constraints on the performance of the instrument and on the geographic locations of the elements. The general categories of constraints, with specific reference to their implication for the VLBA and the CLBA, are:

Resolution

VLBI arrays will generally be designed to have the maximum possible resolution which means the maximum possible length for the longest baseline. The ultimate limit is set by the diameter of the Earth and by the requirement that telescopes at opposite ends of the longest baseline be able to see the same source at the same time. The

---

*The National Radio Astronomy Observatory is operated by Associated Universities, Inc., under contract to the National Science Foundation.

limit is reduced somewhat if the mutual visibility must be long enough to provide more than just a point in the u-v plane. Baselines of 8000 to 10000 km are about the longest that will provide useful coverage of anything but polar sources. The maximum baseline may be further reduced by geographic limitations.

Geographic Region

The geographic boundary conditions under which the configuration search is to be made must be specified. For large countries such as the United States and Canada, a reasonable choice is that the array will be confined to national territory and this is the choice made for both the VLBA and the CLBA. This avoids the problems inherent in an international project and in operations in other countries. It also allows all of the money for the array to be spent in the country building it, an important consideration in some cases. The restriction to national territory, combined with the isolation of northern Canada, has forced the choice of a one dimensional configuration with a maximum baseline of about 5000 km for the CLBA. On the other hand, restriction of the VLBA to U.S. territory still allows reasonable north-south baselines, so a two dimensional configuration that allows observations of low declination sources was specified. Using Hawaii, the maximum baseline of the VLBA is about 8000 km.

Additional geographic constraints are imposed by weather conditions and operations. The currently planned arrays will operate at frequencies as high as 22 or even 43 GHz so it is important to have high, dry sites where possible. For this reason, sites in the southwestern U.S. were favored for the VLBA. The sites must also be easily accessible because tapes must be shipped to the headquarters regularly and major emergency repairs will be done by personnel from the headquarters who fly to the sites when there is a problem.

Range of Baselines

The largest structures that can be mapped with an interferometer are set by the shortest spacing. The impressive results obtained with MERLIN and, at high frequencies, with the VLA over the last few years show that there is much interesting structure at scales (roughly 0.01 to 0.5 arc seconds) that are small for most linked interferometers but large by VLBI standards. A wide range of spacings would allow an array to be used to study some of those structures. A wide range of spacings is also important because it is very difficult to form an accurate image of a source with information on only a small range of spacings if there is significant flux density in structures that are partially resolved by the shortest spacings. However, with a fixed number of telescopes, the range of spacings can only be increased at the expense of the density of coverage of the u-v plane.

Image Quality

One of the most important goals of a VLB array is to produce high quality images of compact radio sources. To do this requires dense coverage of the u-v plane which in turn requires a large number of telescopes. The nature of the u-v plane coverage is an important parameter that influences the quality of the maps and for which the choice is not obvious. Uniform coverage, such as is obtained by most of the linked arrays, makes the fewest assumptions about the nature of the structures to be mapped. However it also limits the range of spacings possible with a given number of telescopes. If the sources to be mapped have relatively simple structure at any given resolution, as seems to be the case with the very high brightness sources studied with VLBI, it is worth sacrificing some uniformity in favor of a wider range of scales that can be mapped. However this should not be carried to extremes. The large holes in the u-v plane that will be required if a very wide range of spacings is to be obtained with a small number of telescopes will seriously degrade the quality of the images that can be made.

Both the VLBA and CLBA projects have chosen to emphasize the short spacings by providing centrally condensed coverage of the u-v plane. The desired coverage is approximately uniform on an logarithmic scale. This provides scale independent coverage within the range of spacings covered, thus making the fewest assumptions about the scales of structures to be mapped. It is also a useful type of coverage to have when trying to obtain maps at different frequencies with the same resolution in order to study spectral indices.

Number of Telescopes

The quality of the maps that can be made with an interferometer increases rapidly with the number of telescopes. The number of baselines, which determines the amount of data available, increases with the square of the number of telescopes. The adaptive calibration techniques (alias self cal, hybrid mapping, etc.) discussed elsewhere in these proceedings rely on a high ratio of baselines to stations, so they work better with a large number of telescopes. The number of telescopes is also an important factor in setting the range of spacings that can be covered. However, the telescopes are by far the most expensive part of an array project so the number must be kept low enough to keep the budget within acceptable bounds. Note that the size (sensitivity) of the each telescope is the other important factor in setting the cost of the array so a compromise must be reached between the sensitivity of each baseline and the number of baselines. The operating costs are also a strong function of the number of telescopes because of the need for personnel at each site. A balance between costs and performance must be found. In practice, perceptions of what constitutes an acceptable budget are a major factor in setting the number of telescopes and, therefore, the performance of the array.

The number of telescopes planned for the VLBA is 10, each 25 m in diameter. These telescopes are to be distributed in two dimensions so that scaled coverage, as described above, is provided for baselines between 200 and 8000 km for sources from -30 to +90 degrees declination. Baselines occur at approximately 10 to 20% increments in u-v space. If the number of telescopes were changed, the density of coverage would probably be kept about constant but the minimum spacing would be adjusted to allow that density.

The number of primary telescopes specified for the CLBA is 8, each 32 m in diameter. The telescopes are placed along an east-west line just north of the U.S. border. There is an additional, smaller telescope in the North to optimize the array for geodesy. The configuration provides scaled coverage, also at about 10% increments, of baselines between 70 and 5000 km. The CLBA obtains a wider range of spacings with fewer telescopes than the VLBA by sacrificing the ability to observe low declination sources.

Interaction with Existing Instruments

Dedicated array projects consisting entirely of new telescopes provide the opportunity to use configurations that are not influenced by the locations of existing facilities. However it is still worth considering the interaction with especially powerful existing instruments. This turns out not to be a major consideration for the CLBA in Canada but, for the VLBA in the United States, there are strong reasons to take into account interaction with the VLA. Baselines to the entire VLA, when used as a phased array, are about five times more sensitive than regular VLBA baselines so it is desirable to provide a wide range of spacings between the VLA and the VLBA elements. Also, if the VLBA configuration is properly designed, a few of the missing spacings between the maximum spacing of the VLA (35 km) and the minimum spacing of the VLBA (200 km) can be filled. With an appropriate configuration, it might be possible to add telescopes later to fill the missing spacings properly. With the planned VLBA configuration, that capability can be provided with three additional telescopes in New Mexico.

Local Factors

Once the basic configuration of an array is chosen, there is still freedom to choose the actual site of each telescope. The extent of that freedom depends on the shortest baselines to the individual site, but ranges from about 10 km to several hundred kilometers. The specific sites need to be chosen to have good access, reasonable interference environments, low horizons, high altitude if possible, and good local support. It is this last factor that provides a bias toward existing astronomy facilities. A significant factor leading to the choice of the proposed VLBA configuration over other configurations of similar performance was that 8 of the 10 sites are at, or are easily supported from, existing radio astronomy facilities.

TESTING ARRAYS

Once the above constraints are specified, the basic characteristics of a configuration are determined. However before proceeding to test and select a specific configuration, it is necessary to decide upon the criteria by which configurations will be compared. This gets into a difficult area for which no good, widely accepted solution has been found. Fortunately many testing schemes give results that are qualitatively similar so the choice of a testing scheme is not extremely important. This fact is probably not accidental because any testing scheme must provide ratings that agree with visual inspection of the u-v coverage or it will be discarded.

An unfortunate aspect of all of this is that there is some ambiguity in what constitutes 'good' coverage. For example, the relative effects of a moderately large hole (say 20% of the u-v distance over half a radian) as opposed to a number of smaller holes may depend on source structure and are not well known in the presence of modern image restoration techniques such as CLEAN and Maximum Entropy. Also the desired distribution of baselines (e.g., uniform vs logarithmic) is a function of preconceived notions of what sources look like. Even if the desired distribution of spacings is decided upon, there are ambiguities such as how to treat the longest spacings where the uniformity generally breaks down or how, in the logarithmic case, to deal with the shortest spacings which cannot be logarithmic for lack of sufficient telescopes. These concerns may seem to be trivial details but the differences between reasonably good configurations are small and such details can determine how they are ranked.

Computer simulations of actual observations with the array might be considered the fairest method of determining array performance because all of the above concerns can be answered empirically. But what do you use for a test source? Many of the above direct concerns about the details of the u-v coverage translate into differences in the types of sources that can be mapped well. Simulations simply disguise the biases built in rather than remove them. The simulations also test the mapping algorithms that are in use today. How do the results relate to algorithms that might be used in the future?

Several testing methods were proposed both before and during the VLBA and CLBA configuration searches:

Dynamic Range

Simulated observations of a test source are used to make maps that are compared with the input source (Cohen 1980; Kellermann 1981; Linfield 1982; Can. Ast. Soc. 1982). The method directly tests the mapping capabilities of the array using current algorithms but is sensitive to the nature of the test source and, perhaps, to the mapping algorithms.

Distance Between Grid Points and Sampled Points

This quality measure is based on measuring the distance from each point on a uniform grid in the u-v plane to the nearest point sampled by the array (Mutel & Gaume 1982). An inverse radial weighting is applied to the points to emphasize the coverage on short spacings and the analysis is performed for a wide range of declinations. The method tests the uniformity of the coverage but is sensitive to edge effects and to the choice of the grid.

Match of Density of Points to Desired Density

This method is an analog of a statistical test known as the Cramer-von Mises test (Schwab 1982). The test measures the discrepancy between the cumulative distribution function of the sampled points in the u-v plane and the desired distribution function. The desired distribution function (e.g., uniform, logarithmic, etc.) is chosen based on the image quality constraint discussed above and may not be well determined.

Number of Sampled cells in a Polar, Logarithmic Grid

This method counts the number of sampled u-v cells in a polar grid in which each cell has a radial width of some fixed percentage of the u-v distance (Walker 1982). The count is made for several declinations to produce an overall measure. It relies on the concept that, since all configurations give about the same total number of samples, an array with big holes will have more redundancy elsewhere and will receive a lower rating. The polar, logarithmic grid was chosen to favor the logarithmic coverage discussed earlier. Counts from a second grid that covers only the central region of the u-v plane and uses even radial spacings were included to to avoid problems in the inner regions where the cells in the main grid are small. The results from the method are sensitive to the choice of the sizes of the grid cells and to edge effects.

The testing method trusted most by persons working on configurations is probably visual examination of the u-v coverage. The other methods all give sufficiently similar results that the one that is easiest to use can be selected. The cell counting method uses very simple calculations and was coded to maximize computation speed when testing large numbers of related configurations. It runs about 100 times faster than the other methods so it became the method of choice in the VLBA configuration search.

REGULAR CONFIGURATIONS

Once the constraints have been specified and the testing tools have been chosen, the search can begin. It would be useful at this point to have some guidance from special geometric patterns that are known to provide good coverage. The general problem of an optimal two-dimensional configuration, even without geographic constraints, has not been solved, especially since the definition of 'optimal' is ambiguous. Special cases, such as providing uniform coverage with a reasonable layout for electronic links and tracks for moving telescopes have been addressed. One dimensional configurations are significantly simpler and a variety of optimal choices exist in the literature depending on the desired coverage.

Possible one dimensional configurations include: even spacing, such as is used by the Westerbork Synthesis Radio Telescope; minimum redundancy arrays, that provide a maximum number of different, regularly spaced baselines with a given number of telescopes (Moffet 1968; Ishiguro 1980); and optimal closure phase arrays, that provide just enough redundancy to bootstrap phases and leave only one unknown antenna phase gain (Morita & Ishiguro 1982). Two dimensional configurations include: instantaneous, non-redundant arrays, that are analogous to minimum redundancy arrays but do not utilize Earth rotation (Klemperer 1974); circles, such as the Culgoora Radioheliograph; Wyes, such as the VLA; and Tees.

It happens that none of the systematic configurations were used for either the VLBA or the CLBA because they all provide uniformly spaced coverage, redundancy, or are incompatible with the available geography. Both projects favored an emphasis on short spacings such as could be provided by the logrithmic coverage described earlier. The advantage of reducing the number of unknown phases led the Canadians to specify an array of the optimal closure phase type in the original CLBA concept (Can. Ast. Soc. 1979). However, the adaptive calibration techniques that have gained acceptance since then do not require the redundancy. In recognition of this development, the CLBA, like the VLBA, will use a non-redundant configuration in order to provide the maximum number of different baselines and, therefore, the maximum amount of information on source structure.

THE CONFIGURATIONS OF THE CLBA AND VLBA

A possible configuration for the CLBA (the final configuration has not been chosen) is shown in Figure 1 and the u-v coverage provided for two declinations is shown in Figure 2. The geographic constraints in Canada forced the choice of a linear configuration of about 5000 km extent. Because of the inaccessibility of some of the relevant parts of eastern Canada, the main concentration of telescopes must go in the western provinces. Beyond these factors, the geographic constraints did not tightly restrict the configuration. Therefore, the possible configurations under study provide about as close an approximation to the desired coverage as possible. Note that the coverage provided by any array that is centrally condensed will not be ideal because it will have groups of closely spaced, long baselines between the farthest out antennas and the close group of antennas that provides the short spacings.

The geographic constraints do tightly restrict the possible configurations of the VLBA. They preclude most of the possible systematic geometries (there is a lot of ocean between Hawaii and the

Figure 1. The dots mark the locations of the antennas of the U.S. Very Long Baseline Array (VLBA). The stars mark the antenna locations for a possible configuration of the Canadian Long Baseline Array (CLBA).

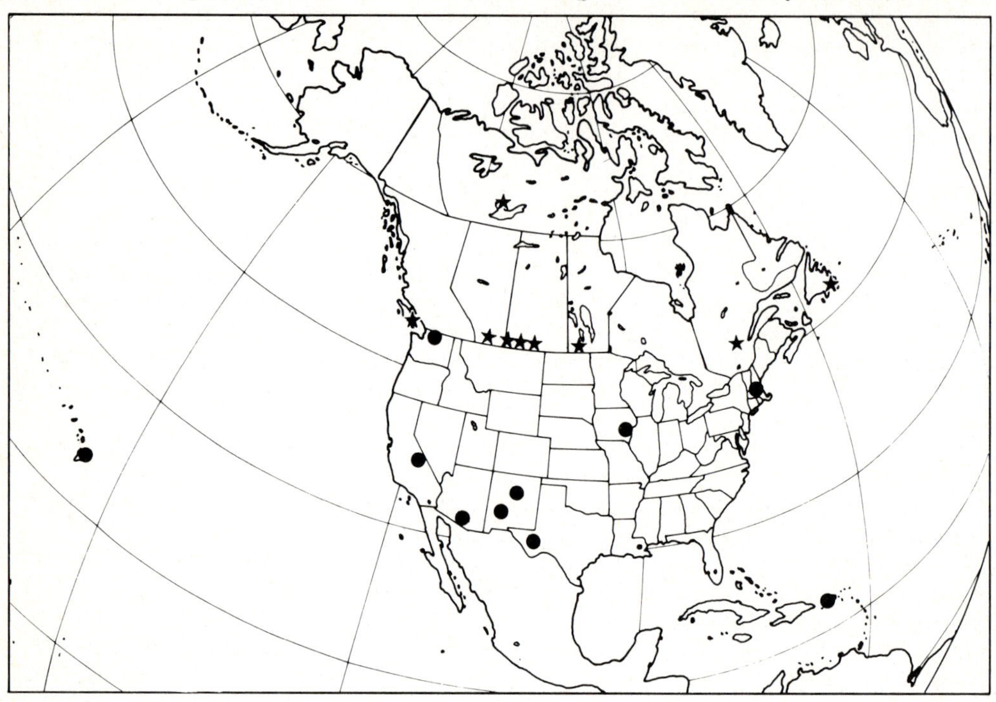

Figure 2. The coverage of the u-v plane for two declinations that is provided by the eight primary antennas of the possible CLBA configuration shown in Figure 1. The scale is in thousands of kilometers with a maximum of 5000 km.

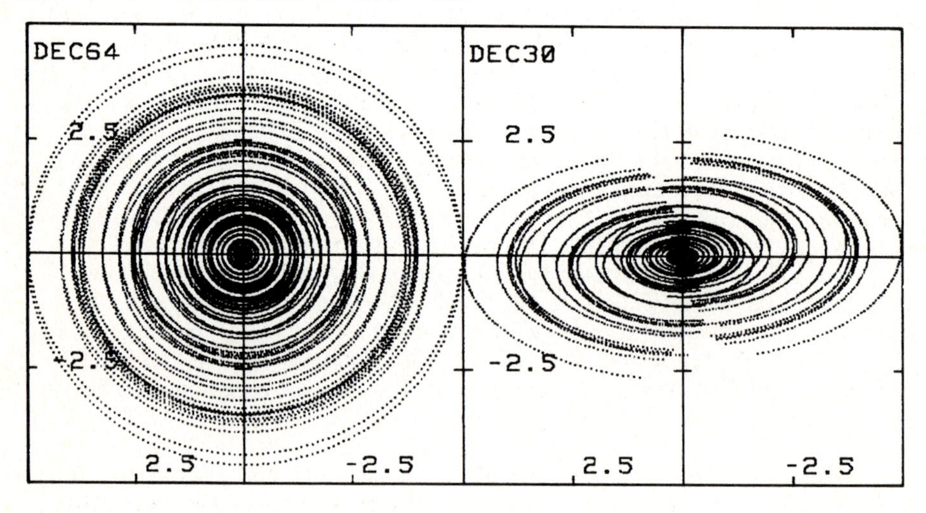

mainland) and set many of the details of the configuration. The following discussion of how the VLBA configuration was found provides a case study in how a configuration can be picked in the presence of strong geographic constraints. It has the advantage of hindsight so it seems rather more orderly than what actually occurred. However the basic steps in the process are accurately reflected.

The procedure used to find the VLBA configuration was to explore the coverage provided by many general classes of configurations using plots of the u-v coverage and then to measure the relative performance of large numbers of variations within each of the promising classes. In this usage, a class of configurations is a group of configurations with sufficiently similar distributions of antennas that there is an identifiable, one-to-one correspondence between each element of one array and some element in each of the other arrays in the class. For example, all members of the class to which the final array belongs have a Northeast site, a Midwest site, a Northwest site, a California site, a Southern Texas site, etc.

With the experience gained during the configuration search, it is clear why an array of the adopted class was chosen. Each of the sites is important for some special aspect of the coverage. Hawaii, along with the East Coast and Puerto Rico provide the longest baselines possible in the U.S. Puerto Rico to New England provides the longest possible north-south baselines available without using Alaska. Alaska is so far north that sources at the southern declinations where the north-south baseline is most important cannot be seen. Intermediate length east-west baselines require stations near the east and west coasts. At least two such baselines with similar lengths but different position angles are needed to avoid holes near zero declination. With New England already specified and the water vapor conditions so poor in the Southeast, the obvious way to get those baselines is with a site in Washington and one in California. Intermediate length north-south baselines are best obtained using a site in southern Texas but that station should not be too near the Gulf Coast where the water vapor content is high. The shortest baseline should be across the VLA in order to provide baselines less than 200 km in length as discussed earlier. The sites for possible future additions to the VLBA that would fill the hole between the VLA and the VLBA should be found as part of the original configuration. With the concentration of telescopes in the Southwest, there is a large hole between the Hawaii-New England baselines and the Hawaii-New Mexico baselines that must be filled with a Midwest site. One more site is needed to complete the coverage of short to intermediate length baselines. It should go somewhere in the Southwest although it is not tightly constrained from first principles.

Once a general class of configurations was identified by the criteria outlined above, numerical quality measures were used to search for the actual location of antenna sites. Such factors as ease of access, existing facilities, climate, etc. were considered in choosing the sites to examine. A strong bias was given to sites with existing radio astronomy activity. In general, at least half of the sites can be picked on grounds other than coverage, as long as they are in the general regions specified by the requirements of the class. The rest of the sites can then be adjusted to give performance almost indistinguishable from that of an array for which all sites were chosen purely for the coverage. This allows considerable freedom to use existing facilities. The number of sites tested was large but was still severely limited by the fact that many thousands of combinations are possible with only a few test sites in each region.

Figure 3. The coverage of the u-v plane provided by the VLBA for four sample declinations. The scale is in thousands of kilometers with a maximum of 8000 km.

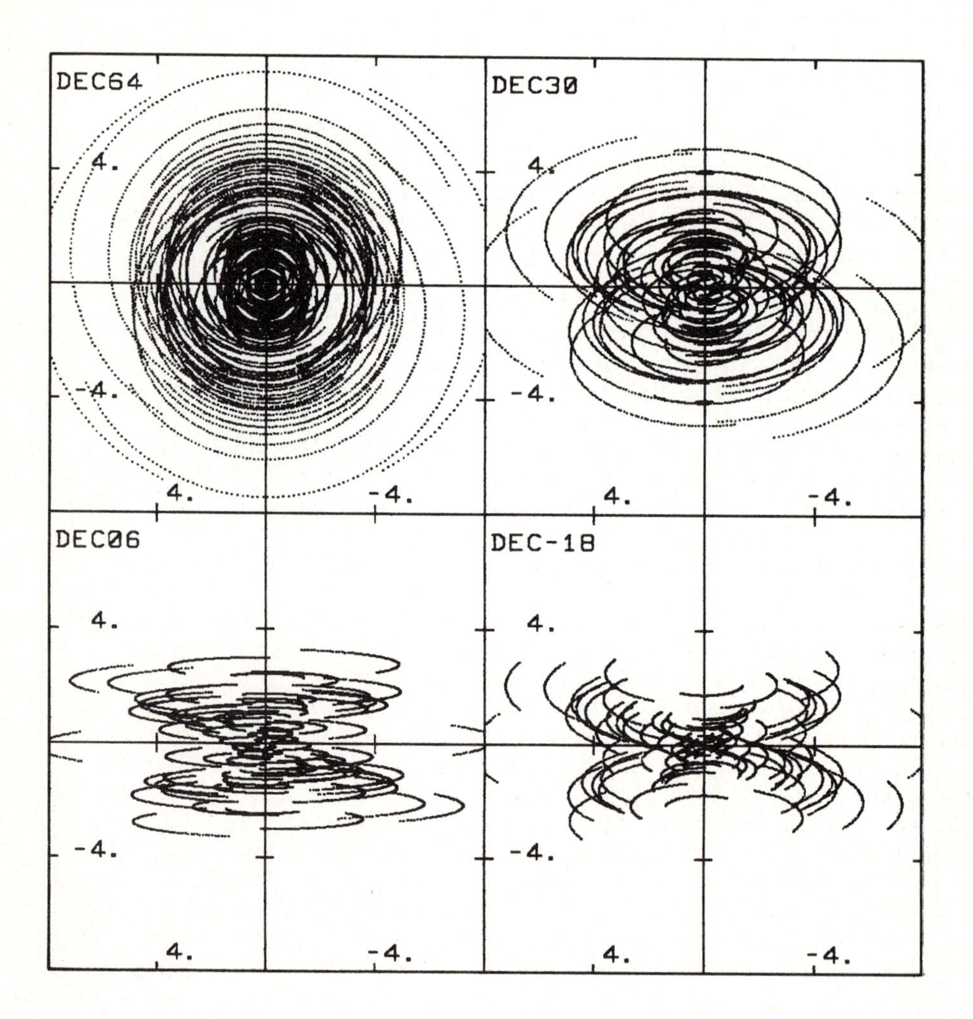

All of the quality measures gave similar results for the ranking of tested arrays and those rankings agreed with the impressions derived from examination of u-v plots. The final configuration is not necessarily the absolute best according to any given quality measure but it is among the top few for which measured differences are as much a result of details of the measuring methods as of real differences in coverage. It has the distinct advantage that it uses a large number of sites with good local support.

The configuration proposed for the VLBA is shown in Figure 1 and the u-v coverage for 4 representative declinations is shown in Figure 3. The plot was made under the assumptions that each telescope has an elevation limit of 10 degrees and that the source is followed for the entire time that it can be seen from any two antennas.

CONCLUSIONS

The steps involved in the design of a VLBI array configuration are: 1.) identify the constraints for the array, 2.) decide on a quality measure, 3.) check possible systematic configurations, 4.) find a basic layout, and 5.) fine tune the configuration for both coverage of the u-v plane and use of satisfactory sites.

The most important step is to specify the constraints, especially the number of telescopes and the nature of the desired coverage. The performance of the array improves very rapidly with the number of telescopes and with the size of each telescope. However those are the most important factors that set the cost of the array so a compromise must be reached between performance and budgetary realities. The nature of the desired coverage is a function of the types of source structure expected in the objects that will be observed. It must be picked on the basis of tradeoffs between uniformity of coverage, spatial dynamic range, and available geography.

The most productive way to find a configuration once the constraints are specified seems to be simply to try many possibilities. With the strong boundary conditions imposed by geography, especially for two dimensional arrays of intercontinental scale, systematic distributions provide little guidance. Tests of many possible configurations will reveal basic patterns in the distribution of telescopes that are required by the available geography and the desired coverage. Once those patterns are identified, many variants on the basic configurations can be tested to find optimal site locations.

The precise configuration chosen is not especially important, as long as it is among the better ones available. The differences in performance between the better configurations are small, compensating somewhat for the lack exact specifications for the desired coverage and the lack of a well determined geometric pattern. The flexibility that is available allows operational factors such as local support and accessibility to play a major role in setting the locations of most of the elements of the array.

REFERENCES

Canadian Astronomical Society, Committee on Radio Astronomy 1979 "A Proposal for a Canadian Very-Long-Baseline Array".

Canadian Astronomical Society 1982, "Canadian Long-Baseline Array".

Cohen, M. C., ed. 1980, "A Transcontinental Radio Telescope", Calif. Inst. of Tech., Pasadena.

Ishiguro, M. 1980, Radio Science, 15, 1163.

Kellermann, K. I., ed. 1981, "The VLB Array Design Study", National Radio Astronomy Observatory, Green Bank.

Klemperer, W. K. 1974, Astron. Astrophys. Suppl. 15, 449.

Linfield, R. 1982, "VLBA Memo No. 49", National Radio Astronomy Observatory, Charlottesville.

Moffet, A. T. 1968, IEEE Trans., AP-16, 172.

Morita, K.-I., & Ishiguro, M. 1982 preprint.

Mutel, R. L., & Gaume, R. A. 1982, "VLBA Memo No. 84", National Radio Astronomy Observatory, Charlottesville.

Schwab, F. 1982, "VLBA Memo No. 100", National Radio Astronomy Observatory, Charlottesville.

Walker, R. C. 1982, "VLBA Memo No. 144", National Radio Astronomy Observatory, Charlottesville.

DISCUSSION

P.E. DEWDNEY

What effect will losing the Arecibo site have on the u-v coverage, and at what frequency will this happen often? (i.e. how much water vapour is there at Arecibo?)

R.C. WALKER

Loss of the Arecibo site results in the loss of the long northwest-southeast baselines but has relatively little effect on the remainder of the u-v plane. The degradation is 'graceful' in the sense that it will give elongated beams but will not cause a large increase in sidelobes. The water-vapour induced delay path is between 2.5 and 5 cm in Puerto Rico. We are now trying to measure the short-time-scale fluctuations in the delay which will affect coherence. We expect problems at 22 and 43 GHz and may be at lower frequencies. Dual frequency observations may help overcome the problems. Successful VLBI observations are now being made at 5 GHz with Arecibo.

J.E. BALDWIN

The choice of a centrally condensed array must presumably be an astronomical one determined by the types of sources which one expects to observe. Can you say what are the expected features which drive one to such a choice and what do you lose by making it?

R.C. WALKER

Uniform coverage maximizes the complexity of structures that can be mapped but minimizes the range of scale sizes of structures that can be seen. A centrally condensed array can map a wider range of scale sizes but sacrifices some capability to map complex sources. Antennas of a VLBI array cannot be moved to provide coverage of different scales so the degree of central condensation must be chosen as a compromise between complexity and range of scale sizes in maps. The few current VLBI experiments that have used as many telescopes as will be available in the VLBA indicate that high brightness sources will not be as complex as the sources often mapped with the VLA but will have structures over a wide range of scales. Therefore moderately strong central condensation has been chosen for both VLB array projects.

J.R. FORSTER

Are there plans to provide simultaneous observation at two frequencies (besides S and X bands) using dichroics on the VLBA?

R.C. WALKER

The feeds will be located so as to allow 15/43 GHz and 10/22 GHz as simultaneous pairs. The dichroics may not be installed until after the instrument is in operation.

2.6 ANTENNA LAYOUT FOR THE AUSTRALIA TELESCOPE

G.T. Poulton
Division of Radiophysics, CSIRO, Sydney, Australia

Abstract. Station locations and antenna positions for the 'compact' Australia Telescope array are designed using a numerical optimization technique. Results are presented for 1.5, 3 and 6 km baselines, using a total of 37 stations. The design also allows 'frequency scaled' observations in two configurations, (1.5, 3) and (3, 6) km.

INTRODUCTION

The 'compact' part of the Australia Telescope will comprise six antennas on an east-west baseline, five movable on a 3 km railtrack and the sixth a further 3 km to the west, also possibly with limited movement. A schematic diagram of the arrangement is shown in Figure 1.

Figure 1. Schematic representation of the 'compact' part of the Australia Telescope.

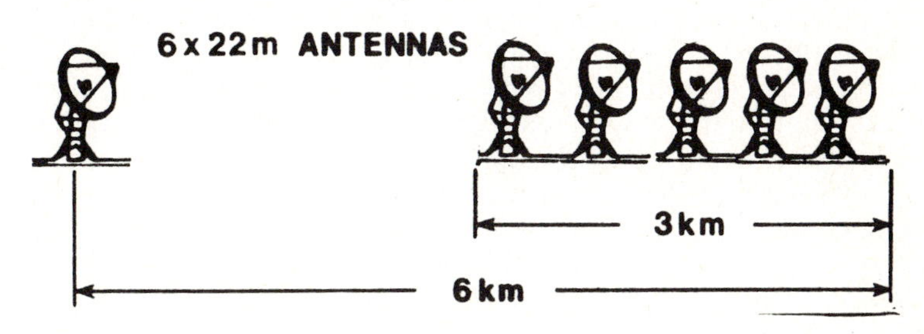

In the course of an observing program each antenna is to be accurately located at one of a number of stations on the railtrack, these locations being changed daily in a manner which will produce a good synthesized beam. Since considerable expense is involved in each station, the number of stations is to be kept to a minimum consistent with having a high-quality and flexible instrument. It is the determination of the location of these stations, and their number, which is the subject of this paper.

As with any synthesis telescope, the aim is to obtain good coverage of the u-v plane for all desired observing programs. Since this part of the Australia Telescope is situated on an east-west baseline the situation is simplified by the fact that all u-v plane tracks due to earth rotation are concentric ellipses, one for each distinct spacing between antennas. In such a case the determination of optimum u-v coverage reduces to the one-dimensional problem of optimizing the distribution of spacings produced by a linear array of antennas. Some of the techniques available for the design of 'minimum redundancy' linear arrays are therefore relevant.

Before proceeding we should mention the problem of grating vs. non-grating arrays. By grating array we mean one which has spacings which are all integral multiples of a unit value. Its advantage is that existing design methods may be used, in contradistinction to non-grating arrays, which have been talked about but not much analysed. Of course, the former possess a grating lobe structure which can limit observations in certain cases, whereas a non-grating array will have a reduced grating response at the expense of an increased overall sidelobe level. There is no reason why design techniques should not be developed for such arrays, but we have decided in the first instance to take advantage of existing design methods.

DESIGN AIMS AND REQUIREMENTS

The diameter of each reflector has been provisionally set at 22 m. The unit spacing should be somewhat smaller than this, with the restriction that the total number of spacings is not too great to be covered in a reasonable time. For this reason a 20 m unit spacing has been chosen. Three array lengths are to be considered: 1.5, 3 and 6 km, having 75, 150 and 300 unit spacings respectively. Since five antennas can give a maximum of 10 spacings per day, it will take a minimum of 8 and 15 days respectively to fill the 1.5 and 3 km arrays. The 6 km array cannot be completely filled without a large number of stations, because of the restricted movement of the distant antenna. The maximum observing program will be taken to be 15 days for this array also. Since the antenna diameter is 22 m, mechanical requirements dictate that the unit spacing (20 m) must be absent from all programs.

In addition, an acceptable design attempt should include 'scaled' arrays for spectral index observations, with a scaling factor of two being the most convenient because the array lengths differ by this factor. Each spacing of a 'scaled' array is twice the corresponding spacing of its 'base' array, allowing identical beams to be produced at frequencies differing by a factor of two.

For all arrays, spacing coverage should increase uniformly from day to day, allowing efficient observation for any number of days. Some constraint must also be put on the total distance moved by each antenna during an observing program. Finally, and perhaps most importantly, cost has to be kept down by minimizing the total number of stations.

Obviously if a numerical optimization is used not all of the above criteria can be satisfied simultaneously, and some compromise is necessary.

METHOD OF CALCULATION

The following process is carried out in an attempt to find a good compromise. The numerical technique used will be discussed later.

(1) A 1.5 km array is designed for near-minimum redundancy, over eight days of observation.

(2) The days are ordered so as to produce a good compromise between (a) even spacing coverage from day to day, and (b) minimum total antenna movement.

(3) A 'scaled' (double-size) version of this 'base' array is formed and juxtaposed on the original so as to minimize the total number of stations by making as many stations as possible coincide.

(4) The composite array is used as a basis for the design of a 3 km, near-minimum redundancy array over 15 days, adding more stations as required.

(5) Day ordering is carried out as for the 1.5 km array to ensure an even 'fill' and minimum antenna movement.

(6) Design of the 6 km array is then accomplished using the existing stations. Additional stations are located near the 6 km point consistent with the requirements of a 6 km 'scaled' array on a 3 km 'base'. Optimum daily antenna locations are then found for the 6 km array to give near-minimum redundancy.

METHOD OF OPTIMIZATION

Minimum redundancy arrays are known for small numbers of elements (Moffet 1968), and crude optimization schemes have been used to obtain low redundancy solutions for larger arrays (Mathur 1969; Ishiguro 1980). For multi-day observation some solutions have been found, again for small numbers of elements (Blum et al. 1975). Our smallest array comprises five elements and eight days, for which no published results exist. We have therefore modified a method of Mathur, and the result may be simply described as follows.

The number of stations is chosen and kept constant throughout the procedure. Initial station locations and antenna positions for each day are allocated at random multiples of the 20 m unit spacing. Each of these values is in turn replaced so as to maximize the spacing coverage. The process converges after several iterations, but there is no guarantee of a global extremum. Nevertheless quite acceptable results have been obtained from this very simple process. By repeating the procedure with different numbers of stations the minimum number which achieves an acceptable design is found.

RESULTS

Application of the above method has produced a design which requires a total of 37 stations, 35 on the 0 to 3 km railtrack and the other 2 near the 6 km point. All desired arrays are present, and their predicted performance is given below.

1.5 km array

Using 20 stations over eight days for the 1.5 km array, 67 out of a possible 74 spacings are produced with a redundancy of 1.19. Figure 2 shows how the aperture is filled from day to day. Daily antenna locations and the paths between them are also given. It can be seen that an even coverage has been obtained, especially after three,

Figure 2. Antenna locations and spacings over eight days for the 1.5 km array. Antenna positions on each day are represented by a cross, and the dashed line shows the antenna movement from day to day. The cumulative set of spacings generated is given each day by an open circle. At the bottom of the diagram the total set of stations used is shown by an asterisk.

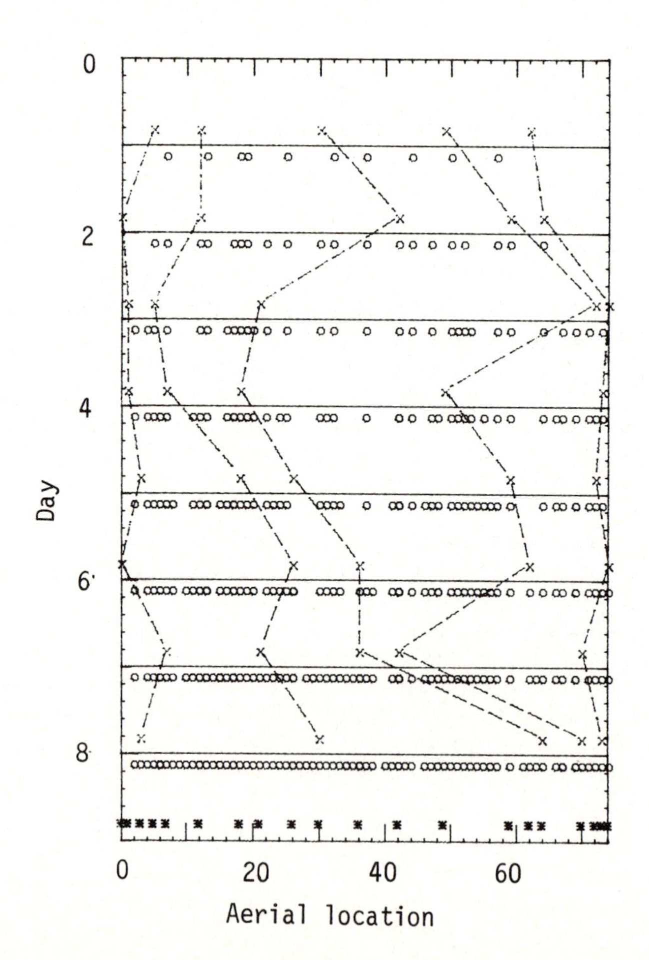

six and eight days. Averaged over eight days, movement per antenna per day is 138 m.

3 km array

Figure 3 shows similarly optimum antenna locations over 15 days for the 3 km array, using all 35 stations. Spacing coverage has been designed to be best at 5, 10 and 15 days, and this can be seen from the diagram. After 15 days 131 out of 149 spacings have been covered, with a redundancy of 1.16. Average antenna movement per day is 294 m.

Figure 3. Antenna locations and spacings over 15 days for the 3 km array. Symbols have the same meanings as in Figure 2.

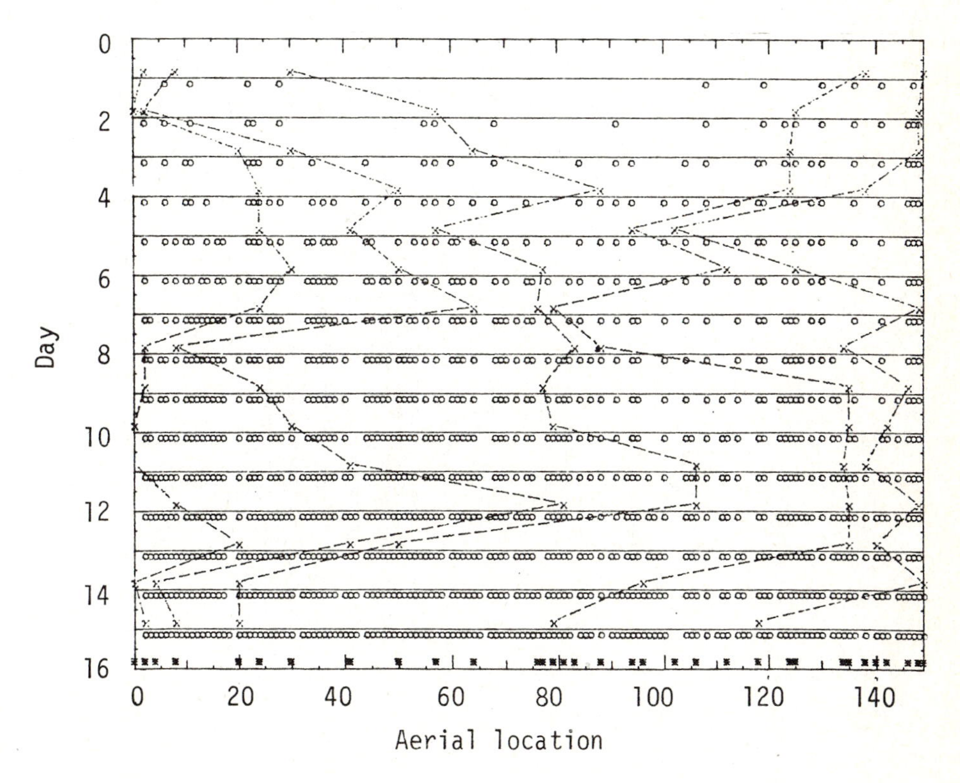

6 km array

Figure 4 gives a 15-day set of antenna locations optimized for the 6 km array, using the 35 railtrack stations plus two near the 6 km point. The expected sparse coverage of the longer spacings is clearly evident. After 15 days 180 out of 300 spacings are present, the redundancy being 1.25.

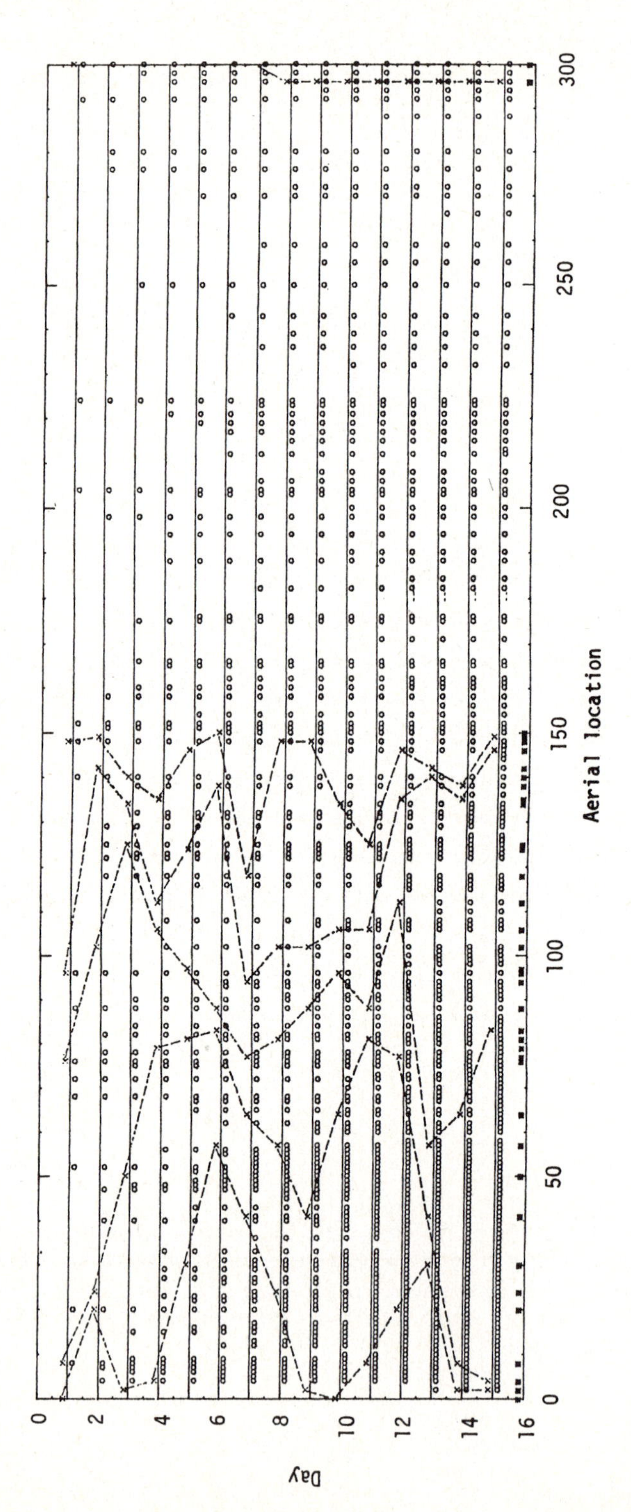

Figure 4. Antenna locations and spacings over 15 days for the 6 km array. Again, symbols are as defined in Figure 2.

Scaled array: (1.5,3) km

Since an entire scaled version of the 1.5 km array is embedded in the 3 km array, any observing program given in Figure 2 may be carried out with a 3 km baseline.

Scaled array: (6,3) km

Here the 'base' is the 1.5 km array with the addition of the two most remote 3 km stations. The corresponding 'scaled' version is the 3 km 'scaled' array plus the two 6 km stations. Figure 5 gives antenna locations over 10 days for the 'base' array, which can be fully duplicated on a 6 km baseline.

Figure 5. Antenna locations and spacings over 10 days for the 'base' 3 km array for which a 6 km 'scaled' replica exists. Symbols are as defined in Figure 2.

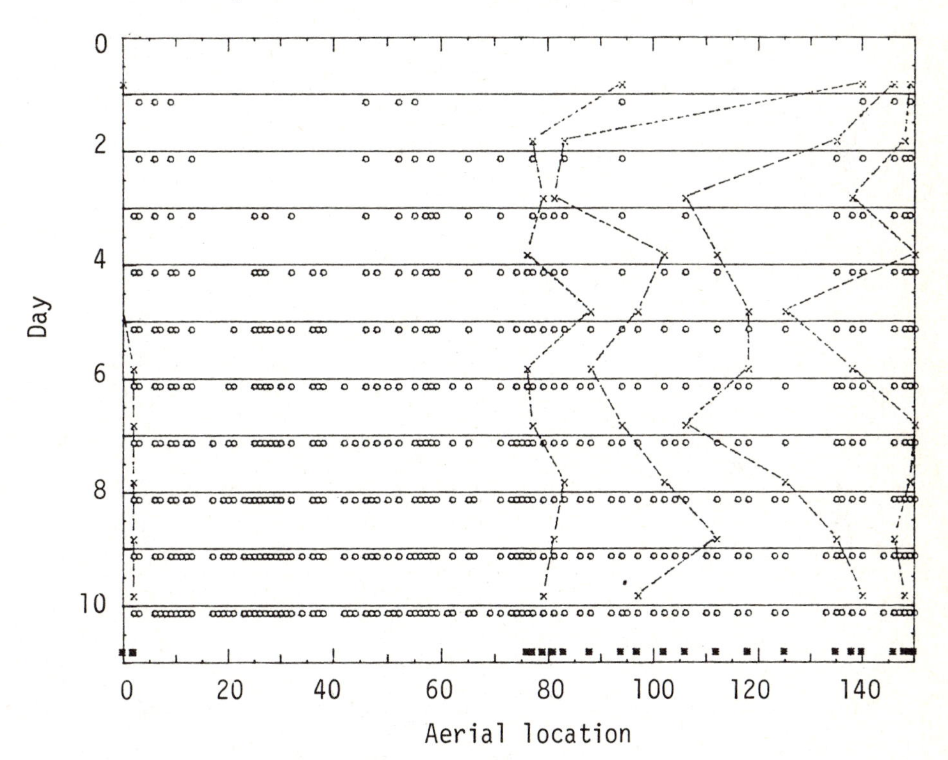

CONCLUSIONS AND FURTHER WORK

A fast and versatile configuration has been designed using a reasonably small number of stations. Evenly distributed sets of spacings exist for different array lengths and periods of observation. In addition, frequency-scaled operation has been achieved for two different configurations.

Since the methods used cannot guarantee a global optimum the search for better solutions is continuing. The area of non-grating arrays has not yet been covered, and this may well prove to be a fruitful field for future investigation.

References

Blum, E.J., Ribes, J.C. & Biraud, F. (1975). Some new possibilities of optimum synthetic linear arrays for radioastronomy. Astron. Astrophys., 41, 409-411.

Ishiguro, M. (1980). Minimum redundancy linear arrays for a large number of antennas. Radio Sci., 15, No. 6, 1163-1170.

Mathur, N.C. (1969). A pseudodynamic programming technique for the design of correlator supersynthesis arrays. Radio Sci., 4, No. 3, 235-244.

Moffet, A.T. (1968). Minimum-redundancy linear arrays. IEEE Trans., AP-16, 172-175.

2.7 A FAN BEAM RADIO TELESCOPE DESIGNED FOR INDIRECT IMAGING

J.M. Durdin,
Astrophysics Dept., University of Sydney, NSW 2006, Australia

M.I. Large,
Astrophysics Dept., University of Sydney, NSW 2006, Australia

A.G. Little
Astrophysics Dept., University of Sydney, NSW 2006, Australia

The Molonglo Observatory Synthesis Telescope (the MOST) is a multiple fan beam radio telescope designed to produce maps by a novel type of rotational synthesis (Mills, 1981). A description of the principle of this method of mapping has been given by Mills (1976) and by Crawford (1983) in this symposium. In the more usual process of "aperture synthesis" in radio astronomy, correlations are recorded between multiple pairs of antenna elements, and the image is formed by a Fourier inversion. By contrast, in the MOST, multiple fan beams are formed in real time and the outputs are added into a map field in the computer. By this means, a 23 arcmin field is covered with full sensitivity. The field size is extended to 1 degree by time sharing.

The circuit networks which form the fan beams perform the necessary Fourier transformation, and no subsequent computer transform is required. The MOST principle is analogous to the principle of medical imaging by tomography, with one important difference: the beam visibility function is tailored to yield a synthesised beam having a spatial frequency content which is substantially uniform out to the limit determined by the physical size of the antennae (see Fig. 1). This is achieved by *multipying* together the signals from the two antennae. The analogue of medical tomography would be to *add* the signals from each antenna, producing a synthesised beam with a visibility function heavily weighted to low spatial frequencies.

Fig. 1. The visibility functions and corresponding beam shapes of the fan beam and the synthesised beam.

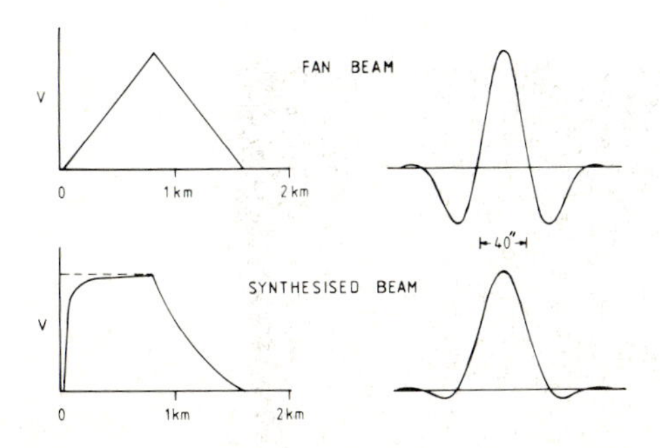

The telescope consists of two colinear cylindrical paraboloid reflectors, 12 m wide, orientated East-West with an overall length of 1570 m and central gap of 15 m. The line feed consists of 7744 circularly polarized elements (Mills & Little 1972). These elements are rotated under computer control to direct the primary beam away from the meridian. Rotation of the two antennae about the East-West axis tilts the beam in the other coordinate (i.e. an alt-alt system). Amplified intermediate frequency signals are returned from 88 equal length sections of the antennae. These signals are converted to digital form (2 bit) and appropriate phase shifts and delays are introduced. They are then converted back to analogue voltages (both in-phase and quadrature components) and combined in resistor arrays (Fig. 2) to form 64 independent fan beams from each of the two antennae. Finally, the corresponding beam signals from each antenna are multiplied together to produce d.c. outputs which are added into the map field in the computer. The 64 beam raw data (averaged over 2 s) are stored on magnetic tape for archive purposes. For outside users, map data are available in FITS form. Further technical details of the MOST are given in Fig. 3 and table 1.

Since the MOST has a very large fully filled aperture, very high sensitivity is obtained in a complete 12 hour synthesis. Furthermore, the redundancy inherent in the filled aperture makes the system reasonably tolerant of random phasing errors in the illumination: high quality maps of good dynamic range are obtained without further processing. However, for observations of complex or extended sources it is desirable to remove the effects of the negative sidelobes of the synthesised beam by using cleaning techniques (Crawford 1983). An example of a 12 hour observation is shown in Fig. 4. This map was plotted directly after the observation, and is uncleaned. The lowest contours shown are 1% of the peak.

Fig. 2. One pathway through the multibeam phasing array. In-phase and quadrature components are added via resistors R_C and R_S to generate the appropriate phase angle ϕ (N, M). N = 1-44, M = 1-64.

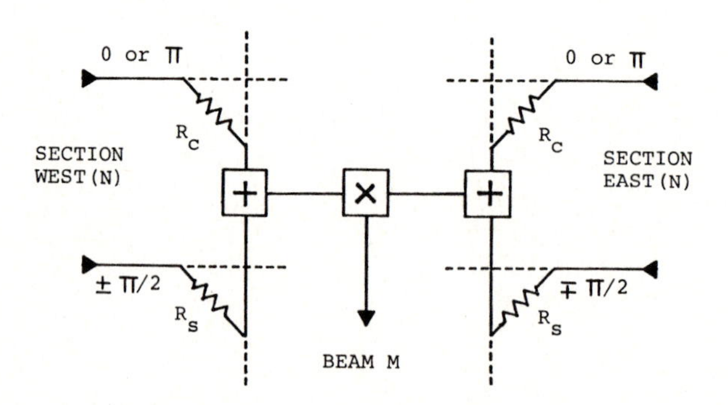

REFERENCES

Crawford, D.F. (1983). Map construction with the Molonglo Observatory Synthesis Telescope. Proceedings of this Symposium.

Mills, B.Y. & Little, A.G. (1972). Future plans for the Molonglo Radio Telescope. Proc. Astro. Soc. Aust., 2, 134.

Mills, B.Y., Little, A.G. & Joss, G.H. (1976). A synthesis telescope at Molonglo. Proc. Astro. Soc. Aust., 3, 33.

Mills, B.Y. (1981). The Molonglo Observatory Synthesis Telescope. Proc. Astro. Soc. Aust., 4, 402, 156.

Table 1. Specifications and performance of the MOST

Centre Frequency 843 MHz

Effective Bandwidth 3 MHz

Time resolution

archive records 24 s

potential ~ 0.5 μs

Maximum Declination Coverage $+18\overset{\circ}{.}5 \rightarrow -90^{\circ}$

Declination Range for full H.A. Coverage $-30^{\circ} \rightarrow -90^{\circ}$

East-West Beam Swing ~ $\pm 60^{\circ}$

Field Size for Full Synthesis 23' x 23' cosec δ

Field Size for Time Shared Synthesis $1\overset{\circ}{.}17 \times 1\overset{\circ}{.}17$ cosec δ ($\delta < -30^{\circ}$)

Synthesised Beam 43" x 43" cosec δ

System Temperature = 110 K

System Noise for very Southern Sources after 12 hr Integration ≈ 0.2 mJy (r.m.s.)

Fig. 3. A block diagram of the MOST.

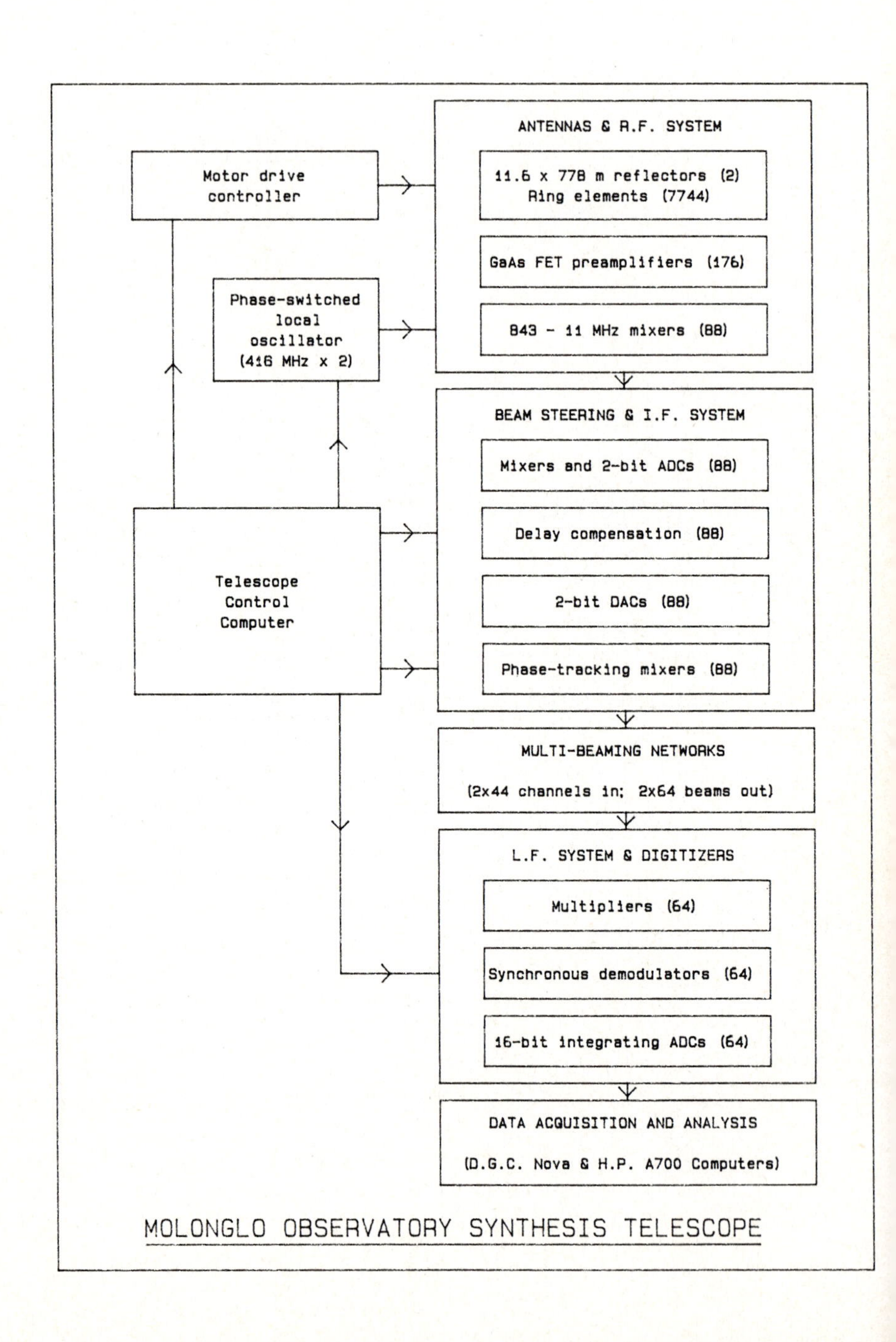

Fig. 4. An uncleaned map of the supernova remnant Puppis A.

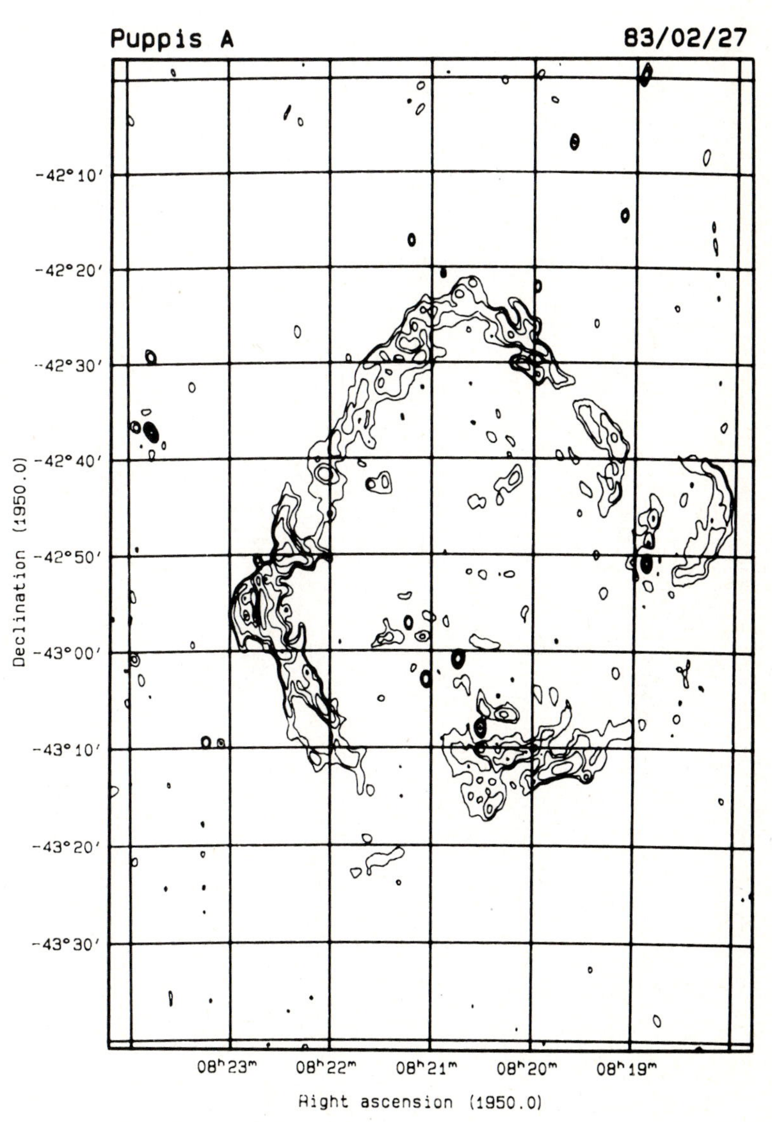

2.8 PHASED ARRAY TELESCOPE WITH LARGE FIELD OF VIEW TO DETECT TRANSIENT RADIO SOURCES

T. Daishido, T. Ohkawa, T. Yokoyama, K. Asuma and H. Kikuchi
Department of Science, School of Education, Waseda University
1-6-1 Nishiwaseda, Shinjuku-ku, Tokyo 160, Japan

K. Nagane and H. Hirabayashi
Nobeyama Radio Observatory, Tokyo Astronomical Observatory, University of Tokyo, Nobeyama, Minamisaku-gun, Nagano-ken, Japan

S. Komatsu
Department of Applied Physics, School of Science and Engineering, Waseda University, Nishiohkubo, Shinjuku-ku, Tokyo 160, Japan

Abstract. A design of large field phased array telescope in future is discussed under the condition to use present technology. In order to detect transient radio sources like Cyg X-3, Cir X-1, SS 433, or supernovae, we need large field radio telescopes with appropriate resolution and sensitivity. The system discussed here is 64 x 64 = 4096 elements filled aperture at 10 GHz with overall size 12.8 m x 12.8 m. 4096 receivers are needed, and the recent developments in GaAs FET/MIC(GaAs Field Effect Transistor/Microwave Integrated Circuit) technology have been able to supply low noise and low price receivers (Tsys = 200 - 300 K). IF (Intermediate Frequency) signals are two dimensionally transformed by FFT (Fast Fourier Transform) processors at 10 - 50 MHz band width, and integrated finally. 4096 points image is recorded every 0.1 sec to optical disk. Whole sky mapping could be done within a week. The field of view is 9 deg x 9 deg, the resolution 9 arc min, and the sensitivity 200 mJy at 10 min integration. The system will be an automatic all sky patrol camera in radio wave region. The results in the test telescope are reported.

INTRODUCTION

A couple of luminous transient radio sources have been discovered since the detection of large outburst in Cyg X-3 (Gregory 1972; Daishido et al. 1974). These transient radio sources are associated with high energy X-ray and/or Gamma ray sources; i.e. Cir X-1 is with Black Hole candidate in X-ray binary (Haynes et al. 1979), SS 433 is with precessing jet object, and LSI+61°303 is probably with COS-B Gamma-ray source. In Cyg X-3, ultra high energy Gamma-ray (10^{12} and 10^{15} eV) light curves with 4.8 hour period have also been observed. Although there should be many more sources like these objects, we cannot detect them because of their abrupt flaring and our poor positional information. The situation is the same in the talk in the URSI meeting in Washington

D.C. (Sullivan 1981). He discussed the possibility of discovering the next supernova in our Galaxy by the radio telescope. According to him, the anticipated flux densities of the next supernova are 200 - 2000 Jy even if it appears on the other side of our Galaxy. In order to detect these transient radio sources, we need large field radio telescopes with appropriate resolution and sensitivity.

A 64 x 64 elements array of filled aperture type telescope at 10 GHz was proposed (Daishido 1978; Daishido et al. 1981). It has large field (9 deg x 9 deg), appropriate resolution (8 arc min) and sensitivity (200 mJy at 10 min integration). 4096 beams are simultaneously synthesized by FFT processors (Chikada 1981, Chikada et al. 1983) at 10 - 15 MHz band width. 4096 RF receivers are made by the GaAs FET MIC technology and they are available as satellite TV receivers at low price. The system temperature is now 300 K at 10 GHz, and will be 200 K within a few years. The array of 4096 horns of 16 cm x 20 cm has 12.8 m x 12.8 m aperture and allows mapping of the sky 4096 faster than the same aperture single dish. For the technical test and the development, eight elements telescope has been built at Waseda University in 1980. The results are discussed.

IMAGE FORMING

Fourier synthesis telescope is a tool of the two dimensional spectrum analysis in spatial frequency. Signals simultaneously received by the array are superposed harmonic waves from various directions. Sampled complex signals (0 and $\pi/2$ shifted) are transformed, for example, by complex FFT processors into a set of Fourier components. Their amplitudes and spatial frequencies determine the intensity and the position of the sources respectively.

We consider here an N points one dimensional array at equal intervals of Δx. From the sampling theorem, the band width of spatial frequencies are limited. The complete set of Fourier components that can be analysed by this system are

$$E(x,m) = E(m)\exp(i\,\frac{2\pi x}{N\Delta x}\,m), \quad (m = 0, \pm 1, \ldots, \pm N/2) \tag{1}$$

in which $N\Delta x/m$ is equal to the projected wave length of $\lambda/\sin\theta_m$, and the projected phase velocity is $c/\sin\theta_m$, where λ is true wave length (Fig. 1). θ_m is defined by

$$\sin\theta_m = m\,\frac{\lambda}{N\Delta x} = m \sin\theta_1, \tag{2}$$

for large field mapping. When $\theta_m \ll 1$, it is reduced to $\theta_m = m\theta_1$. Thus the equation (1) is interpreted as the amplitude and phase of the radiation at x illuminated from θ_m with the intensity of $E^2(m)$. The total signal received by the element of the array at $x = j\Delta x$ is given by summing the equation (1) over all directions of θ_m, as

$$\hat{E}(j) = \sum_{m=-N/2}^{N/2} E(x,m) = \sum_{m=-N/2}^{N/2} E(m)W^{jm}, \quad (j = 0, 1, \ldots, N-1) \qquad (3)$$

where $W = \exp(i2\pi/N)$. This means that a wave from θ_m generates a phase gradient W^m along the array. Multiplying phase gradient compensator of $W^{-jn} = \exp(-i2\pi j\Delta x \sin\theta_n/\lambda)$ by equation (3) and summing over j, we can adjust the phase to the source at θ_n as follows,

$$\sum_{j=0}^{N-1} \hat{E}(j)W^{-jn} = \sum_{j=0}^{N-1} \sum_{m=-N/2}^{N/2} E(m)W^{j(m-n)}$$

$$= \sum_{m=-N/2}^{N/2} E(m)\delta_{mn} = E(n). \qquad (4)$$

Thus the brightness distribution $E^2(n)$ is obtained from a set of the sampled signal $\hat{E}(j)$. Several analogue and digital methods exist or have been proposed to perform this Fourier transformation (Cole 1979). Recently, CMOS (Complementary Metal Oxide Semiconductor) FFT butterfly chips of 10 MHz band width have been developed (Chikada 1981; Chikada et al. 1983), and we have tested Butler Matrix as we discuss lately. Excellent result in the matrix obtained by Rogers & Craggs (1977). Butler Matrix is mathematically equivalent to FFT algorithm. In N elements filled aperture, the number of complex butterfly processors is $N \log_2 N$, and that of complex correlators is $N(N-2)/2$. This means that the FFT type is more economic than the correlator type when the number of elements is large. In the correlator type, raw data are visibility functions, and in the FFT type they are $\hat{E}(j)$. Fourier analysis telescope may be a more appropriate term for the latter.

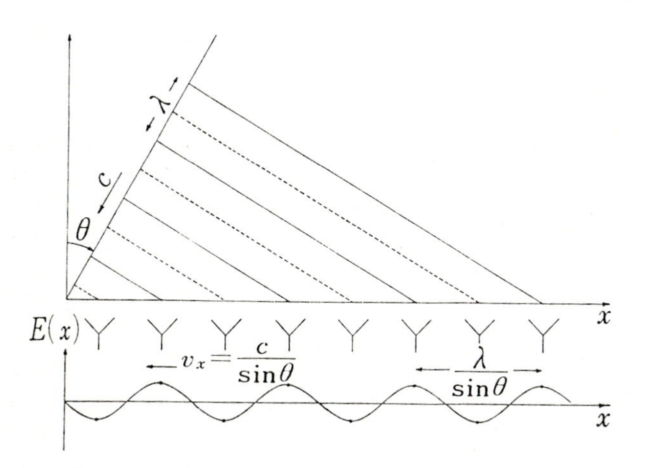

Fig. 1. Array and projected wave length.

Can we use FWT (Fast Walsh Transform)? A given function can be expanded precisely enough to both Walsh and Fourier series. They are similar, and Walsh transform is much simpler and faster. However, Walsh transform is not invariant to phase shift, and this is the case in Fourier transform. So, the results from FWT applying to random signals as in radio astronomy are not true images. But it may be useful to find the change of the sky.

SENSITIVITY

Three questions arise on sensitivity. (a) When many (4096) receivers are used, don't they increase the total system temperature and decrease the sensitivity? On the contrary, if the number of elements increases, does the system become more sensitive when the total aperture is constant? (b) Is there much difference in the sensitivity of filled aperture between the FFT type and the correlator type? (c) Since there are so many redundant Fourier components in filled aperture, can most of the elements be removed?

These are frequently asked questions. Conclusions are; (a) the sensitivity does not decrease nor increase, if the total aperture is constant. It does not depend on the number of the receivers. It depends on the total aperture. (b) Neglecting small factors, the sensitivities are nearly equal both in the FFT type and in the correlator type. (c) Removing a redundant pair of elements does not affect the resolution nor the view. However, it decreases the sensitivity. And FFT type is only possible in filled aperture.

The above conclusions are derived from the difference of the laws of superposition, in which random signals and coherent signals obey respectively. Let N be the number of antenna elements, T_n the noise temperature of receivers, and T_A the antenna temperature of each antenna element. Since the radiation from radio sources is random noise which is coherent to all the antenna elements, the summed amplitude is $N\sqrt{T_A}$. On the other hand, each receiver generates the rms amplitude of noise $\sqrt{T_n}$ incoherently, and the summed amplitude is $\sqrt{N}\,\sqrt{T_n}$. Thus the power ratio of signal to noise is NT_A/T_n. Since NT_A is proportional to the total aperture, we confirm the conclusion of (a). The sensitivity if correlators are used is obtained by the standard manipulation. The conclusion (b) is confirmed by this calculation. Conclusion (c) is apparent by the above consideration.

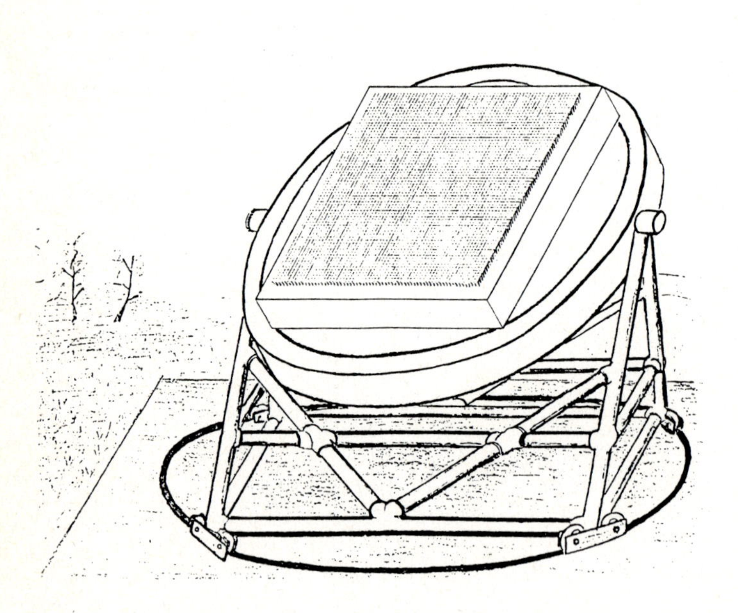

Fig. 2. Large field radio patrol camera

Fig. 2 is the 64 x 64 array image. According to face on tracking, no delay compensators are used. Output data of 4096 points are recorded every 0.1 sec with

Fig. 3. Eight element test telescope at Waseda University

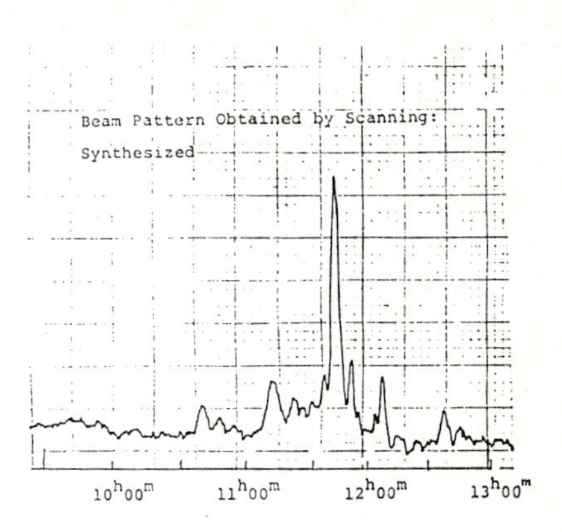

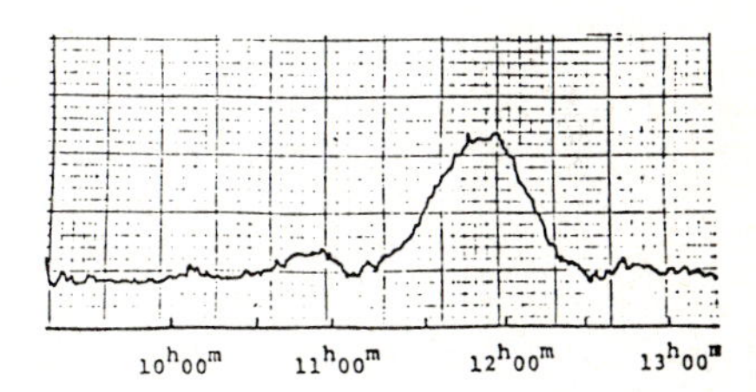

Fig. 4. Synthesized beam (above) and the pattern of single horn

8 bit dynamic range into optical disk. Running cost of the disk will be A$3,000/year.

RESULTS OF TEST TELESCOPE

Eight elements phased array telescope at 10.6 GHz has been built in 1980 at Waseda University for technical test and development. GaAs FET MIC receivers are used. Noise temperature of the receivers are 300 K, and the total system temperatures including isolators and IF amplifiers are 400 K. Beams are synthesized by Butler matrix. In this test telescope, we have obtained eight beams of 1 deg HPWB in 7 deg field of view (Fig. 3, Fig. 4). (a) Receivers: RF signals (10.6 - 10.7 GHz) are amplified 18 dB by two stage GaAs FET at first, and converted to IF signals (1.0 - 1.1 GHz). Total conversion gain of the head amplifier is 50 dB. Supplied powers and frequency of local signals to each receiver are +10 dBm and 9.6 GHz respectively. Reliability of MIC receivers is very important, when 4096 receivers are used. We have no experience of receiver trouble for these three years, except the initial period. In future systems, local signal will be supplied between the 1st and 2nd stage of FET. This will reduce 10 dB in local power supply. (b) Butler matrix: delay lines of the Butler matrix are printed on an alumina substrate. Eight beams are synthesised by the matrix in this experiment. It operates at 1.05 GHz IF, and the band width is 100 MHz. Butler matrix is mathematically equivalent to FFT. It is convenient and useful in small systems. However, there exist some problems also in practical use. (1) If the matrix is used at IF stage, band widths are limited in the synthesis of a lot of beams. (2) No tunable points exist in this analogue system. So, the print and try system. (3) Loss in power depends on the length of delay line.

We have detected solar bursts using the system. The largest is the event on 6th November 1980.

REFERENCES

Chikada, Y. (1981). Paper CDJ2 presented at URSI General Assembly (Washington D.C.), Abstract pp 151; and Nobeyama Radio Observatory Technical Rep. No. 8.

Chikada, Y. et al. (1983). in this issue.

Cole, T.W. (1979). Analogue processing methods for Synthesis Observations, in Image Formation from Coherence Functions in Astronomy, ed. C. van Schooneveld, pp. 123-141. Dordrecht: Reidel

Daishido, T. et al. (1974). Nature, 251, 36.

Daishido, T. (1978). Paper 34 presented at Annual Meeting of Astronomical Society of Japan (Sendai).

Daishido, T. et al. (1981). Paper J2-4 presented at URSI General Assembly (Washington D.C.).

Gregory, P.C. et al. (1972). Nature, 239, 440.

Haynes, R.F. et al. (1979). Aust. J. Phys. 32, 43.

Rogers, P.G. & Craggs, I.S. (1977). IREE 16th International Conf., 378.

Sullivan, W.T. (1981). Paper J2-4 presented at URSI General Assembly (Washington D.C.).

DISCUSSION

R.H. FRATER

Have you considered using an optical processor for the transform required in your telescope?

T. DAISHIDO

Not yet. For automatic observations, an optical processor seems to be one of the most promising devices in future.

R.N. MANCHESTER

When will the full 4096-beam system be completed? What declination range will the observations cover?

T. DAISHIDO

The full system is technologically possible at present. On the budget, I cannot say anything. Total construction cost is A$$1.5 \times 10^7$. The telescope should cover all declinations.

R.H.T. BATES

Have you thought about the possibility of dips in the gain pattern of the array (at certain angles) due to mutual coupling (in particular pseudo-surface-waves on the array)? These become more pronounced as the array gets bigger.

T. DAISHIDO

Horn aperture is 16 cm $\times$ 20 cm, and λ = 3 cm. So there is probably no problem.

3

Optical imaging

3.1 OPTICAL ASTRONOMICAL SPECKLE IMAGING

R.H.T. Bates
Electrical & Electronic Engineering Department
University of Canterbury, Christchurch, New Zealand

Abstract. The theoretical basis of speckle imaging is reviewed. The best known techniques are described and are grouped according to the "order" of their interferometric processing. Their capabilities for imaging, as opposed to merely resolving, astronomical objects are assessed.

Introduction

In astronomical contexts the term "interferometry" tends to imply the reception of signals by separated telescopes. This is unnecessarily restrictive conceptually. It is useful, in particular, to think of a single imaging instrument of large aperture (i.e. pupil) as an interferometer when it is viewing an object through a highly distorting medium. The point is that the degradation due to the medium is manifested by patterns of "fringes" across the image. When a large optical telescope (pupil diameter of 1m or more) is operated under normal seeing conditions, the unwanted interference is so severe that the effective diameter of the instantaneous image of an unresolvable object is an order of magnitude or more larger than that of the Airy disc. This instantaneous image is also broken up into hundreds, sometimes thousands, of "speckles".

Astronomical speckle imaging has its genesis in Labeyrie's (1970) speckle interferometry. The latter is strictly a method of "resolving" objects, as opposed to forming true images of them. The more recently developed speckle techniques are concerned with reconstructing actual images, as faithfully as possible.

Labeyrie's method of processing short exposures (typically 10 ms), called speckle images (sometimes called specklegrams), discards the visibility phase, so that it appears, at first sight at least, useful only for resolving, as opposed to forming true images of celestial objects. Phase retrieval is, however, fast becoming a fully practical possibility (refer to the papers in this book by Fienup and by Bates, Fright and Norton) implying that speckle interferometric data may soon prove a sufficient basis for image reconstruction. One might then question the point of studying other speckle imaging methods, especially since they tend on the whole to be less robust than speckle interferometry. There are three general reasons why it is worth doing so.

The first reason for investigating different speckle techniques is that they may provide initial estimates of the visibility phase which permit accurate phase retrieval to be carried out more efficiently. The second reason is that one has more faith in a result if it is obtained in several different ways. The third reason is perhaps the most cogent. Some speckle imaging techniques can be expected to exhibit a better signal-to-noise ratio than speckle interferometry.

This paper draws heavily on a comprehensive review article (Bates, 1982) which, besides describing in detail the material summarised here, covers its background and lists the observational results that have been obtained with the aid of speckle imaging techniques and also discusses practical matters, such as optical bandwidths and exposure times of speckle images (sometimes called specklegrams), sizes of seeing cells and seeing discs, numbers of speckle images needed to form useful resolved images, etc. It is to be understood that justification for anything stated without qualification below can be found in this review article.

Preliminaries

Observation, under statistically stationary seeing conditions, of an unresolvable object (whose image, under ideal seeing conditions, would be an Airy disc) provides a sequence of "reference speckle images", the mth being

$$som(x,y) = hom(x,y) + com(x,y) \tag{1}$$

where x,y are Cartesian coordinates in the image plane, $hom(x,y)$ is the point spread function (psf) characterising the isoplanatic (i.e. point spread invariant - the blurring is the same everywhere) part of the distortion due to the atmosphere above the telescope during the mth short exposure (the interval during which the mth speckle image is recorded) and $com(x,y)$ is the contamination (which represents all measurement imperfections and includes the non-isoplanatic part of the distortion). It is worth emphasising that speckle imaging techniques fail when the amplitudes of the psfs are negligible compared to those of their respective contaminations. On the other hand it is remarkable how tolerant speckle imaging is to high contamination levels. When observing faint objects it is vital to make use of all of the received light. It is explained in §4.4 of Bates (1982) how this can be made compatible with the assumption (adopted here) that all of the psfs are statistically independent.

The mth "reference speckle visibility" $Som(u,v)$ is defined by

$$som(x,y) \leftrightarrow Som(u,v) = Hom(u,v) + Com(u,v) \tag{2}$$

where u,v are Cartesian coordinates in the Fourier plane, $\leftrightarrow$ identifies a two-dimensional Fourier transform pair and the symbols H and C represent optical transfer functions (OTF) and spectral contaminations respectively. The "reference speckle transfer function"

$To(u,v,\alpha,\beta)$ is defined by

$$To(u,v,\alpha,\beta) = \langle Som^*(u,v)\ Som(u+\alpha,v+\beta)\rangle \qquad (3)$$

where the asterisk denotes the complex conjugate, α and β are non-negative real constants and the angular brackets denote the average over the available sequence of speckle visibilities.

It is here appropriate to define the "size" of an image - this is the smallest convex area of the image plane within which the image has significant intensity.

Observation of a resolvable object under seeing conditions statistically similar to thos under which the unresolvable object is observed provides a sequence of "resolved speckle images", the mth being

$$sm(x,y) = f(x,y) \odot hm(x,y) + cm(x,y) \qquad (4)$$

where $\odot$ denotes two-dimensional convolution and where the h and c are of course different from the similar quantities introduced in *(1)* but they have the same respective connotations. The quantity $f(x,y)$ is the "true image" of the object (i.e. it is the image that would be obtained under ideal seeing conditions with a perfectly figured telescope having the same pupil as the telescope which is actually used). It is here understood that the size of $f(x,y)$ is much smaller than that of any of the $hm(x,y)$ - this is the situation of most interest in current astronomical speckle imaging contexts - it implies that the "image" of the object obtained by operating the telescope in the conventional manner (i.e. forming a long exposure while tracking the object) is virtually indistinguishable from the "seeing disc" $sd(x,y)$, which is defined by

$$sd(x,y) = \langle som(x,y)\rangle = \langle sm(x,y)\rangle \qquad (5)$$

where the total brightnesses of the unresolvable and resolvable objects have been normalised for convenience to be equal. This of course means that all detail in the true image is destroyed when the telescope is operated conventionally (it merely acts as a "light bucket"). It is also understood here that the fluctuating distortions characterised by the psfs are significantly more severe than the fixed aberrations of the telescope, thereby permitting the latter to be compensated by speckle imaging (Dainty, 1973; Cady and Bates, 1980). The mth "resolved speckle visibility" and the "resolved speckle transfer function" are defined respectively by

$$Sm(u,v) = F(u,v)\ Hm(u,v) + Cm(u,v) \quad \text{and} \qquad (6)$$

$$T(u,v,\alpha,\beta) = \langle Sm^*(u,v)\ Sm(u+\alpha,v+\beta)\rangle \qquad (7)$$

where $f(x,y) \leftrightarrow F(u,v)$, $hm(x,y) \leftrightarrow Hm(u,v)$ and $cm(x,y) \leftrightarrow Cm(u,v)$.

Because of the aforementioned "stationarity" and "similarity" of the statistics of the seeing conditions, it is assumed that

$$\langle Hm^*(u,v)\, Hm(u+\alpha, v+\beta)\rangle = To(u,v,\alpha,\beta) \tag{8}$$

The "brightest" points in the mth reference and resolved speckle images are denoted by $(\xi om, \eta om)$ and $(\xi m, \eta m)$ respectively, i.e.

$$som(\xi om, \eta om) = \max_{x,y} som(x,y) \text{ and } sm(\xi m, \eta m) = \max_{x,y} sm(x,y) \tag{9}$$

Note that the psfs, contaminations and true image are non-negative real because all astronomical sources are here taken to be spatially incoherent.

It is convenient to recall the concept of the "order" of an interferometric processing scheme (Bates and Gough, 1975). It is important to distinguish this from the order of a correlation detector. The practical significance of the "interferometric order" is that it is much more directly related to "signal-to-noise" performance. When the processing can be symbolised by $[\{s\}n]$, where the square brackets denote an averaging operation and $\{s\}n$ represents an integral operator acting on n speckle images, then n is the order of the "interferometry". If the processing of the sequence of resolved speckle images could be perfect (i.e. the effects of the psfs could be perfectly compensated) then

$$[\{s\}n] = [\{f\}n] + [\{c\}n] \tag{10}$$

where f and c symbolically identify the true image and the contamination respectively. Many interesting astrophysical objects are very faint, implying that the contamination tends to dominate (but not completely of course) the psf in each resolved speckle image. Consequently, provided all else is equal (which it never is, as is explained below), the overall signal-to-noise ratio can be expected to increase spectacularly as n decreases.

When one is required to recover $g(x,y)$ from

$$a(x,y) = g(x,y) \odot b(x,y) + c(x,y) \tag{11}$$

where $b(x,y)$ is given but $c(x,y)$ is unknown apart from a few statistical descriptors, the Fourier transform of $g(x,y)$ can often be usefully approximated by

$$G(u,v) = A(u,v)\, W(u,v) \tag{12}$$

where $a(x,y) \leftrightarrow A(u,v)$ and the "Wiener filter" $W(u,v)$ is defined by

$$W(u,v) = B^*(u,v)/\{|B(u,v)|^2 + \Phi(u,v)\} \tag{13}$$

where $b(x,y) \leftrightarrow B(u,v)$ and $\Phi(u,v)$ is the best available estimate of the ratio of contamination-to-signal intensity. The operation on RHS (12) is sometimes called "multiplicative deconvolution" (for obvious reasons). It is occasionally more appropriate to carry out "subtractive deconvolution" in the image plane. The two types of deconvolution are theoretically equivalent (Bates, McKinnon and Bates, 1982). Whichever type is invoked, it is helpful to describe the recovery procedure as deconvolution of $a(x,y)$ by $b(x,y)$ to provide an estimate of $g(x,y)$.

Speckle imaging methods

The better known methods are grouped below according to the order of their interferometry. It is noteworthy that the method which in theory restores true images most faithfully is of third order and therefore exhibits the worst ratio of signal-to-noise. The first order methods are of course much less sensitive to noise but they are not always able to produce faithful images.

First-order interferometry

The notion that a blurred image is least distorted where it is brightest (Bates, 1976) cannot be justified theoretically but it is nevertheless often useful in practice. It suggests that the brightest point in the blurred image can be expected to correspond reasonably faithfully to the brightest point in the true image. This leads to the shift-and-add principle (Bates and Cady, 1980) which asserts that, if each speckle image in a particular sequence is shifted until its brightest point is at the origin of the image plane, and if all speckle images are then averaged, the true image is revealed superimposed upon a smooth background. The "simple" shift-and-add image is defined by

$$fsa(x,y) = \langle sm(x+\xi m, y+\eta m) \rangle \tag{14}$$

It is the simplest member of a class of first-order interferometric images, the other members of which are obtained by, for instance, either appropriately weighting each speckle image (to compensate for the tendency of contamination levels to be largely independent of the brightnesses of the brightest points) or effecting multiple superpositions of individual speckle images (Lynds, Worden and Harvey, 1976).

Since a convolution remains a convolution after both weighting (i.e. multiplication by a constant) and shifting, any member, $fg(x,y)$ say, of the abovementioned class of images can be written in the form

$$fg(x,y) = f(x,y) \odot hg(x,y) + cg(x,y) \tag{15}$$

where $hg(x,y)$ and $cg(x,y)$ are the "generalised" psf and contamination respectively. It is now convenient to denote by $fog(x,y)$ the image obtained from the sequence of reference speckle images by the same processing as $fg(x,y)$ is obtained from the sequence of resolved speckle images, i.e.

$$fog(x,y) = hog(x,y) + cog(x,y) \tag{16}$$

If the brightest region in $f(x,y)$ is only just resolved and is much brighter than the rest of $f(x,y)$ then $hg(x,y)$ is usually quite similar to $hog(x,y)$, so that $fog(x,y)$ deconvolves a faithful version of $f(x,y)$ from $fg(x,y)$ provided the level of the contamination $cog(x,y)$ is low enough.

When $f(x,y)$ does not contain a single "much brightest" point of small size the image obtained by deconvolving $fg(x,y)$ with $fog(x,y)$ is seldom a faithful version of $f(x,y)$, but its size is usually much smaller than that of any of the $hm(x,y)$. This means that $fg(x,y)$ virtually always possesses a much larger spatial frequency content than the seeing disc. The imaging capability has thus been improved as compared with conventional telescope operation and yet the signal-to-noise ratio has not been degraded, as it unavoidably is for other types of speckle imaging.

Second-order interferometry

The assumption inherent in *(8)* is much superior in general to the assumption that the forms of $hg(x,y)$ and $hog(x,y)$ are similar, because the truth of *(8)* is virtually independent of the form of $f(x,y)$. Second-order speckle imaging is based on deconvolving $T(u,v,\alpha,\beta)$ with $To(u,v,\alpha,\beta)$ to provide an estimate of the quantity

$$\Psi(u,v,\alpha,\beta) = F^*(u,v)\, F(u+\alpha,v+\beta) \qquad (17)$$

This processing is only useful if α and β are both small enough that there is sensible phase coherence on the average between the two speckle visibilities inside the angular brackets on RHS *(3)*, or RHS *(7)*. This means that α and β must each be about one quarter of the effective radius of the Fourier transform of the seeing disc. If they are appreciably greater than this, the estimate of $\Psi(u,v,\alpha,\beta)$ tends to be far too noisy to be useful. The celebrated algorithm of Knox and Thompson (1974) permits $F(u,v)$ to be evaluated recursively at points $(j\alpha,k\beta)$ in the Fourier plane, where j and k are integers. The above-mentioned constraints on α and β ensure that the mesh of the rectangular grid at which $F(u,v)$ is found is small enough to avoid any aliasing errors. The recursive evaluation of $F(u,v)$ leads to a build-up of phase errors with increasing radius in the Fourier plane, although the redundant processing recently developed by Noyes, Stachnik and Nisenson (1981) compensates significantly for this and is producing useful results.

The noise sensitivity of the processing is least when

$$\alpha = \beta = 0 \qquad (18)$$

which corresponds to Labeyrie's (1970) speckle interferometry, which technique has several theoretical equivalences to intensity interferometry (Hanbury Brown, 1974). The latter is of course also a species of second-order interferometry. The visibility phase is discarded, but this may soon be of little account (as noted in the

Introduction). Speckle interferometry is now producing many useful astrophysical results (cf. §9 of Bates, 1982).

Several variants of second-order processing have been suggested (cf. §8.4 of Bates, 1982), Walker's (1982) technique being particularly interesting.

Third-order interferometry

Weigelt (1983) has devised a third-order processing technique that recovers the phase of $F(u,v)$ without having to rely on the "coherence assumption" inherent in the Knox-Thompson method (i.e. that $To(u,v,\alpha,\beta)$ and $T(u,v,\alpha,\beta)$ are not over-noisy when α and β are large enough for the processing not to be over-protracted). What has yet to be demonstrated is whether a useful signal-to-noise ratio can be obtained from sequences of speckle images of faint objects.

Discussion

Of all speckle imaging techniques, Labeyrie's speckle interferometry is undoubtedly the most robust, using this term in its statistical sense, implying comparative insensitivity both to uncertainties concerning the a priori information and to noise in the data. It relies on fewer underlying assumptions in order to be able to recover the information it sets out to gather. The astronomical community is not, however, completely satisfied with the performance of the technique in the observatory. Several comments were passed at the conference on "Image reconstruction from astronomical speckle interferometry" (held from 10 - 13 April at the University of Arizona Conference Center in Oracle) on the difficulty of obtaining accurate versions of visibility intensities. There are many possible reasons for this, but the most pertinent may be that speckle interferometry (being of second order) is akin in its signal-to-noise performance to intensity interferometry (cf. Hanbury Brown, 1974). It makes sense therefore to inquire whether a better estimate of $|F(u,v)|$ might be obtained by first-order interferometry. The Fourier transform of *(15)* is

$$FG(u,v) = F(u,v)\ HG(u,v) + CG(u,v) \tag{19}$$

where $fg(x,y) \leftrightarrow FG(u,v)$, $hg(x,y) \leftrightarrow HG(u,v)$ and $cg(x,y) \leftrightarrow CG(u,v)$. If it was to transpire that $|HG(u,v)|$ is a slowly varying function of u and v (so that $HG(u,v)$ introduces mainly phase distortion) for a wide class of true images then $|FG(u,v)|$ could be expected to be a more faithful estimate of $|F(u,v)|$ than that provided by deconvolving $T(u,v,0,0)$ with $To(u,v,0,0)$.

The above question is probably much too difficult to be answered by mathematical analysis. What is needed is exhaustive computational experimentation followed by investigations in the optical laboratory, with observational testing as the final step.

References

Bates, J.H.T., McKinnon, A.E. & Bates, R.H.T. (1982). Subtractive image restoration. II: comparison with multiplicative deconvolution. Optik, 62, no.1, 1-14.

Bates, R.H.T. (1976). A stochastic image restoration procedure. Optics Communications, 19, no.2, 240-4.

Bates, R.H.T. (1982). Astronomical speckle imaging, Physics Reports, 90, no.4, 203-97.

Bates, R.H.T. & Cady, F.M. (1980). Towards true imaging by wideband speckle interferometry. Optics Communications, 32, no.3, 365-9.

Bates, R.H.T. & Gough, P.T. (1975). New outlook on processing radiation received from objects viewed through randomly fluctuating media. IEEE Transactions, C-24, no.4, 449-56.

Cady, F.M. & Bates, R.H.T. (1980). Speckle processing gives diffraction-limited true images from severely aberrated instruments. Optics Letters, 5, no.10, 438-40.

Dainty, J.C. (1973). Diffraction-limited imaging of stellar objects using telescopes of low optical quality. Optics Communications, 7, no.2, 129-34.

Hanbury Brown, R. (1974). Intensity Interferometer. London: Taylor & Francis.

Knox, K.T. & Thompson, B.J. (1974). Recovery of images from atmospherically-degraded short-exposure photographs. Astrophysical J., 193, no.1, L 45-8.

Labeyrie, A. (1970). Attainment of diffraction-limited resolution in large telescopes by Fourier analysing speckle patterns in star images. Astronomy & Astrophysics, 6, no.1, 85-7.

Lynds, C.R., Worden, S.P. & Harvey, J.W. (1976). Digital image reconstruction applied to Alpha Orionis. Astrophysical J., 207, no.1, 174-80.

Noyes, R.W., Stachnik, R.V. & Nisenson, P. (1981). Speckle image reconstruction of solar features. Report # AFGL-TR-0155 (Air Force Geophysics Laboratory). Cambridge, MA: Harvard College Observatory.

Walker, J.G. (1982). Computer simulation of a method for object reconstruction from stellar speckle interferometry data. Applied Optics, 21, no.17, 3132-7.

Weigelt, G.P. & Wirnitzer, B. (1983). Image reconstruction by the speckle masking method. Optics Letters, 8, no.7, 389-91.

3.2 EXPERIMENTAL EVIDENCE OF THE UNIQUENESS OF PHASE RETRIEVAL FROM INTENSITY DATA

J.R. Fienup
Environmental Research Institute of Michigan
P.O. Box 8618, Ann Arbor, Michigan 48107, USA

Summary. An increasing body of theory indicates that the phase retrieval problem usually has a unique solution for 2-D objects. In this paper experimental reconstruction results that support the uniqueness theory are shown.

1 INTRODUCTION

In both optical and radio astronomy, sometimes one can accurately obtain the modulus of the Fourier transform (i.e., the magnitude of the complex visibility function) of an image, but not the Fourier phase. In order to obtain an image it then becomes necessary to retrieve the Fourier phase. Since the autocorrelation function can be computed as the inverse Fourier transform of the squared Fourier modulus, the problem is equivalent to reconstructing an image from its autocorrelation.

In this paper we are concerned with the phase retrieval problem in optical astronomy, in which case one cannot rely on such aids as closure phase (Jennison 1958). However the results shown here do have relevance to radio astronomy as well.

Several methods have been put forward for solving the phase retrieval problem (Liu & Lohmann 1973; Napier & Bates 1974; Frieden & Currie 1976; Baldwin & Warner 1978; Fienup 1978, 1979, 1982; Bates et al. 1982a). In addition there are a number of reconstruction techniques that depend on the specific method of data collection, for example, astronomical speckle interferometry (Bates 1982b). Of the methods that would work for the most general case, the iterative input-output Fourier transform algorithm (Fienup 1978, 1979, 1982) appears to be the most practical.

When any of the reconstruction algorithms finds a solution, the question remains: is it the only (unique) solution or is it one of many possible (ambiguous) solutions? In Section 2 the theory of the uniqueness will be briefly reviewed. Then in Section 3 experimental reconstruction results will be shown that are consistent with the theory that the 2-D case is usually unique. In addition, experimental reconstruction results are shown that indicate that noise in the Fourier modulus data does not radically change the uniqueness of the solution.

2 UNIQUENESS THEORY

When we speak of the reconstruction being unique or ambiguous, we ignore translations and 180° rotations since neither of these operations affects the Fourier modulus. Here we are also assuming that the object has a finite spatial (or angular) extent.

The one-dimensional (1-D) phase retrieval problem has long been known to be highly ambiguous (Walther 1963). Only for the special cases of objects known to consist of sufficiently separated parts or nonnegative objects having sufficiently separated parts is the 1-D phase retrieval problem usually unique (Greenaway 1977; Crimmins & Fienup 1983).

The 2-D case is quite different. This can best be understood from the theory developed by Bruck and Sodin (1979). They considered the special case of an object sampled on a rectangular lattice. For the 1-D case the Fourier transform can then be expressed as a polynomial of order M of a single complex variable, and such a polynomial can always be factored into M irreducible factors (by the fundamental theorem of algebra). They showed that this implies that in the 1-D case there are 2^{M-1} possible solutions, although not all of those solutions would satisfy a nonnegativity constraint (Bates 1969). On the other hand, polynomials of two complex variables having arbitrary coefficients are only rarely factorable. Consequently the 2-D case is usually unique. Although the 2-D theory for continuous functions has not yet been fully developed, it is likely that a similar result will hold.

Of course one can always fabricate 2-D examples that are not unique. An example is an object formed by convolving two nonnegative functions. A second object, formed by convolving the first nonnegative function with an inverted (i.e., rotated by 180°) version of the second nonnegative function, has the same Fourier modulus as the first object. Another method of synthesizing ambiguous cases was given by Huiser and van Toorn (1980). However, these fabricated ambiguous objects are very special cases--most 2-D objects do not fit into these categories.

There are also a number of classes of objects for which the phase retrieval problem is known to be unique (as opposed to just being usually unique). For example, if the object includes an unresolved (delta-function-like) point far enough away from the rest of the object, then the autocorrelation includes the rest of the object as one of its terms (Liu & Lohmann 1973), analogous to holography. It has also been recently discovered that for objects having a special support there is a unique reconstruction even if the reference points are very close to the rest of the object (Fiddy et al. 1983). The support of an object is the set of points over which it is nonzero, i.e., its shape. Also using latent reference points it can be shown that these and other objects having certain supports can be uniquely reconstructed from their Fourier modulus (Fienup 1983a). These recent results point to the importance of the support of an object in determining whether the object can be uniquely reconstructed from its Fourier modulus. Methods for reconstructing support information without resorting to a complete reconstruction are also being investigated (Fienup et al. 1982).

3 UNIQUENESS EXPERIMENTS

3.1 Iterative reconstruction algorithm

One approach to determining whether most objects of interest are uniquely reconstructable from their Fourier modulus is to perform a number of reconstruction experiments. This is now possible due to the existence of a practical reconstruction algorithm, the iterative Fourier transform algorithm (Fienup 1978, 1979, 1982).

The iterative Fourier transform algorithm uses all the available measurements and a priori information to arrive at a solution. In the Fourier domain one has the measured Fourier modulus data, which is an estimate of the true modulus of the Fourier transform of the object. In the object domain one has the a priori constraint that the object's spatial (or angular) brightness distribution is a nonnegative function. From the Fourier modulus data one can compute an estimate of the object's autocorrelation function. From the autocorrelation one can place upper bounds on the diameter of the object (only in special cases can the support of the object be readily determined from the support of its autocorrelation) (Fienup et al. 1982).

The iterative Fourier transform algorithm is a modification of the Gerchberg-Saxton (1972) algorithm that has been used in electron microscopy and for other applications (Fienup 1983b). The simplest version of the iterative algorithm consists of the four following steps. (1) An estimate of the object (an input image) is Fourier transformed. (2) The resulting Fourier-domain function is forced to conform to the measurements by replacing the computed Fourier modulus with the measured Fourier modulus. (3) The result is inverse Fourier transformed, yielding an output image. (4) A new input image is formed by forcing the output image to conform to the object-domain constraints, i.e., it is set equal to zero where it is negative or where is exceeds the known diameter (i.e., the support constraint). This algorithm, which we call the error-reduction algorithm, can be proven to converge in the sense that the error at the kth iteration is always less than or equal to the error at the $(k - 1)^{th}$ iteration. Here the error is defined as the amount by which the computed Fourier modulus differs from the measured Fourier modulus or as the amount by which the output image violates the object-domain constraints. However, in practice the error-reduction algorithm usually converges so slowly that it is impractical for this application (Fienup 1982).

Fortunately there exist a number of accelerated versions of the algorithm which converge in a reasonable number of iterations. To date the fastest version of the algorithm is the hybrid input-output algorithm. Its first three steps are identical to those of the error-reduction algorithm described above. The fourth step of the hybrid input-output algorithm consists of forming a new input image that is equal to the output image wherever the output image satisfies the constraints, and is equal to the previous input image minus a constant factor times the output image wherever the output image violates the constraints. Any value between 0.5 and 1.0 works well for constant factor, which is similar to a negative feedback parameter.

In one series of trials, the algorithm was run on a fabricated Fourier modulus which was known to have two solutions. One of the two solutions was reconstructed in about half of the trials and the other solution was reconstructed in the other half of the trials. Which of the two solutions was obtained depended on the array of random numbers used as the initial input to the algorithm. Therefore we believe that if there are multiple solutions, then the algorithm is equally likely to find any one of them (if the initial input is sufficiently random and unbiased), and if run enough times with different initial inputs, it will probably find all of them. In a practical reconstruction situation in which the solution is not known beforehand, if one were to run the algorithm two or three times, each time using a different array of random numbers for the initial input, and if the reconstructed images were the same each time, then one would be highly confident that one had found the solution and that it is unique (Fienup 1979).

A problem with experimental reconstruction experiments is that there is no guarantee that the iterative algorithm will converge to <u>any</u> solution, even when an accelerated version of the algorithm is used. One can think of the reconstruction algorithm as an iterative search through an N^2-dimensional parameter space (each dimension or parameter corresponding to the value of one of the pixels of the image), seeking to minimize the error of the estimate. While searching for the global minimum of the error, the algorithm could stagnate at a local minimum of the error in that N^2-dimensional space. The likelihood of stagnation and the success of the algorithm depend on the N^2-dimensional topography of the error function, which varies from one type of object to another. Therefore, for particularly difficult objects, i.e., ones for which the error has many local minima, one may not be able to test for uniqueness since the reconstruction algorithm fails. Fortunately such a problem has occurred only occasionally for the types of objects examined.

One particular convergence problem has occurred on several occasions. Sometimes the algorithm stagnates at a deep local minimum at which the output image resembles the original object but with a pattern of stripes superimposed. A similar phenomenon has occurred in other reconstruction situations (Cornwell 1983). In most cases the stripes are of low contrast, superimposed on an otherwise excellent reconstructed image, and are of little concern. In other cases the stripes are of high enough contrast to be objectionable. When the Fourier modulus data is sufficiently noisy, then the stripes do not appear (Feldkamp & Fienup 1980). The nature of the stripes is as yet not fully understood and methods of avoiding them remain to be developed.

An example of the stripes phenomenon is shown in Figure 1. Figure 1(a) shows the original object and Figure 1(b) shows a reconstructed image, which appears to be quite faithful. Figure 1(c) shows the same reconstructed image, but heavily overexposed, in order to emphasize the low-contrast vertical stripes that are present, although difficult to discern, in the image. Figures 1(d-f) show the overexposed reconstructed images resulting from three other trials of the algorithm, each of which was initialized with a different array of random numbers. In each of these three cases the reconstructed image contains a more easily discernable pattern of stripes, but the spatial frequencies and orientations of the stripes are different in each case. The stripes extend throughout image space (although they are weaker away from the support of the object), and therefore by inspection of the reconstructed images it is possible to determine that the stripes are an artifact rather than a true feature of the object. Furthermore, it is possible to discern the true image from the stripes since the stripes change from one reconstruction to the next, but the true features of the object are present in all the reconstructed images.

3.2 Experimental uniqueness results for various objects

The iterative reconstruction algorithm was used to reconstruct a number of different objects from their Fourier modulus. The objects examined are of a very practical and interesting class: digitized photographs of satellites. They also share a feature that we suspect makes them "good" objects to reconstruct: they have interesting (i.e., complicated) shapes.

A typical result is shown in Figure 2, in which (a) is the original object and (b) is the reconstructed image (Fienup 1981). For this and almost all of the cases examined, the reconstructed image looks much like the original object except for differences that could be attributed to stripes. For example, horizontal stripes are evident over portions of the reconstructed image shown in Figure 2(b). Therefore, except for the presence of the stripes artifact which we believe is a characteristic of a local minimum rather than an inherent ambiguity, most objects of this type are uniquely related to their Fourier modulus.

There are exceptions, however. Figure 3 shows one case that worked particularly poorly. The object shown in Figure 3(a) is nearly centrosymmetric. Figure 3(b) shows the reconstructed image, which is not very faithful. This particular case has similarities with the ambiguous case fabricated by Huiser and van Toorn (1980). From this we see that, although ambiguous cases may be unusual, they are by no means nonexistent in the real world.

3.3 Experimental uniqueness in the presence of noise

As with any reconstruction method the sensitivity of the algorithm to noise is a major point of concern. Reconstruction results using noisy Fourier modulus data have shown that the iterative Fourier

Figure 1. (a) Original object; (b) image reconstructed from Fourier modulus using iterative algorithm; (c)-(f) four images reconstructed using different starting inputs--these pictures were intentionally overexposed in order to emphasize the stripes.

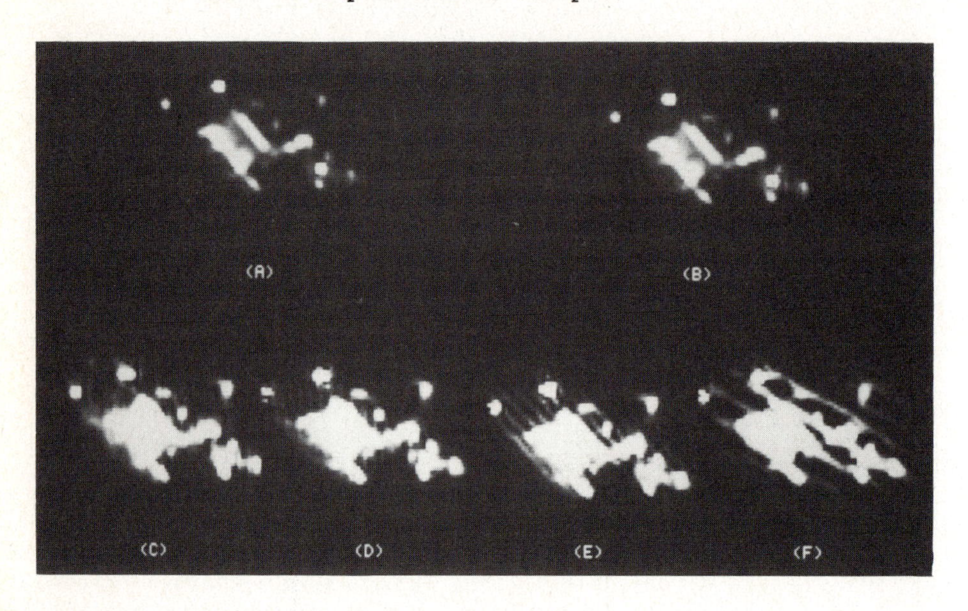

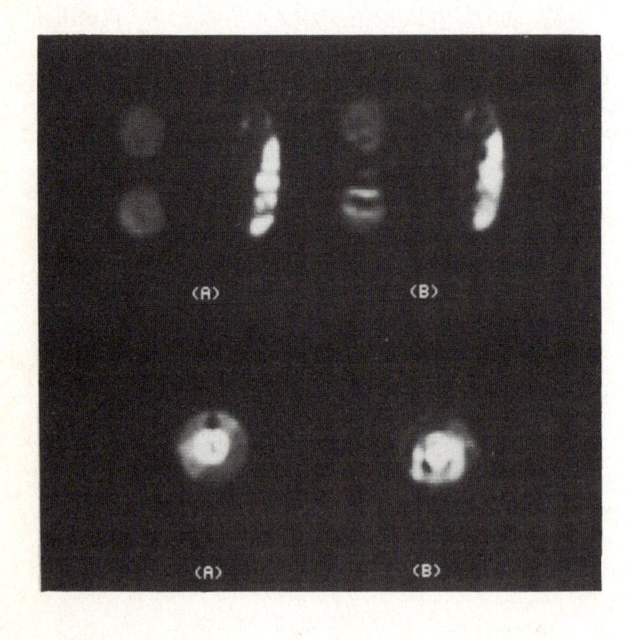

Figure 2. (a) A typical object; (b) image reconstructed from Fourier modulus.

Figure 3. (a) An atypical object, for which the reconstructed image (b) does not resemble the object.

transform algorithm is not highly sensitive to noise (Fienup 1978, 1979). In this section the results of a systematic study of the noise sensitivity of the reconstruction (Feldkamp & Fienup 1980) are summarized.

When noise is present in the Fourier modulus data, then there is generally no solution that is completely consistent with both the measured data and the constraints. For example, an autocorrelation function computed from a noisy Fourier modulus would be very likely to have some negative values for the largest separations. Obviously no nonnegative object can have an autocorrelation having negatives; therefore there could be no nonnegative object consistent with the noisy Fourier modulus. Nevertheless the algorithm searches for a solution that is most consistent with the measured data and constraints, and in doing so it can arrive at a useful image.

Fourier modulus data was simulated to have the type of noise that would be present in astronomical speckle interferometry. The object shown in Figure 1(a) was convolved with 156 different point-spread functions to produce 156 different blurred images. Each of the point-spread functions represents a different realization of the blurring due to the turbulent atmosphere. The widths of the point-spread functions were comparable to the width of the object. The blurred images were then subjected to a Poisson noise process to simulate the effects of photon noise. The degraded images were then processed to produce a noisy Fourier modulus estimate by Labeyrie's (1970) method, as modified by Goodman and Belsher (1976) to eliminate the bias noise term from the squared Fourier modulus.

Figure 4 shows a noise-free Fourier modulus (a) and three examples of the simulated noisy Fourier modulus estimates (b)-(d) with increasing noise. Figure 5 shows the original undegraded object (a) and three images (b)-(d) reconstructed from the respective noisy Fourier modulus estimates of Figure 4. For the case shown in Figures 4(b) and 5(b), which represent a realistic amount of noise for this situation, the normalized rms error of the Fourier modulus estimate was 2.9% and the reconstructed image is very good. For the case shown in Figures 4(c) and 5(c), only 1/50 as many photons were assumed to be available, and the rms error of the Fourier modulus estimate is a very poor 32%; nevertheless the reconstructed image still retains some recognizable features. In the case shown in Figures 4(d) and 5(d), an extreme amount of noise was present, and the rms error of the Fourier modulus estimate is near 100%; since this Fourier modulus estimate does not resemble the true Fourier modulus, then, as one would expect, the reconstructed image does not resemble the original object.

4. CONCLUSIONS

Theory, which points toward the conclusion that a 2-D object of finite extent is ordinarily uniquely related to the modulus of its Fourier transform, has been supported by experimental reconstruction re-

Figure 4. Fourier modulus estimates with noise, having rms error (a) 0%, (b) 2.9%, (c) 32%, (d) ~100%.

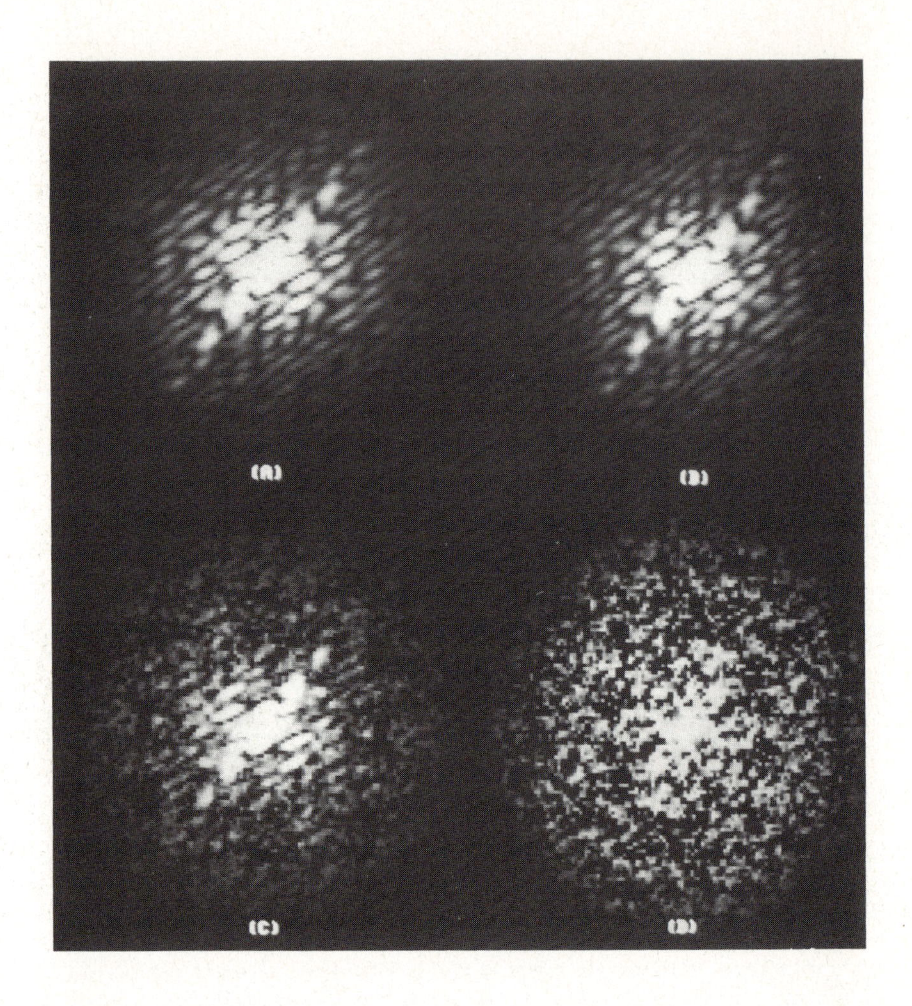

Figure 5. Images reconstructed from noisy Fourier modulus estimates shown in Figure 4.

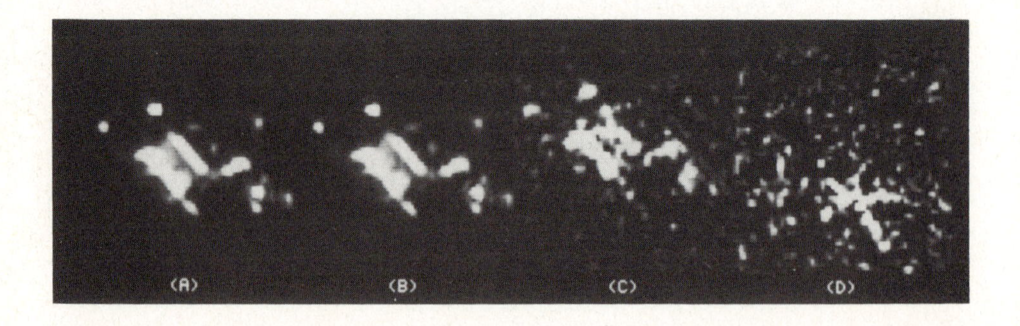

sults. The vast majority of reconstructed images of satellites resemble the original objects from which the Fourier modulus was computed. Furthermore, contrary to some predictions (Huiser & van Toorn 1980), the uniqueness properties do not change radically when noise is present. Rather, as more noise is introduced into the Fourier modulus estimate, the reconstructed image simply becomes correspondingly noisier, and degrades in a gradual manner.

ACKNOWLEDGEMENT

This research was supported by the U.S. Air Force Office of Scientific Research and Rome Air Development Center.

REFERENCES

Baldwin, J.E. & Warner, P.J. (1978). Phaseless aperture synthesis. Mon. Not. R. Astr. Soc. 182, 411-422.

Bates, R.H.T. (1969). Contributions to the theory of intensity interferometry. Mon. Not. R. Astr. Soc. 142, 413-428.

Bates, R.H.T., et al. (1982a). Fourier phase problems are uniquely solvable in more than one dimension. I: Underlying theory. Optik 61, 247-262; ... II: One-dimensional considerations. Optik 62, 131-142; ... III: Computational examples for two dimensions. Optik 62, 219-230.

Bates, R.H.T. (1982b). Astronomical speckle imaging. Physics Reports (Reviews Sect. of Phys. Lett.) 90, 203-297.

Bruck, Yu.M. & Sodin, L.G. (1979). On the ambiguity of the image reconstruction problem. Opt. Commun. 30, 304-308.

Cornwell, T.J. (1983). A method of stabilizing the CLEAN algorithm. submitted to Astron. Astrophys.

Crimmins, T.R. & Fienup, J.R. (1983). Uniqueness of phase retrieval for functions with sufficiently disconnected support. J. Opt. Soc. Am. 73, 218-221.

Feldkamp, G.B. & Fienup, J.R. (1980). Noise properties of images reconstructed from Fourier modulus. In 1980 International Optical Computing Conference. Proc. SPIE 231, 84-93.

Fiddy, M.A., Brames, B.J., & Dainty, J.C. (1983). Enforcing irreducibility for phase retrieval in two dimensions. Opt. Lett. 8, 96-98.

Fienup, J.R. (1978). Reconstruction of an object from the modulus of its Fourier transform. Opt. Lett. 3, 27-29.

Fienup, J.R. (1979). Space object imaging through the turbulent atmosphere. Opt. Eng. 18, 529-534.

Fienup, J.R. (1981). Fourier modulus image construction. Rept. RADC-TR-81-63.

Fienup, J.R. (1982). Phase retrieval algorithms: a comparison. Appl. Opt. 21, 2758-2769.

Fienup, J.R. (1983a). Reconstruction of objects having latent reference points. To appear in J. Opt. Soc. Am. 73, November.

Fienup, J.R. (1983b). Reconstruction and synthesis applications of an iterative algorithm. In Transformations in Optical Signal Processing, ed. W.T. Rhodes, J.R. Fienup & B.E.A. Saleh. Bellingham, Washington: SPIE.

Fienup, J.R., Crimmins, T.R., & Holsztynski, W. (1982). Reconstruction of the support of an object from the support of its autocorrelation. J. Opt. Soc. Am. 72, 610-624.

Frieden, B.R. & Currie, D.G. (1976). On unfolding the autocorrelation function. J. Opt. Soc. Am. 66, 1111 (Abstract).

Gerchberg, R.W. & Saxton, W.O. (1972). A practical algorithm for the determination of phase from image and diffraction plane pictures. Optik 35, 237-246.

Goodman, J.W. & Belsher, J.F. (1976). Fundamental limitations in linearly invariant restoration of atmospherically degraded images. In Imaging through the atmosphere. Proc. SPIE 75, 141-154.

Greenaway, A.H. (1977). Proposal for phase recovery from a single intensity distribution. Opt. Lett. 1, 10-12.

Huiser, A.M.J. & van Toorn, P. (1980). Ambiguity of the phase-reconstruction problem. Opt. Lett. 5, 499-501.

Jennison, R.G. (1958). Mon. Not. Roy. Astr. Soc. 165, 25.

Labeyrie, A. (1970). Attainment of diffraction limited resolution in large telescopes by Fourier analysing speckle patterns in star images. Astron. and Astrophys. 6, 85-87.

Liu, C.Y.C. & Lohmann, A.W. (1973). High resolution image formation through the turbulent atmosphere. Opt. Commun. 8, 372-377.

Napier, P.J. & Bates, R.H.T. (1974). Inferring phase information from modulus information in two-dimensional aperture synthesis. Astron. Astrophs. Suppl. 15, 427-430.

Walther, A. (1963). The question of phase retrieval in optics. Optica Acta 10, 41-49.

DISCUSSION

J.A. ROBERTS

When you added noise did you still require the transform of the reconstruction to match the data exactly?

J.R. FIENUP

The reconstructed images shown have exactly the measured noisy Fourier modulus. When noise is present there is no solution that is perfectly consistent, but the algorithm seeks the solution most consistent with both the noisy data and the a priori constraints.

P.E. DEWDNEY

I notice that many of your test images are very contrasty (i.e. have sharp edges). How does this property affect your reconstruction?

J.R. FIENUP

Objects having sharp edges cause a greater degree of interference within the Fourier transform (have more prominent high-frequency components) and are easier to reconstruct than smoother objects. We have successfully reconstructed photographs of satellites that were low-pass filtered, so sharp edges are not required for success. Of course the majority of interesting astronomical objects are fairly contrasty (have moderately large dynamic range) and they should be easy to reconstruct. For objects that have a high background level, methods such as Bates' defogging should help.

J. NOORDAM

Both transform-planes seem to have the same number of points, whereas in radio-astronomy we have many more map-points than measured u,v-points. This should influence the retrieval-process, because the degree of freedom in the map is much larger.

J.R. FIENUP

These reconstruction algorithms will work less well for partially-filled u,v apertures than for fully-filled apertures. Experiments will have to be performed to determine the effects.

J.P. HAMAKER

I am inclined to think that iterative techniques are a brute-force approach which one uses when one has not been clever enough to find a more subtle method. I wonder if other more 'intelligent' approaches exist?

J.R. FIENUP

The problem can be stated in terms of solving N^2 nonlinear simultaneous equations having M^2 unknowns, where both N^2 and M^2 are very large numbers. Unfortunately no one has been able to solve this nasty problem in an elegant way. Standard methods including least squares (Fieden and Currie) and gradient search have been used, but the iterative Fourier transform algorithm has proven to be the most efficient so far.

3.3 AMBIGUITIES IN SPECKLE RECONSTRUCTIONS - SOME WAYS OF AVOIDING THEM

A H Greenaway, J G Walker and J A G Coombs
Royal Signals & Radar Establishment, Malvern, WR14 3PS, UK

Abstract. Examples of two dimensional objects, which present uniqueness problems for phaseless reconstruction methods, are presented and used to illustrate the need for methods which avoid such ambiguities. Two such methods are briefly discussed. One, based on exponential apodization, is suitable for restoration of diffraction-limited images of bright and faint objects from an ensemble of turbulence degraded images. The other method, based on phase closure, requires a special interferometer, but for bright objects permits diffraction-limited image reconstruction from a single frame of turbulence-degraded data.

1 INTRODUCTION

A number of methods have been proposed for mitigating the degrading effects of atmospheric turbulence in high-angular resolution astronomy (van Schooneveld, 1979). None-the-less, relatively few "diffraction-limited" image reconstructions from experimental data have been published. One reason for this is the difficulty to obtain the very high quality data required but a further problem is the question of uniqueness of the solutions obtained from phaseless reconstructions using algorithms like the Maximum Entropy (Gull & Daniell, 1978) and the Fienup (1978) algorithms. Whilst neither of these methods is restricted to modulus only reconstructions, the frequently made assertion (Barakat & Newsam, 1983) that the two-dimensional phaseless reconstruction problem is, or is almost always, unique means that phaseless applications are still the methods most commonly tried.

This statement regarding the uniqueness of two-dimensional phaseless reconstruction rests on the fact that ambiguities can arise only when the Fourier transform of the object may be factorized into two or more analytic factors. Since such factorizability is rather exceptional in two-dimensions, it follows that ambiguity in two-dimensional problems is unlikely. This applies to perfect data. Real data is truncated and subject to error. In consequence, the measured data is not strictly compatible with object constraints such as reality, positively and finite support. Thus one no longer seeks a solution which exactly matches the measured power spectrum and known object constraints, but one that approximates these within some predetermined error. It has been supposed that noise on the data will lead to "noisy reconstructions" which form a compact set in the solution space. Such a conclusion has been supported by the results of computer simulations (Fienup, 1979). However, relaxation of the precision with which the solution must fit the data, necessitated by the presence of error, increases the chances that a factorizable function may be found within the set of functions which give an acceptable fit to measured data. If such a factorizable function can be found the risk of ambiguity will increase and the set of "solutions" may become disjoint in the solution space, ie the algorithms may produce a number of dissimilar "solutions". This hypothesis has been tested (van Toorn et al, 1983) by producing a power spectrum which corresponds to an object whose Fourier transform is factorizable. In the example shown in figure 1, the factorizability was chosen to give a four-fold ambiguity. Using a noise-free but truncated power spectrum, a Fienup algorithm was then used to reconstruct the object from the power spectrum. The algorithm minimizes the energy in negative excursions or beyond the geometric bounds of the object. The results of figure 1 show that, for this example, the hypothesis was correct. The algorithm produced solutions not only from the theoretically correct set of simple, high contrast objects but also several stable pseudo-solutions not contained within the set of correct solutions. Some progress has been made toward understanding the type of object structures which will guarantee the uniqueness of phaseless reconstructions (Fiddy et al, 1983) but it would seem that unless ones "reconstructions" fit into this class (in which case repeated, randomly seeded starts will yield at most one solution and its mirror image) one should make maximum use of any other information which may be available. If repeated trials do not yield a single solution one must try an alternative approach.

2 ALTERNATIVE APPROACHES

2.1 Exponential Filtering

This method (Walker, 1982) is similar to the Fienup algorithm in implementation but guarantees to exclude ambiguity by using the turbulence degraded data to produce two functions of the object. One is the modulus of the Fourier transform of the object, which is measured by speckle interferometry processing. The other is the modulus of the Fourier transform of the product of the object and an exponential, which is obtained by multiplying each of the images by an exponential and then using speckle interferometry processing. These two functions of the object together contain all the information to reconstruct the object. Unlike Fienup's algorithm, the exponential filtering algorithm can diverge, a solution is not guaranteed, but if found that solution will closely resemble the object. In practice such divergence seldom occurs. Further, it should be noted, the exponential filter method guarantees an unambiguous solution even for the one-dimensional case, in which ambiguities are considerably more likely even with perfect data. A computer comparison of the exponential filtering algorithm with Fienup's algorithm is given in figure 2.

2.2 Phase Closure

This method (Greenaway, 1982; Arnot, 1983) offers a way of obtaining a unique solution from a single frame of turbulence-degraded data and may be understood from the following.

Suppose that an image of some object, $o(\underline{r})$, is recorded through the matrix of apertures shown in figure 3a. The Fourier transform of the image of a self-luminous object is the Fourier transform of the object itself multiplied by the autocorrelation of the aperture system through which the image was formed (shown in 3b). If $\underline{\xi}_{AF}$ is the spatial frequency corresponding to the displacement of pinhole F from pinhole A, if the Fourier transform of the object is $O(\underline{\xi})e^{i\theta(\underline{\xi})}$ and if the effect of the atmosphere is to introduce constant random phase changes, ϕ_j, over each of the apertures ϕ_A, ϕ_B etc, then at the point labelled AF in figure 3b the log of the Fourier transform of the image has the value

$$M(\underline{\xi}_{AF}) = \ln\left|O(\underline{\xi}_{AF})\right| + i[\theta(\underline{\xi}_{AF}) + \phi_A - \phi_F] \qquad (1)$$

Now, because of the position occupied by pinhole 1 prior to its displacement, one has that

$$\underline{\xi}_{AF} \equiv \underline{\xi}_{1G} \quad , \quad \underline{\xi}_{AE} \equiv \underline{\xi}_{1F} \quad , \quad \underline{\xi}_{BD} \equiv \underline{\xi}_{1B} \qquad (2)$$

etc. Also, one has available 3 disposable parameters. Arbitrarily setting $\theta(\underline{\xi})$ to zero at two $\underline{\xi}$'s, $\underline{\xi}_{AE}$ and $\underline{\xi}_{1A}$ say, is equivalent to displacing the object on the sky without changing its morphology. Arbitrarily setting ϕ_1, say, to zero does not affect the imaging process (only phase differences matter). Combining equations (1) in pairs as indicated by equations (2) one has, for example,

$$M(\underline{\xi}_{AF}) - M(\underline{\xi}_{1G}) = \phi_A - \phi_F + \phi_G \qquad (3)$$

giving a set of equations which may be solved for all ϕ's using Guass elimination. The ϕ values thus obtained may be used in equations (1) to solve for $O(\underline{\xi})$.

This method suffers from the disadvantages that it requires the construction of a specially designed interferometer and will only work on bright objects.

ACKNOWLDGEMENTS

The authors wish to acknowledge discussion with N Arnot.

REFERENCES

Arnot N, 1983, to appear in Optics Commun.
Barakat R & Newsam G, 1983, to appear in Optics Commun.
Fiddy M A, et al, 1983, Optics Letters 8, 96
Fienup J R, 1978, Optics Letters 3, 27
Fienup J R, 1979, Optical Engineering 18, 529.
Greenaway A H, 1982, Optics Commun 42, 157
Gull S F & Daniell G J, 1978, Nature 272, 686
Van Schooneveld C, 1979, "Image Formation from Coherence Functions in Astronomy", Reidel.
Van Toorn P, et al, 1983, to be submitted to Optics Commun.
Walker J G, 1982, Applied Opt, 21, 3132.

Figure 1. The pictures on the top line show four objects having identical power spectra. These four objects, and their twin images, are the only admissible solutions. Twelve randomly seeded starts of a Fienup algorithm gave the results shown on the lower three rows. Numbering left to right, starting top left, 9 and 16 represent admissible solutions and fit the data to within 2% rms error. 6, 10 and 15 fit the data to within 3% and 7, 8, 11, 12 and 14 to 4% error. 13 has failed to converge (~8%) and 5 is converging only slowly (~5%). The apparent failure to find solution like 3 or 4 would seem to be exceptional. Evidently one requires very accurate data for successful phaseless reconstructions.

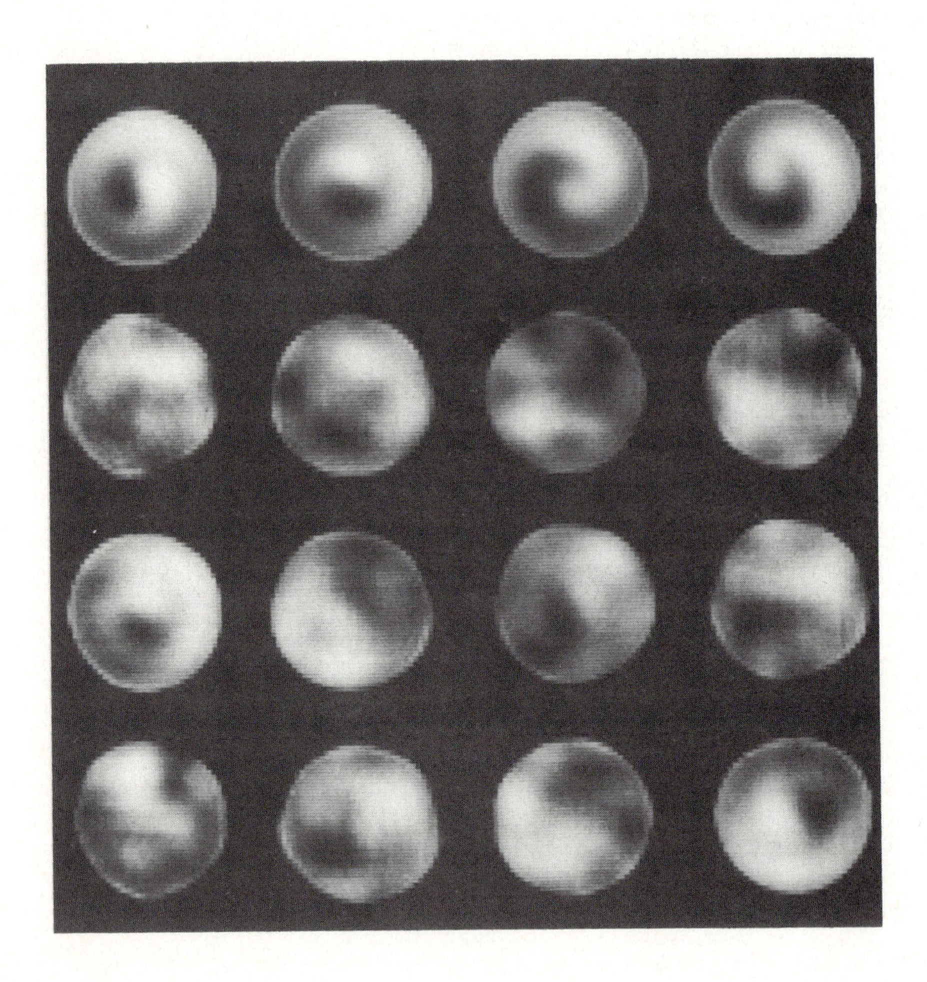

Figure 2. A comparison of the Fienup and Exponential filtering methods of processing stellar speckle interferometry data. The top row shows, from left to right, the test object; a randomly selected high light level short exposure image; a randomly chosen high light level image of a point source; the twin image of the object. To obtain the data for the reconstructions on the next two rows 200 images of a point spread function were generated. 100 of them were convolved with the test object to produce simulated short exposure images of the object, the other 100 were used to evaluate the short exposure atmospheric transfer function.
Repeated attempts were made to reconstruct this object using a randomly seeded Fienup algorithm, constrained to fit the given power spectrum and yield a positive solution within the geometric image with zero outside. The second row shows the four solutions corresponding to the best fits to the data, all of which are within the known error bound.
Solutions were then obtained from the same data using the exponential filter method (Walker, 1982); the last row shows the four solutions corresponding to the best fits to the data.

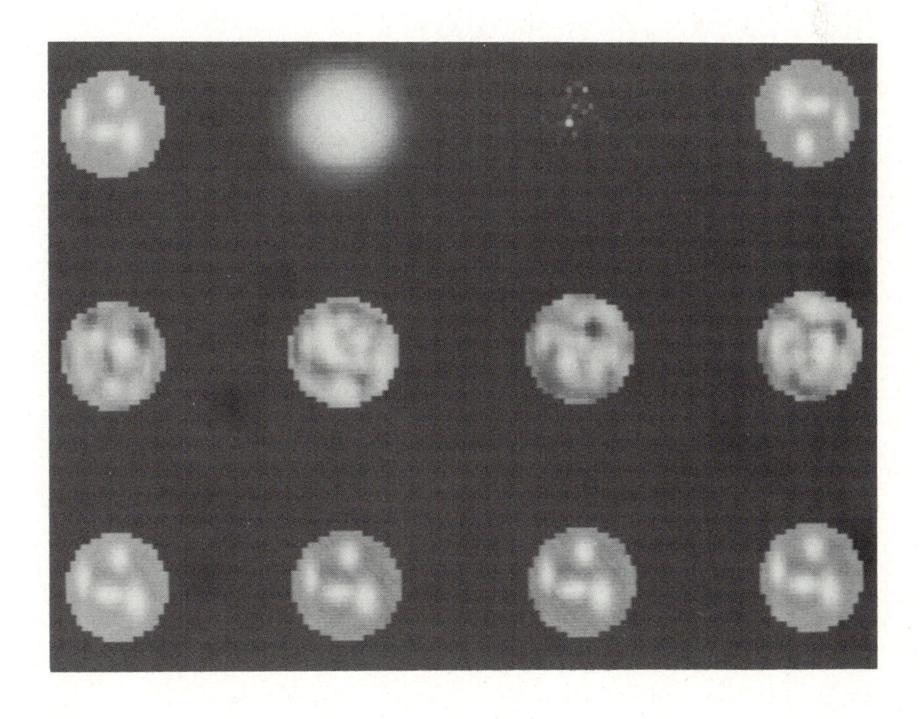

Figure 3.

3a. The matrix of apertures used in the phase closure method. The interferometer must displace the reference aperture from the dotted position to position 1 using a periscope mirror arrangement.

3b. The autocorrelation of aperture assembly 3a. The heavier circles represent cross-correlations between aperture 1 and the T. Some of the other circles are labelled to indicate from which cross-correlation they arise. The axes indicate the origin in Fourier space. The method directly yields the object spectrum at all spatial frequencies within the dotted line, with the exception of those shown cross-hatched. These latter may, in principle, be deduced from the values obtained.

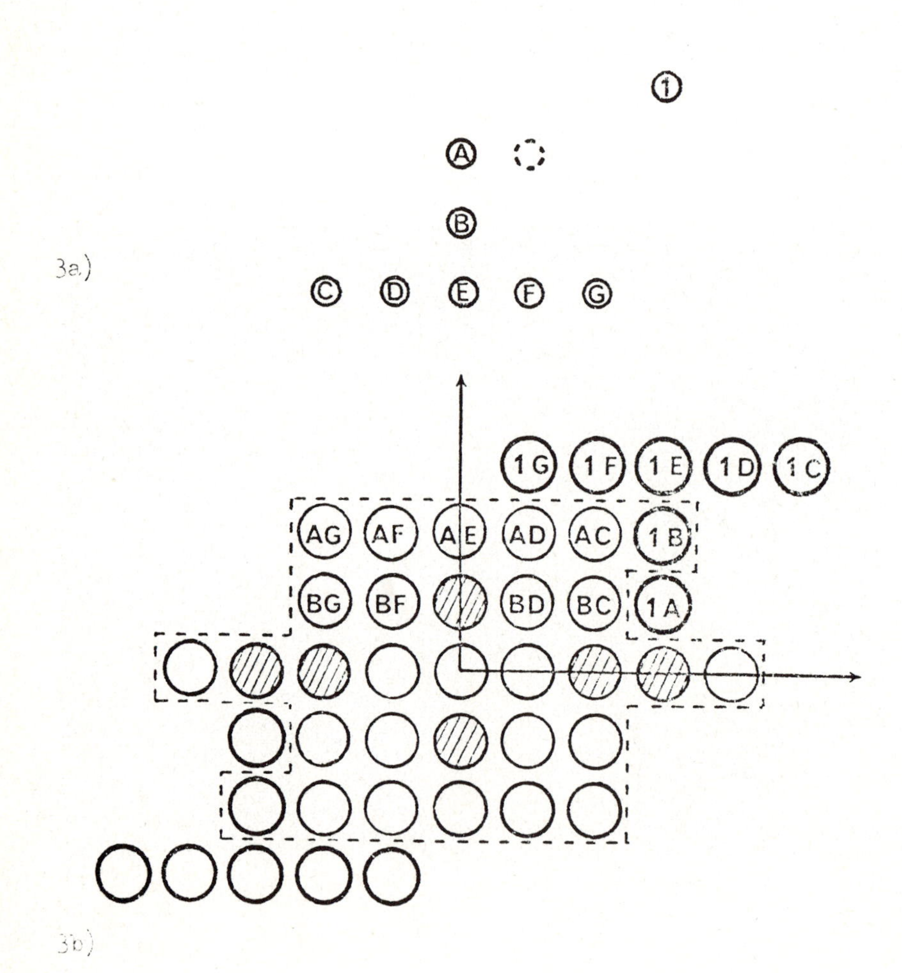

DISCUSSION

J.R. FIENUP

It should be noted that in no case was an absolutely true solution found. 'Convergence' was declared to be achieved (as is customary) when the residual rms error seemed to be low, and slow progress was being made from one iteration to the next. In the examples shown, when the residual rms error was driven to the lowest levels (i.e. closest to a solution), then the reconstructed image was quite good. The poor reconstructions shown had larger residual errors. Therefore the seeming lack of uniqueness for those poorer reconstructions were really just cases of incomplete convergence, which hopefully could be corrected by a larger number of iterations or by a more clever reconstruction algorithm.

J.E. BALDWIN

The author's belief is that only one of the reconstructions had failed to converge. Some of the solutions were only very slightly worse in fit (3% rms) than those corresponding to the expected ambiguous solutions (2% rms).

3.4 PHASE RESTORATION IS SUCCESSFUL IN THE OPTICAL AS WELL AS THE COMPUTATIONAL LABORATORY

R.H.T. Bates
University of Canterbury, Christchurch, New Zealand

W.R.Fright
University of Canterbury, Christchurch, New Zealand

W.A.Norton
University of Canterbury, Christchurch, New Zealand

Abstract. The current status of the two-dimensional astronomical Fourier phase problem is assessed. Experience of phase retrieval from Fourier intensities measured in our optical laboratory is summarised. Results of phase retrieval from measured Fourier intensities of spatially coherent sources are presented. Four problem areas most urgently in need of further research are identified and discussed, and promising lines of investigation are outlined.

Introduction

High-resolution astronomical imaging - using either radio aperture synthesis (Thompson et al, 1980) or optical speckle imaging (Bates, 1982b) - is concerned with forming a diffraction-limited, non-negative real image $f(x,y)$ of a region of the celestial sphere from measurements of its complex visibility (i.e. Fourier transform) $F(u,v)$. Note that x,y and u,v are Cartesian co-ordinates in the image and Fourier planes respectively. The intensity of the visibility can usually be directly inferred from measurement appreciably more accurately than its phase. When the latter is completely lost, it is impossible to determine exactly where the image lies on the celestial sphere. Also the mirror image is as compatible with the visibility intensity as is the image itself. However, an object can be recognised when viewed in a mirror and its form is not altered merely by changing its position. Accordingly, $f(x-\xi 1,y-\eta 1)$ and $f(-x-\xi 2,-y-\eta 2)$ are said to possess the same "image-form" as $f(x,y)$, where $\xi 1$, $\eta 1$, $\xi 2$ and $\eta 2$ are arbitrary constants (Bates, 1982a; Fright and Bates, 1982).

We pose the astronomical phase problem as: given $|F(u,v)|$, at points spaced sufficiently closely in the Fourier plane that the auto-correlation of $f(x,y)$ can be immediately reconstructed accurately with the aid of the FFT algorithm, recover the image-form of $f(x,y)$.

Following a decade of hints that the Fourier phase problem is "much less non-unique" in two dimensions than in one, it has finally been made clear to the optical/image-processing community by Bruck and Sodin (1979) why it is that one can almost always (in the mathematical sense) expect to be able to recover the image-form from the visibility intensity. Even though some active workers (cf. Fiddy, Brames and Dainty, 1983) remain concerned about potentially ambiguous image-forms, the experience of Fienup (1982) and ourselves (Bates and Fright, 1983a) strongly suggests that non-uniqueness only occurs in contrived situations.

It is worth emphasising that phase retrieval is likely to become essential in astronomy when (and if) optical synthesis telescopes are constructed, because it is most improbable that visibility phases will be directly inferable from the measurements. In current speckle interferometric practice there are several ways of estimating phase to a useful accuracy (cf. §8 of Bates, 1982b).

We comment below on our reconstructions of image-forms from measured Fourier intensities and we conclude by summarising our ideas on what needs to be done most urgently to make phase retrieval a practical astronomical reality.

Phase retrieval in the optical laboratory

Something experimenters know well is that the ways in which real-world data become contaminated often bear little resemblance to the "noise models" programmed into computer simulations. This is as puzzling as it is true, and in our experience is much more noticeable for measurements of fields (e.g. in the optical and ultrasonic laboratories or radio anechoic chambers) than of signals (whose corruptions tend to be much more "Gaussian") - perhaps this is yet another manifestation of a change in the number of dimensions being a matter of kind rather than merely of degree. What is certain is that much electromagnetic and acoustic propagation distortion is non-isoplanatic, so that one tends to be sceptical about most theoretical approaches to allowing for data contamination. The moral of this homily is that the efficacy of a phase retrieval algorithm cannot be properly assessed merely by testing it on computer generated data.

We have successfully reconstructed image-forms from visibility intensities measured in our optical laboratory, even when we have made sure that the data are "well contaminated" by dust on the lenses, careless registration of the electronic recording camera, etc. This has been made possible by incorporating various "tricks" (Bates and Fright, 1983b) into our composite algorithm (Bates and Fright, 1983a). Our opinion is that existing phase retrieval algorithms cannot operate effectively on most measured data unless they are extended in some such ways. It is worth listing here our sequence of operations, described in the afore-mentioned papers: defogging, pre-filtering, crude phase estimation, image boxing, Fienup cycling (i.e. employing Fienup's error-reduction and hybrid-input-output algorithms in appropriate combinations to reduce differences between successive iterates of the image-form to below a pre-chosen threshold) and refogging.

Our previous report of phase retrieval from measured data related to a laboratory simulation of astronomical speckle interferometry

Fig.1. True (left) and reconstructed (right) images.

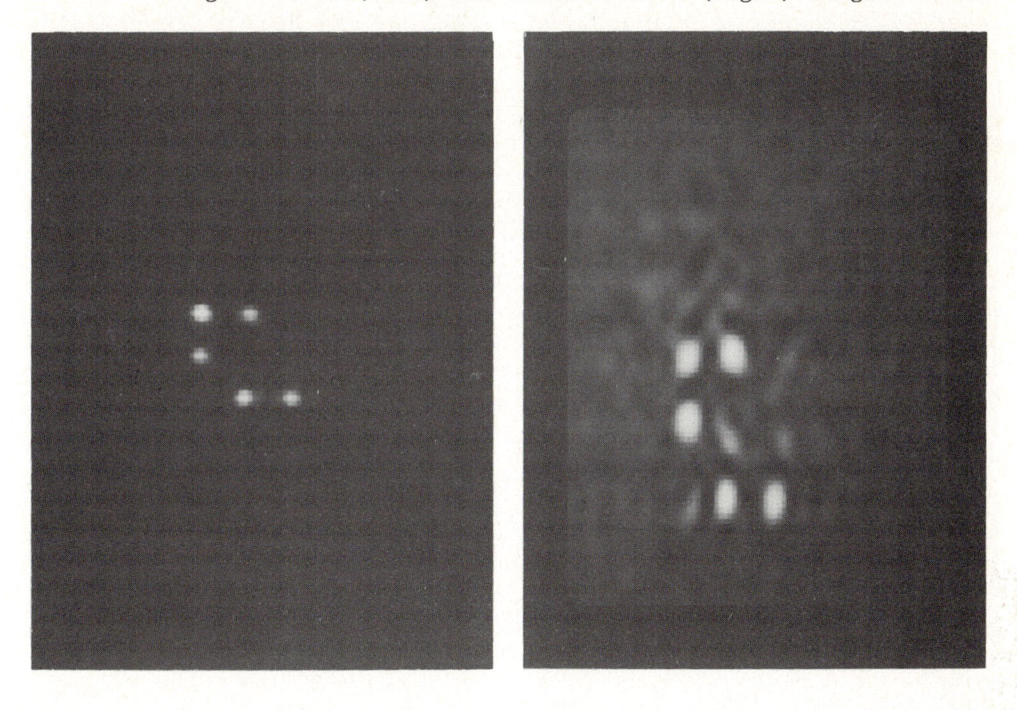

Fig.2. More (left) and less (right) carefully measured visibility intensities.

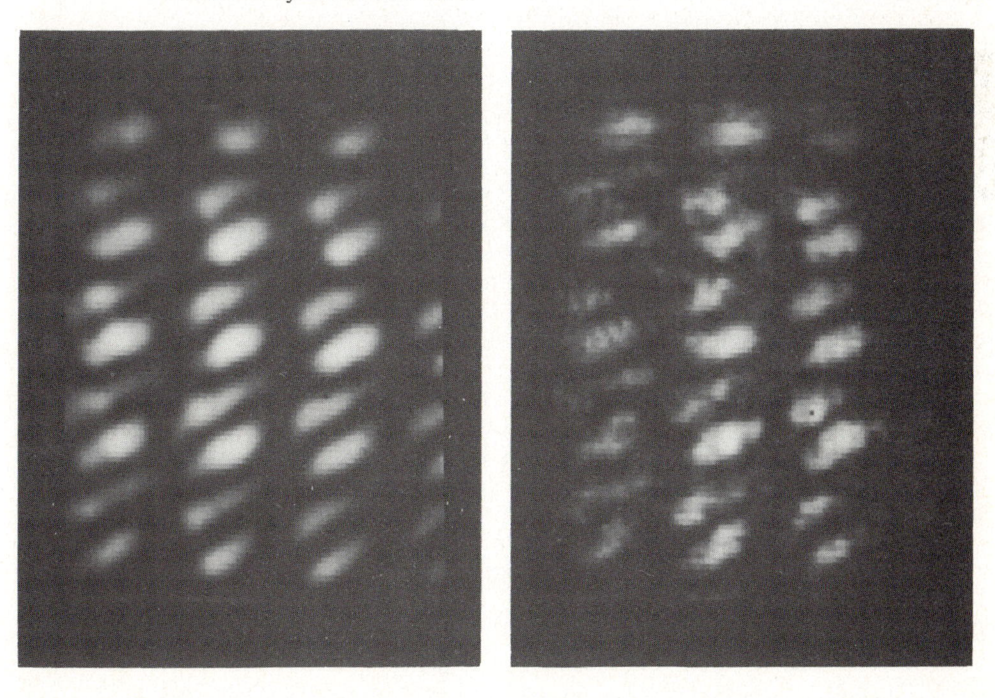

(Bates and Fright, 1983b), for which the sources (i.e. the simulated stars) were spatially incoherent. Such sources of course arise in almost all astronomical contexts. Spatially coherent sources are, however, of general scientific and technical interest. So, we think it is appropriate to complement our earlier results by presenting an image reconstructed from the intensity of the far-field, or Fraunhofer, pattern (this is effectively the same thing as the complex visibility) of a spatially coherent object. The latter consisted of a brass plate, with five holes of roughly 0.5mm diameter drilled in it, illuminated from behind with collimated light from a 632.8 nm laser. On the left in Fig.1 is a focused image of the object. This is to be thought of as the true image. The left hand image in Fig.2 shows the visibility intensity measured as carefully as possible (this led to an encouragingly faithful reconstructed image). We feel that the record of the visibility intensity shown on the right in Fig.2 is more realistic from a practical point of view, because the precautions taken when measuring the visibility intensity shown on the left in Fig.2 are likely to be impracticable in the real world. On the right in Fig.1 is the reconstructed image obtained (from the visibility intensity shown on the right in Fig.2) after pre-filtering, crude phase estimation and 100 Fienup iterations. This reconstruction is appreciably contaminated, which is representative of the differences between the two visibility intensities shown in Fig.2, but it clearly reveals the form of the true image.

While we are sure we are definitely on the right track, we feel that much more research is needed before phase retrieval can be routinely applied in the field and the observatory.

Conclusions and suggestions for future work

Considerations of "uniqueness" are always cogent of course, so that it would be worthwhile being able to categorise the class of image-forms whose Fourier intensities are ambiguous. It would also be interesting to know how "close to this class" an image not in this class has to be before (given the type of data contamination and its level) there is a prescribable "probability of ambiguity". This will be enormously difficult to do properly and, in our opinion, is far less urgent at present than the types of investigation outlined below. Furthermore, there is very little evidence that ambiguities are likely to arise with real-world data.

What is really important is to refine phase retrieval algorithms to the point where image-forms can always be reconstructed in practice. It is now clear that no simple, single algorithm will operate successfully on noisy, measured data, especially when the interesting detail in the image is faint compared with its general background level (we say such an image exhibits "low contrast"). In the following four paragraphs we discuss what we think are the most pertinent problem areas.

If the phase is lost, one must obviously make every effort to measure the visibility intensity as accurately as possible. The relevance of this to optical speckle imaging is discussed in the paper by Bates in this book. If optical "telescopes" having milliarcsecond resolution are ever built, it may well be better to invoke the radio astronomers' synthesis principles than attempt to extend conventional designs. There would then be at least a reasonable chance of measuring visibility intensities with fair precision.

Our experience of Fienup's algorithms is that their efficiencies increase markedly if an improved initial estimate of the visibility phase is available. When the data contain absolutely no phase information, the only known alternative to a pseudo-random initial phase is our "crude phase estimate", which we have shown to be efficacious (Bates and Fright, 1983a,b). Given sequences of speckle images, however, one may be able to construct even better initial estimates by invoking the several extensions of speckle interferometry (cf. §8 of Bates, 1982b).

Our "defogging" technique (Bates and Fright, 1983a,b) is a kind of image enhancement by high-pass filtering in the spatial frequency domain, because it attenuates the "central lobe" of the visibility intensity to prevent it over-shadowing whatever small "interference fringes" may be present in the remainder of the visibility intensity. It has recently occurred to us that a useful extension of the technique may be to maximise the fringe depth in the following sense. We chose a non-negative real "magnifying function" $M(u,v)$ which is multiplied to the visibility magnitude $|F(u,v)|$ to give the "processed magnitude"

$$P(u,v) = |F(u,v)| \, M(u,v) \qquad (1)$$

where $M(u,v)$ appropriately "magnifies" the difference between the "peaks" and the "valleys" of $|F(u,v)|$. It is not immediately obvious how to do this optimally because, for instance, "hard limiting" would obliterate any small fringes manifested as "ripples" superimposed upon larger "lobes". We think it will be necessary to experiment with several strategies. There is another point to be considered. On denoting the Fourier transform of $|M(u,v)|^2$ by $\psi(x,y)$ and the auto-correlation of $f(x,y)$ by $ff(x,y)$ we see from *(1)* that the Fourier transform $pp(x,y)$ of $|P(u,v)|^2$ is given by

$$pp(x,y) = ff(x,y) \odot \psi(x,y) \qquad (2)$$

where $\odot$ is the convolution operator. An essential step in our approach to phase retrieval is to recognise the "support" (i.e. the "extent" within the image plane) of the Fourier transform of the "defogged Fourier intensity" (Bates and Fright, 1983b) - this support is of course the support of $pp(x,y)$. If $\psi(x,y)$ is very "diffuse" (in whatever sense eventually proves to be apposite) it may become impossible to recognise this support. Consequently, it is going to be necessary to adopt some compromise between maximising fringe depth in the Fourier plane, while minimising the "spread" of $pp(x,y)$ in the image plane. We anticipate, however, that even very simple-minded strategies are likely to be useful, considering the remarkable effectiveness of defogging in its present rather primitive form.

After each iteration, during Fienup cycling, we can assess the "error" according to two criteria. There is an "image plane criterion", introduced by Fienup (1982), which measures the mean change in the image from one iteration to the next. There is also a "Fourier plane criterion" which is perhaps more objective because it relates directly to the data (Bates and Fright, 1983b). We find the convergence of all existing phase retrieval algorithms to be erratic, in the sense that both of the above-mentioned "errors", when plotted against the number of iterations, tend to exhibit multiple minima. What we must devise are strategies for modifying the algorithms whenever the errors start to increase. This is probably the most urgent problem area for phase retrieval, and it is one that is almost certainly far too difficult to be resolved by mathematical argument. Intuition and exhaustive numerical experimentation are called for.

References

Bates, R.H.T. (1982a). Fourier phase problems are uniquely solvable in more than one dimension. I: underlying theory. Optik, 61, no.3, 247-62.

Bates, R.H.T. (1982b). Astronomical speckle imaging. Physics Reports, 90, no.4, 203-97.

Bates, R.H.T. & Fright, W.R. (1983a). Composite two-dimensional phase-restoration procedure. J. Optical Soc. America, 73, no.3, 358-65.

Bates, R.H.T. & Fright, W.R. (1983b). Reconstructing images from their Fourier intensities. In Advances in Computer Vision & Image Processing, ed. T.S. Huang. Vol.1, in press.

Bruck, Y.M. & Sodin, L.G. (1979). On the ambiguity of the image reconstruction problem. Optics Communications, 30, no.3, 304-8.

Fiddy, M.A., Brames, B.J. & Dainty, J.C. (1983). Enforcing irreducibility for phase retrieval in two dimensions. Optics Letters, 8, no.2, 96-8.

Fienup, J.R. (1982). Phase retrieval algorithms: a comparison. Applied Optics, 21, no.15, 2758-69.

Fright, W.R. & Bates, R.H.T. (1982). Fourier phase problems are uniquely solvable in more than one dimension. III: computational examples for two dimensions. Optik, 62, no.3, 219-30.

Thompson, A.R., Clark, B.G., Wade, C.M. & Napier, P.J. (1980). The very large array. Astrophysical J. Supplement Series, 44, no.1, 151-67.

3.5 LONG BASELINE OPTICAL INTERFEROMETRY

John Davis
Chatterton Astronomy Department, School of Physics,
University of Sydney, N.S.W. 2006, Australia.

Abstract The development and current status of optical interferometry employing long baselines, meaning baselines in excess of telescope aperture diameters, are reviewed. The prototype modern Michelson interferometer being developed by the University of Sydney is described in some detail to illustrate the problems involved in long-baseline optical interferometry through the earth's atmosphere. The problems and prospects for imaging at optical wavelengths using long-baseline interferometry are discussed.

1 INTRODUCTION

Spatial interferometry at optical wavelengths employing baselines in excess of telescope aperture diameters has developed very slowly since it was first applied to astronomy by Michelson and Pease (1921). This is in sharp contrast with the relatively rapid progress made in radio interferometry since it was first applied to solar observations by Ryle & Vonberg (1946) and which has now been developed to the stage where images of regions of the sky can be produced with angular resolutions corresponding to inter-continental baselines.

There are two main areas of difficulty that have inhibited the development of optical interferometry. One is associated with the severe requirements imposed on the mechanical stability and guidance of the instrument by the wavelength of optical radiation and the other is the problem of making accurate measurements in the presence of the disruptive effects of atmospheric turbulence on the incoming wavefronts. These difficulties are circumvented by an intensity interferometer but with only relatively poor sensitivity compared with amplitude interferometers. The problems imposed by the atmosphere can be avoided by an interferometer in space and several workers have discussed this possibility (see for example Labeyrie 1978; Traub & Gursky 1980; Labeyrie et al. 1980; Stachnik 1982). However, the advent of the laser, and modern control, detection and data-handling techniques now offer the possibility of overcoming the mechanical and atmospheric problems in a ground-based amplitude interferometer. The space proposals are still at an early stage of development and the present discussion will be restricted to ground-based instruments operating at optical wavelengths.

The ultimate target of long-baseline optical interferometry must be to follow the example of radio interferometry and produce images but the technical difficulties are formidable. They are such that workers in the field are currently devoting their efforts to solving the mechanical and atmospheric problems in the simpler context of a two aperture instrument measuring only the modulus of the complex degree of coherence $|\gamma_{12}(0)|$ (Born & Wolf, 1964). In what follows $|\gamma_{12}(0)|$ will be represented by $|\gamma|$. There is a wide range of astronomical programmes to be tackled at optical wavelengths for which measurement of $|\gamma|$ at selected spatial frequencies without the determination of phase is sufficient (Labeyrie 1978; Davis 1979a; McAlister 1979). These programmes basically involve the measurement of angular sizes and separations and they are the current driving force for the development of long-baseline optical interferometry.

Once the basic problems of mechanical stability, optical path matching and atmospheric turbulence have been solved attention will undoubtedly turn towards imaging but there will still be major technical difficulties to overcome. Some of the difficulties will be discussed in Section 3 but first the development and current status of long-baseline optical interferometry will be reviewed.

2 LONG-BASELINE OPTICAL INTERFEROMETERS

2.1 Historical background

The major application of long-baseline optical interferometry has been to the determination of angular diameters of stars and its history and achievements have been reviewed by Hanbury Brown (1974). Only a brief outline will be given here.

The first determination of the angular diameter of a star was made by Michelson & Pease (1921) when they measured α Ori with a 20 ft (6.1 m) Michelson stellar interferometer mounted on the 100 inch (2.5 m) telescope. Subsequently the angular diameters of several stars were measured and the results for 7 stars were published by Pease (1931). The original interferometer was limited in resolution by the length of beam that could be attached to the telescope and so in the 1920's, the 50 ft (15.2 m) Michelson stellar interferometer was constructed (Pease 1931). However, it proved difficult to operate and no final results were published. In both instruments the interference fringes were observed visually and the visibility of the fringes was observed to decrease with increasing baseline at a rate which was seeing dependent even for unresolved sources (Pease 1931). There was no possibility of accurately determining the fringe amplitude, let alone measuring fringe phase, at that time. Consequently a measurement consisted of a search for the first zero in the visibility curve with the attendant difficulties of ensuring that the instrument remained in perfect adjustment even when no fringes could be seen.

Stellar interferometry remained at a standstill after the failure of the 50 ft Michelson instrument until the intensity interferometer was developed in the 1950's. The measurement of α CMa (Sirius) by Hanbury Brown & Twiss (1956) was the first angular diameter determination for a main sequence star and led to the design and construction of the stellar intensity interferometer at Narrabri Observatory (Hanbury Brown et al. 1967). The principles and practical problems of intensity interferometry and the achievements of the extensive observational programme carried out with the Narrabri instrument between 1963 and 1972 have been discussed in detail by Hanbury Brown (1974).

In an intensity interferometer the measured correlation between the intensity fluctuations in the light received at the two apertures of the interferometer is proportional to the square of the modulus of the complex degree of coherence $|\gamma|^2$ (Hanbury Brown 1974) or, in other words, the square of the fringe visibility as observed with Michelson's stellar interferometer. Thus the fringe phase is not determined and in the interpretation of the measurements symmetry in the brightness distribution across the source is assumed.

The main advantages of an intensity interferometer are that it is possible to use long baselines with the mechanical tolerances relaxed by several orders of magnitude compared with an amplitude interferometer and it is essentially unaffected by atmospheric seeing and scintillation. It has the serious disadvantage of being relatively insensitive. The Narrabri stellar intensity interferometer had a limiting blue magnitude B(limit) of only ~ +2.5 (Hanbury Brown et al. 1974) and a possible design for a very large intensity interferometer with ~ 80 times this sensitivity has been discussed by Hanbury Brown (1974) and Davis (1975). There is no doubt that it could be built and that it would work but a modern amplitude interferometer offers the possibility of greater sensitivity at lower cost. If reasonable parameters are assumed it can be shown that it would be necessary to use apertures with diameters of the order of 30 m in an intensity interferometer in order to achieve the same sensitivity as an amplitude interferometer with 100 mm diameter apertures.

Gamo (1961) has suggested the use of a laser as a local oscillator in a heterodyne interferometer but this is not a viable proposition at optical wavelengths (Twiss 1969). The signal to noise ratio is improved by a factor $2/|\gamma|$ compared with an intensity interferometer but the optical constraints are the same as for an amplitude interferometer and apertures would have to be limited to ~ 100 mm diameter because of the effects of atmospheric turbulence. The advantage of using large, relatively crude light collectors is lost and the sensitivity is much less than that of even the Narrabri stellar intensity interferometer.

2.2 Modern long-baseline optical interferometers

The large sensitivity advantage of an amplitude interferometer makes it the most attractive proposition providing the mechanical and atmospheric difficulties can be overcome and the long-baseline optical interferometers currently under development are all amplitude instruments. Small aperture amplitude interferometers are currently being developed at the Universities of Sydney (Davis 1979b) and Maryland (Liewer 1979) and a large aperture amplitude interferometer is being developed at CERGA (Labeyrie 1978). While it is not appropriate to describe all three projects in detail, in order to illustrate the problems faced in a long-baseline instrument for the optical region of the spectrum, the small aperture approach will be outlined and the prototype modern Michelson interferometer being developed at the University of Sydney will be described in some detail.

The basic features of a small aperture amplitude interferometer. Figure 1 shows the basic features of a small aperture amplitude interferometer where the mirrors C reflect the light from a star horizontally to a central station. For baselines exceeding ~ 10 m it is not feasible to mount the whole instrument on a steerable platform and the mirrors C are coelostats located on stable piers situated symmetrically about the centre of the instrument. When the star is not directly above the instrument the coelostats are steered to maintain the horizontal direction of the reflected light and the retro-reflectors R are moved to equalise the light paths from the star to the beamsplitter B where the two beams are combined.

Figure 1. Simplified diagram of a small aperture amplitude interferometer. The broken lines represent the paths of light from a star.

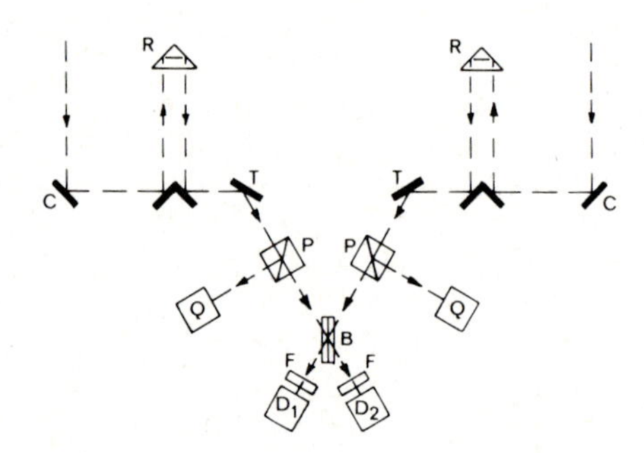

The effect of atmospheric turbulence is to produce time dependent deformations of the wavefronts reaching the coelostats. In the approach proposed by Twiss (Twiss 1965; Tango & Twiss 1980) and demonstrated in the Monteporzio 2 m amplitude interferometer (Tango 1979b) the atmospheric difficulties are circumvented by selecting 'flat' sections of the incoming wavefront, making them parallel to each other where they are combined at the beamsplitter B by means of active optics, and by sampling the outputs of the photodetectors D_1 and D_2 with a time interval short compared with the period of amplitude and phase fluctuations of incoming waves. 'Flat' sections of wavefront are selected by restricting the effective apertures of the coelostats to $\lesssim r_o$, Fried's coherence diameter

(Fried & Mevers 1974). Polarising beamsplitters P pass one plane of polarization to the beamsplitter B and direct the orthogonal plane to quadrant detectors Q which provide signals for driving the tilting mirrors T to remove atmospheric turbulence-induced tilts in the incoming wavefronts. Thus the wavefronts are arranged to interfere at nominally zero angle at the beamsplitter B and, in the absence of aberrations, the complimentary exit pupils of the interferometer will be uniformly illuminated. The optical bandwidth of the instrument is defined by the filters F and the photon-counting detectors D_1 and D_2 measure the total flux in each beam in a sample time typically in the range 1 to 10 ms. In the absence of aberrations the signals from the two detectors will be proportional to $(1 + |\gamma| \cos\phi)$ and $(1 - |\gamma| \cos\phi)$ where ϕ is a randomly varying phase angle resulting from atmospheric optical path length fluctuations and path compensation errors. The sampling time τ of the two detectors D_1 and D_2 is made short so that changes in ϕ are not significant during τ and the photocounts registered by each detector in τ are processed by a correlator which basically measures the square of the difference between the two signals. The data are integrated for a total observing time T and analysed to yield $2|\gamma|^2\langle\cos^2\phi\rangle$, where the brackets indicate an average over T. If the phase is a uniform random variable $2|\gamma|^2\langle\cos^2\phi\rangle$ becomes $|\gamma|^2$, the square of the fringe visibility.

The University of Sydney prototype modern Michelson interferometer. The design of the University of Sydney prototype interferometer incorporates the ideas outlined above. It is being built to test the viability of this approach and to establish the accuracy that can be achieved in measurements of $|\gamma|^2$. The target is ±2% (Davis 1976) and every effort is being made to keep individual sources of error to less than 1% in $|\gamma|^2$. The prototype is also intended to provide the foundations for the design of a major instrument. It has not been designed with the aim of measuring new angular diameters but simply to use selected stars as test objects. The length of the fixed North-South horizontal baseline has been set at 11.4 m to give an effective baseline for the star αCMa corresponding to approximately 0.3 for $|\gamma|^2$.

The prototype instrument is being installed in the grounds of the Australian National Measurement Laboratory at West Lindfield, near Sydney, and Figure 2 shows a North-South cross-section in elevation. The component parts of the interferometer are all mounted on reinforced concrete plinths anchored in a monolithic layer of sandstone approximately 1 m below ground level. A coelostat is mounted on a 1.35 m high plinth at each end of the 11.4 m baseline and each coelostat directs starlight via flat mirrors arranged as periscopes to the interferometer proper which is housed in a central laboratory. For the major part of the optical path from the coelostats to the laboratory the light beams pass through pipes sealed with optical windows at each end. The pipes are thermally insulated with provision for evacuation. The coelostat and the periscope mirrors are all 150 mm diameter Zerodur flats and the 45° inclination of the periscope mirrors limits the maximum diameter of the primary beam to 106 mm.

Figure 2. Cross-section of the University of Sydney prototype interferometer looking towards the east.

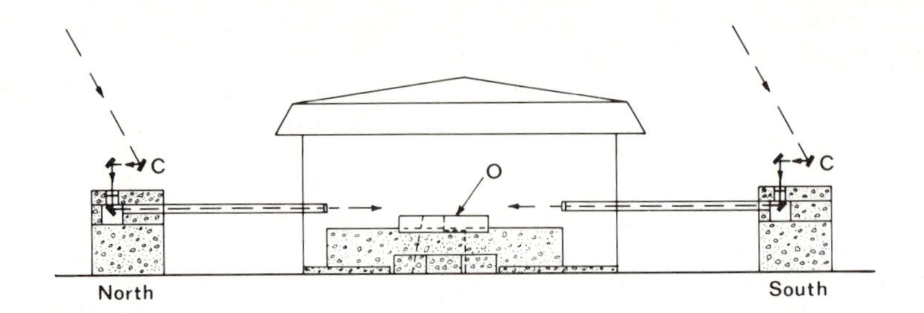

The optical layout of the central part of the interferometer is shown in simplified form in Figure 3. The key components in the basic layout of Figure 1 have been given the same letters in Figure 3 so that they may be identified. The two sides of the instrument are symmetrical so it is only necessary to describe one side. The diameter of the incoming beam is reduced by a factor of 2.5 by a beam reducing telescope (BRT) which consists of two off-axis paraboloidal mirror segments with focal lengths in the ratio of 2.5:1. The reduced diameter beam passes via the optical path length compensator (OPLC), which will be described later, the beamsplitter G, the mirrors S and T, and the polarizing beamsplitter P to the beamsplitter B where the wavefronts from the two sides of the instrument interfere at nominally zero angle. Matched prism spectrometers in the output beams from B select the mean wavelength and spectral bandwidth for observations. For ease of alignment of the spectral bandwidths in the two signal channels of the interferometer, the wavelength of the HeCd laser (441.8 nm) has been adopted as the mean operating wavelength for the prototype instrument. The spectral bandwidth is a variable parameter and will generally be in the range 0.1 to 1 nm. Photon counting detectors D_1 and D_2 measure the total flux in each beam within the sampling time τ (~ 1-10 ms).

A small fraction (~ 5%) of the light incident on the beamsplitter G is reflected to a long focal length lens which forms a diffraction limited image of the star on the quadrant detector Q_g. Error signals from Q_g are used to correct the pointing of the appropriate coelostat so that the beam transmitted by G is accurately aligned with the defined optical axis of the instrument except for the rapid variations in tilt due to seeing (of order 1-2" arc rms). The polarized component of light reflected by the polarizing beamsplitter P is also focussed by a long focal length lens onto a quadrant detector Q. The error signals in this case are used to reduce the seeing-induced tilts via the piezoelectrically-actuated tilting mirror T. The optimisation of the bandwidth of this servo and the compensation for losses in $|\gamma|^2$ due to failure of the servo to completely correct the tilts (by means of the noise power of the error signal) has been discussed by Tango (1979a) and by Tango & Twiss (1980). The limiting sensitivity of the interferometer is set by the tilt-correcting servo at V(limit) ~ +8 (Tango 1979a).

Figure 3. Diagrammatic layout of the main optical components of the University of Sydney prototype interferometer.

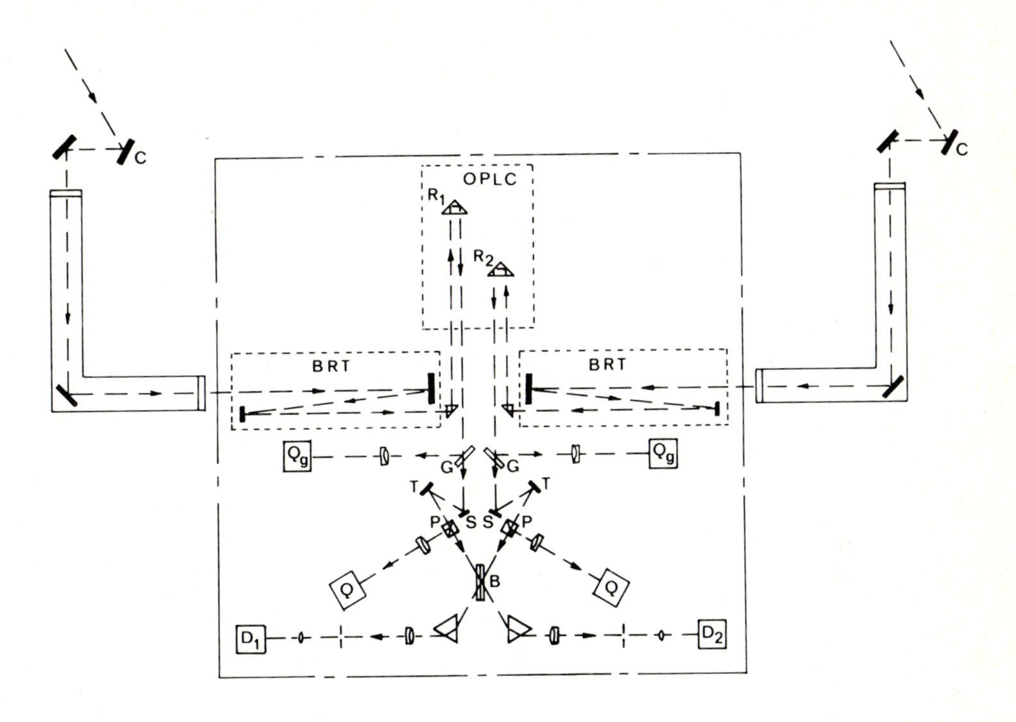

The mirror S is mounted on a piezoelectric transducer, so that the mirror may be translated in a direction normal to its surface. This enables a path difference of $\lambda/4$ to be introduced between the two interfering beams in successive sample periods so that, providing the sample period is short compared with the timescale of changes in the relative phase ϕ of the interfering wavefronts, the analysis gives the apparent correlation equal to $2|\gamma|^2\langle\cos^2\phi + \sin^2\phi\rangle$ instead of $2|\gamma|^2\langle\cos^2\phi$ and it no longer requires the assumption that the phase is a uniform random variable to obtain $|\gamma|^2$.

The optical path length compensation system (OPLC) compensates for the difference in the time of arrival of starlight at the two coelostats. As shown in Figure 3 it consists of two parallel optical delay paths, one in each arm of the interferometer. The retro-reflectors R_1 and R_2 are each mounted on a carriage rolling on a precision track. The positions of R_1 and R_2, relative to their positions for equality of the internal paths in the two arms of the interferometer, are monitored by a laser interferometric system which is not shown in Figure 3. In operation the internal paths of the two arms of the interferometer will be equalised by introducing an artificial star as in the Monteporzio amplitude interferometer (Tango 1979b), tilting the coelostats to an auto-collimating position to form a Twyman-Green interferometer and then adjusting the relative positions of R_1 and R_2 until the zero order white light fringe is observed. To observe a star the optical path difference at transit will be compensated by moving R_1 to the appropriate position on its track (~ 4.5 m long) with the laser system monitoring the actual distance moved. R_1 will remain fixed in position during an observation and the continuously varying component of the optical path difference will be compensated by moving R_2 with its position and velocity controlled via the monitoring laser system.

R_1 and R_2 are simple roof prisms rather than corner cubes, partly to reduce cost, but also to avoid difficulty with polarization. The parallelism of the incident and reflected beams is maintained by an automatic levelling system on the carriage carrying R_1 and by accurately levelled tracks for the much shorter (~ 1 m) arm of the OPLC involving R_2.

It can be shown that the maximum allowable difference in the optical paths via the two arms of the interferometer Δs, assuming a spectral bandpass of Gaussian profile and width at half maximum of $\Delta\lambda$, is given by

$$\Delta s \quad < \quad \frac{3.7 \times 10^{-2}\ \lambda^2}{\Delta\lambda} \qquad (1)$$

if losses in $|\gamma|^2$ are to be less than 1%. If there is an uncertainty of $\Delta\theta$ arcseconds in the angle between the baseline orientation and the direction of the star there will be a corresponding uncertainty in the computed paths to the beamsplitter B in Figure 3 given by

$$\Delta s \quad \simeq \quad d.\Delta\theta/206265 \qquad (2)$$

where d is the effective baseline of the interferometer. Combination of the above equations leads to

$$d.\Delta\theta \quad < \quad \frac{7.7 \times 10^{3}\ \lambda^2}{\Delta\lambda} \qquad (3)$$

if losses in $|\gamma|^2$ are to be less than 1%.

In the prototype interferometer a single optical bandwidth of the order $\Delta\lambda = 0.2$ nm will be used and, to keep losses in $|\gamma|^2 \leq 1\%$, it will only be necessary to match the optical paths to $\lesssim 40$ μm. Once the orientation of the baseline has been established by stellar observations the relative positions of R_1 and R_2 can be computed to within the tolerance and it will not be necessary to use a fringe tracking system at this stage. However, as equation (3) shows, for longer baselines and/or wider bandwidths it will become necessary to introduce fringe tracking. For example, if we take $\Delta\theta = 0.2$ arcseconds and $\lambda = 442$ nm then fringe tracking will be necessary for $d \gtrsim 38$ m unless $\Delta\lambda$ is decreased from 0.2 nm. Shao & Staelin (1980) have developed a fringe tracking system for their 1.5 m baseline prototype astrometric stellar interferometer and this will be relatively easy to incorporate into the interferometer in the future.

The OPLC system has to meet a severe requirement regarding the smoothness of the effective motion of the continuously moving carriage whether it relies solely on the carriage movement or has an additional stage involving a piezoelectrically driven element for high speed correction. Since the relative phase of the beams combined at B in Figure 3 will change if R_2 is not moving at the correct velocity, any error or irregularity in the velocity in the sampling time τ will result in a loss of coherence and an error in the measured $|\gamma|$. Phase errors must be kept to $\Delta\phi(\mathrm{rms}) \lesssim 0.1$ rad to keep losses in $|\gamma|^2 \leq 1\%$ (Tango & Twiss 1980).

The OPLC is the most critical component of the interferometer and considerable effort has been expended in developing an 'on-board' drive system that will allow extension of the OPLC as will be required for longer baselines. Tests on the system are very encouraging although not yet compplete. A mechanical vibration isolation platform to carry R_2 has been incorporated on the moving carriage, the Zeeman difference frequency from the monitoring laser has been stabilized and the Doppler shifted frequency of the laser beam from the moving retroreflector is phase locked to a signal generated by a frequency synthesizer under computer control.

The prototype interferometer is nearing completion and it will be used to establish the accuracy and reliability that can be achieved in measurements of $|\gamma|$ through the atmosphere. This programme is the foundation for the major high angular resolution interferometer that we plan to build. The preliminary specification includes a limiting magnitude of $V(\mathrm{limit}) \gtrsim +8$ and baselines extending to a kilometer or more. This would enable angular sizes down to $\sim 5 \times 10^{-5}$ arcseconds to be measured.

Intercomparison of amplitude interferometer programmes. While the University of Sydney programme has been described in some detail in order to illustrate the problems involved in long-baseline optical interferometry further mention should be made of the other two major programmes to develop this field.

The group led by Currie at the University of Maryland is developing a small aperture ($< r_o$) prototype interferometer (Liewer 1979) based on their amplitude interferometer (Currie et al. 1974). It has a baseline of ~ 3 m oriented parallel to the Earth's axis to simplify the path compensation problem and is designed to work with baselines up to 50 m when installed at a permanent observing site. It includes most of the features of the basic small aperture interferometer outlined earlier although it does not include tilt-correcting active optics at this stage.

Labeyrie's group at CERGA is developing a large aperture ($> r_o$) interferometer (Labeyrie 1978). The Coudé images from two telescopes are relayed to a central station where fringes appear in the speckles of the superimposed images providing the optical paths are matched within the tolerance set by the coherence length of the light. A prototype instrument using 250 mm diameter aperture telescopes has been successfully used out to baselines of ~ 40 m to determine $|\gamma|^2$ for bright stars ($V < +3.1$). Work is in progress to develop telescopes with aperture diameters of 1.52 m (Labeyrie 1978) for use in this type of instrument. Labeyrie has discussed long-term plans to build an array of large apertures (discussed further in Section 3.2) but initially a two element interferometer is to be built with baselines up to 60 m.

The relative merits of the large aperture and small aperture approaches to a modern Michelson interferometer have been outlined by Davis (1976). The large aperture approach is expected to have greater sensitivity but a small aperture amplitude interferometer should be capable of achieving greater accuracy.

3 IMAGING WITH LONG-BASELINE OPTICAL INTERFEROMETERS

Imaging with long-baseline optical interferometers has not been attempted to date because of the severe technical difficulties involved. In this section some of the difficulties will be outlined.

3.1 The measurement of phase

The measurement of the phase of the complex degree of coherence, which is identical with the spatial phase of the interference fringe pattern relative to that produced by a point source equidistant from the apertures of the interferometer, is in principle straightforward. Possible methods for determining the phase have been discussed by Rogstad (1968), Goodman (1970) and recently by Mertz (1983). In principle the instantaneous phase could be determined by one of these methods or, indeed, with the interferometer described in Section 2.2 if the appropriate data analysis was performed. However, in practice it is not straightforward. Atmospheric turbulence introduces a differential phase shift between the light waves received at the two ends of the baseline and this is a random function of time with an amplitude of a few wavelengths and a spectrum extending to a few hundred Hertz. A determination of the instantaneous phase would therefore have to be completed in a time short compared with the averaging effect of the turbulence. This enforced limitation of the measurement time to the order of 1-10 ms would restrict the measurement of instantaneous phase to bright stars. Walkup & Goodman (1973) have determined the limitations of fringe phase detection at low light levels and their results can be used to estimate the limiting magnitude for a phase measurement. For example, assuming the measurement is made with τ = 5 ms, an optical passband $\Delta\lambda$ = 1 nm at the baseline corresponding to a fringe visibility of 0.5 and that the rms phase uncertainty should be less than 0.25 radians, the limiting magnitude for the small-aperture interferometer described in Section 2.2 is V(limit) $\sim$ +3. If $\Delta\lambda$ is increased to 50 nm V(limit) is lowered to $\sim$ +7 but neither of these estimates takes into account optical losses in the instrument which will make matters worse.

Although the instantaneous fringe phase can be measured with an amplitude interferometer, albeit with low sensitivity, it is not possible to separate the visibility phase from the atmospheric contribution unless measurements are made simultaneously in two colours (Currie 1979; Shao & Staelin 1980).

Gamo (1963) has pointed out the possibility of measuring phase with a three aperture intensity interferometer but, as shown by Twiss (1969), the sensitivity of a triple intensity interferometer for phase measurements at optical wavelengths is so low that it can be completely ruled out.

3.2 Extension of long-baseline optical interferometry to more than two apertures

The simplest extension of the two aperture interferometer shown in Figure 3 is to a linear three aperture or 'triple' interferometer. This is shown in diagrammatic form in Figure 4 and it is based on Goodman's (1970) adaptation of Jennison's (1958) triple interferometer for radio wavelengths. As shown in Figure 4 the light received at the three coelostats C can be combined to correspond to three baselines. The actual physical layout, bearing in mind the mechanical stability requirements, is difficult to envisage but in addition the following factors have to be considered:

a) Each baseline requires its own OPLC system, beam combining optical system and detectors. In other words the boxes marked I in Figure 4 each contain the equivalent of the interferometer shown in Figure 3.

b) Since phase-coherent amplifiers are not available at optical wavelengths it is necessary to divide the light received by each coelostat between baselines. For the triple interferometer each coelostat beam is divided in two at the beamsplitters B shown in Figure 4. For a small-aperture interferometer the S/N is proportional to N_o, the mean combined counting rate from both detectors (D_1 and D_2 in Figure 3), near the limiting sensitivity (Tango 1979a) and so dividing the light by two loses a similar factor in S/N. Losses in the beamsplitters and other additional optics will decrease the S/N still further.

Figure 4. Linear 'triple' interferometer (after Jennison (1958) and Goodman (1970)). The labelled components are coelostats (C), mirrors (M) and beamsplitters (B) and each of the boxes (I) represents a complete path-compensated interferometer (see text).

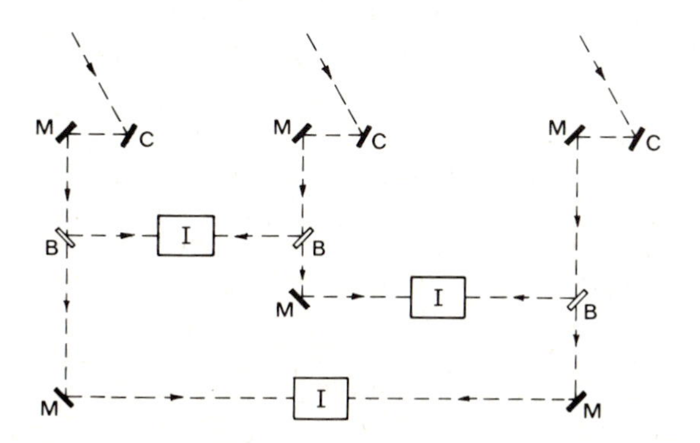

c) If it is intended to measure the instantaneous phase at each baseline so that phase-closure techniques might be implemented it would be necessary to complete a set of phase measurements in ~ 1-10 ms so the same limitations will be imposed as discussed in Section 3.1 with a further reduction due to the beam division discussed above.

d) One baseline in Figure 4 would most likely remain unchanged in practice and be used as a reference. However, if the other baselines are changed there is a major difficulty. Each interferometer I should ideally lie at the centre of its baseline. In principle they could remain fixed and the asymmetry in the post-coelostat optical paths produced by changing the baseline could be accommodated by adjustment of the OPLC system. However, there is a limit to the asymmetry in the paths that can be accepted since there will be a loss of coherence due to differential diffraction effects in the post-coelostat optical beams (Tango & Twiss 1974).

The problems outlined above are daunting and an alternative approach has been suggested by Labeyrie (1978). He proposes that a ring-shaped array of telescopes might use a single fixed central station if the telescopes are moved along radial tracks during observations. No optical delay lines for path-length equalization would be needed if zero path difference is maintained by positioning the telescopes along an ellipse, the shape of which varies during observations. As Labeyrie has pointed out, this approach would be difficult to implement with coelostats because of the attendant complexity of beam manipulation. Instead, Labeyrie proposes using telescopes of conventional Cassegrain-Coudé design to provide a Coudé beam of reduced diameter for transmission to the central station. Labeyrie (1978) has described a new form of telescope mount that he has developed and which would form one element of the type of array he proposes to build.

Labeyrie's approach certainly simplifies the layout problem but it does not remove the severe constraints on path matching and phase tracking which are common to all amplitude interferometers. Fine tuning can be achieved with piezoelectrically-actuated optical components but it will still be necessary to move the individual telescopes, each weighing several tonnes, very smoothly and with high precision. It remains to be seen whether this can be done.

If the beams from an array are combined simultaneously with their paths matched to within the tolerances set by the coherence length then fringes will appear in the superimposed images with orientations corresponding to each baseline present in the array. Simulated examples have been shown by Labeyrie (1974) and Oneto (1980). If fringe visibility only is to be determined then high sensitivity has been predicted with V(limit) $\lesssim +17$ (Labeyrie 1978). In principle phase-closure could be used by determining the relative phases of the fringes within a single speckle. However, the limiting magnitude will not differ significantly from that estimated for small-aperture interferometers since the speckle pattern must be sampled with the same short measurement time. The phases from speckle to speckle will be randomly distributed so each would have to be analysed separately - the limiting magnitude would not be changed significantly but many measurements, equal to the number of speckles per sample time, would be made in parallel.

3.3 Prospects for imaging with long-baseline optical interferometry

In the preceding sections it has been seen that while the instantaneous phase can be measured and that it is possible, at least in principle, to apply phase closure techniques they appear to be decidedly unattractive at optical wavelengths. This is because of the complexity introduced into the optical system and the severe mechanical problems that accompany it, and also because the sensitivity is severely limited by the need to complete each closure measurement in the order of a very few milliseconds.

At the current state of development of long-baseline optical interferometry it is hard to envisage significant progress in true imaging in the immediate future.

4 CONCLUDING REMARKS

This contribution has been written with the emphasis on imaging aspects of long-baseline optical interferometry since that is the theme of the meeting. While it appears that the problems which have delayed the development of long-baseline two aperture amplitude interferometers to measure $|\gamma|$ reliably through the atmosphere are near solution the extension to multi-aperture instruments is difficult. This is particularly true for an imaging instrument in which phase measurements are required. The sensitivity limits imposed on phase measurements by the atmosphere are discouraging.

The best prospects for increasing the imaging resolution lie with the next generation of large optical telescopes and the use of speckle-imaging techniques. The increase in resolution will be relatively modest but it can be predicted with confidence based on the experience gained with existing telescopes and particularly with the MMT.

The emphasis on the imaging aspects of long-baseline optical interferometry and the attendant difficulties might give the quite erroneous impression that long-baseline optical interferometry has little to offer astronomy in the foreseeable future. On the contrary there is a wide range of important astronomical programmes (Labeyrie 1978; Davis 1979a; McAlister 1979) to be tackled with instruments that measure $|\gamma|$ at selected spatial frequencies. The next decade will undoubtedly see signficant progress in this area.

ACKNOWLEDGEMENTS

It is a pleasure to acknowledge my colleagues Professor R. Hanbury Brown, Dr. L.R. Allen and Dr. W.J. Tango with whom I share the development of the University of Sydney prototype interferometer that I have described. It is also a pleasure to acknowledge the support given to this research project by the Australian Research Grants Scheme, the University of Sydney Research Grants Committee and the Science Foundation for Physics within the University of Sydney.

REFERENCES

Borne, M. & Wolf, E. (1964). Principles of Optics. Oxford: Pergamon Press.

Currie, D.G. (1979). Two color refractometry, precision stellar catalogs and the role of anomalous refraction. In Refractional Influences in Astrometry and Geodesy (I.A.U. Symposium No. 89), ed. E. Tengström & G. Teleki, pp.131-155. Dordrecht: Reidel.

Currie, D.G., Knapp, S.L. & Liewer, K.M. (1974). Four stellar-diameter measurements by a new technique: amplitude interferometry. Astrophys. J., 187, 131-134.

Davis, J. (1975). A proposed successor to the Narrabri stellar intensity interferometer. In Multicolor Photometry and the Theoretical HR Diagram, ed. A.G.D. Philip & D.S. Hayes, Dudley Observatory Report No. 9, pp.199-214.

Davis, J. (1976). High angular resolution stellar interferometry. Proc. A.S.A., 3, 26-32.

Davis, J. (1979a). The application of high angular resolution stellar interferometry to the study of single objects in the visual region of the spectrum. In High Angular Resolution Stellar Interferometry (I.A.U. Colloquium No. 50), ed. J. Davis & W.J. Tango, pp.1.1 - 1.27. Chatterton Astronomy Department, University of Sydney.

Davis, J. (1979b). A prototype 11 metre modern Michelson stellar interferometer. In High Angular Resolution Stellar Interferometry (I.A.U. Colloquium No. 50), ed. J. Davis & W.J. Tango, pp.14.1 - 14.12. Chatterton Astronomy Department, University of Sydney.

Fried, D.L. & Mevers, G.E. (1974). Evaluation of r_o for propagation down through the atmosphere. Applied Optics, 13, 2620-2622. (Correction in Applied Optics, 14, 2567).

Gamo, H. (1961). On the intensity interferometer with coherence background. In Advances in Quantum Electronics, ed. J.R. Singer, pp.252-266. New York: Columbia University Press.

Gamo, H. (1963). Triple correlator of photoelectric fluctuations as a spectroscopic tool. J. Appl. Phys., 34, 875-876.

Goodman, J.W. (1970). Synthetic-aperture Optics. In Progress in Optics VIII, ed. E. Wolf, pp.1-50. Amsterdam: North Holland.

Hanbury Brown, R. (1974). The Intensity Interferometer. London: Taylor & Francis.

Hanbury Brown, R., Davis, J. & Allen, L.R. (1967). The stellar interferometer at Narrabri Observatory - I. Mon. Not. R. astr. Soc., 137, 375-392.

Hanbury Brown, R., Davis, J. & Allen, L.R. (1974). The angular diameters of 32 stars. Mon. Not. R. astr. Soc., 167, 121-136.

Hanbury Brown R., & Twiss, R.Q. (1956). A test of a new type of stellar interferometer on Sirius. Nature, 178, 1046-1048.

Jennison, R.C. (1958). A phase sensitive interferometer technique for the measurement of the Fourier transforms of spatial brightness distributions of small angular extent. Mon. Not. R. astr. Soc., 118, 276-284.

Labeyrie, A. (1974). Observations interférométriques au Mont Palomar. Nouv. Rev. Optique, 5, 141-151.

Labeyrie, A. (1978). Stellar interferometry methods. Ann. Rev. Astron. Astrophys., 16, 77-102.

Labeyrie, A., Paderie, F., Steinberg, J., Vatoux, S. & Wonters, F. (1980). "Flute" a long-baseline optical interferometer in space. In Optical and Infrared Telescopes for the 1990's, ed. A. Hewitt, pp.1020-1026. Tucson: Kitt Peak National Observatory.

Liewer, K.M. (1979). The prototype very long baseline amplitude interferometer. In High Angular Resolution Stellar Interferometry (I.A.U. Colloquium No. 50), ed. J. Davis & W.J. Tango, pp.8.1 - 8.14. Chatterton Astronomy Department, University of Sydney.

McAlister, H.A. (1979). High angular resolution binary star interferometry. In High Angular Resolution Stellar Interferometry (I.A.U. Colloquium No. 50), ed. J. Davis & W.J. Tango, pp.3.1 - 3.16. Chatterton Astronomy Department, University of Sydney.

Mertz, L. (1983). Complex interferometry. Applied Optics, 22, 1530-1534.

Michelson, A.A. & Pease, F.G. (1921). Measurement of the diameter of α Orionis with the interferometer. Astrophys. J., 53, 249-259.

Oneto, J. (1980). Microprocessors and large telescope arrays. In Optical and Infrared Telescopes for the 1990's, ed. A. Hewitt, pp.976-983. Tucson: Kitt Peak National Observatory.

Pease, F.G. (1931). Interferometer methods in astronomy. Ergabnisse der Exacten Naturwissenschaften, 10, 84-96.

Rogstad, D.H. (1968). A technique for measuring visibility phase with an optical interferometer in the presence of atmospheric seeing. Applied Optics, 7, 585-588.

Ryle, M. & Vonberg, D.D. (1946). Solar radiation on 175 Mc/s. Nature, 158, 339-340.

Shao, M. & Staelin, D.H. (1980). First fringe measurements with a phase-tracking stellar interferometer. Applied Optics, 19, 1519-1522.

Stachnik, R.V. (1982). A multiple spacecraft Michelson spatial interferometer. Transactions of the International Astronomical Union, XVIII B, 107-108.

Tango, W.J. (1979a). The limiting sensitivity and visibility loss in a small aperture amplitude interferometer. In High Angular Resolution Stellar Interferometry (I.A.U. Colloquium No. 50), ed. J. Davis, & W.J. Tango, pp 12.1 - 12.10. Chatterton Astronomy Department, University of Sydney.

Tango, W.J. (1979b). The Monteporzio two metre amplitude interferometer. In High Angular Resolution Stellar Interferometry (I.A.U. Colloquium No. 50), ed. J. Davis, & W.J. Tango, pp.13.1 - 13.7. Chatterton Astronomy Department, University of Sydney.

Tango, W.J. & Twiss, R.Q. (1974). Diffraction effects in long path interferometers. Applied Optics, 13, 1814-1819.

Tango, W.J. & Twiss, R.Q. (1980). Michelson stellar interferometry. In Progress in Optics XVII, ed. E. Wolf, pp.239-277. Amsterdam: North Holland.

Traub, W. & Gursky, H. (1980). Coherent arrays for optical astronomy in space. In Optical and Infrared Telescopes for the 1990's, ed. A. Hewitt, pp.250-262. Tucson: Kitt Peak National Observatory.

Twiss, R.Q. (1965). Reported in Observatory, 85, 138-139.

Twiss, R.Q. (1969). Applications of intensity interferometry in physics and astronomy. Optica Acta, 16, 423-451.

Walkup, J.F. & Goodman, J.W. (1973). Limitations of fringe-parameter estimation at low light levels. J.O.S.A., 63, 399-407.

DISCUSSION

R.H.T. BATES

Please compare the probable difficulties to be expected with your instrument, the MMT and Labeyrie's proposed instrument.

J. DAVIS

The major difficulty common to all forms of amplitude interferometry is in matching the optical paths and in ensuring that variations in phase due to the atmosphere, mechanical instabilities and irregularities in tracking are not significant during the sampling time which is of the order 1-10 ms. In the MMT the guidance and stability of a single structure are involved; Labeyrie proposes to move his telescopes continuously whereas we are moving relatively small retroreflectors. I cannot comment further on the other two projects but as far as our project is concerned we believe that we have the problem under control although it still remains to be fully tested.

R.D. EKERS

Is there a fundamental reason why the small aperture approach is expected to be more accurate?

J. DAVIS

In the small aperture approach, as being implemented in our instrument, we are eliminating atmospheric effects in real time from the measurement of visibility. The aperture size and sampling time are restricted to minimize the effects of curvature and phase fluctuations, and wavefront tilts are removed by active optics. Residual curvature and tilt will be monitored and corrected for and, as indicated, we are aiming for $\lesssim \pm 2\%$ in $|\gamma|^2$. In a large aperture instrument the measured visibility is always less than the true value due to the statistical nature of the speckle phenomenom. It is therefore necessary to calibrate the visibility by observing unresolved sources with uncertainty introduced when the equipment configuration is changed to point in a different direction. Experience with Labeyrie's 250 mm aperture interferometer shows a large spread in calibration measurements as a function of time and direction.

J.R. FIENUP

Regarding Ron Ekers question about the accuracy of the Michelson stellar interferometer versus speckle interferometry: I agree with John Davis that the Michelson interferometer can more accurately measure the visibility magnitude. The reason is that in speckle interferometry we must compensate for the modulation transfer function (MTF) of the process (which can be determined from a reference star measured through an atmosphere having the same statistics as the atmosphere through which the object was measured). However, not only is the atmosphere changing very rapidly, but also the statistics of the fluctuations vary with time and space. Therefore it has not been possible to calibrate out the speckle MTF as accurately as we would like.

3.6 A LARGE FORMAT, PHOTON COUNTING IMAGING SYSTEM FOR THE STARLAB TELESCOPE

T.I. Hobbs, D.A. Carden, R.A. Gorham, A.W. Rodgers, T.E. Stapinski

Mount Stromlo and Siding Spring Observatories
Research School of Physical Sciences
The Australian National University
Private Bag WODEN P.O. A.C.T. 2606
AUSTRALIA

ABSTRACT

The two dimensional image detectors to be used in STARLAB are based on the Photon Counting Array (PCA) developed at MSSSO. STARLAB is a one metre, UV-visible telescope to be placed in earth orbit and will incorporate several 90mm versions of the PCA system as the detector for the on board spectrograph and direct imager. The performance of scaled down prototypes is described.

The latest PCA system comprises a micro channel plate based intensifier with sufficient gain to permit photon detection and counting, fibre optic conduit to dissect the image and couple each sub-image to a self scanned charge coupled device (CCD) image array. A digital processor centres the events for storage in an accumulating memory.

The overall system performance is described in several sections corresponding to various requirements placed on the PCA. The amplitude distribution of events from the image tube comprise photon responses, internal ion events and photocathode dark noise. The way in which these event categories limit the dynamic range and image quality is described. Detector resolution and related parameters are discussed. Final image quality is controlled by fibre optic defects and CCD blemishes. Image pixel signal variance indicates the photon counting performance as a function of count rate. Detector linearity is governed by coincidence losses at high count rates. Graphs of the pixel response versus incoming photon arrival rate for point, line and flat field images are presented. Finally, flat field calibration accuracy under environmental influences and image stability are discussed.

The results of these studies with the PCA system make the authors confident that the design goals for the STARLAB detector can be realised.

This paper will be published in full in the Proceedings for Instrumentation in Astronomy V by SPIE - The International Society for Optical Engineering.

DISCUSSION

J. NOORDAM

Why is the image divided up into 64 fibre bundles, and are any spatial rearrangements of these bundles considered?

D.A. CARDEN

The 90 mm diameter image is too large to be processed all at once. In particular the CCD devices used have a 1 cm^2 imaging area and this determines the sub-image area and hence the cross-sectional area of the fibre bundle.

Use is made of the flexible nature of the bundles to physically locate the CCDs at convenient positions in the instrument package. There is no spatial rearrangement of the image.

4

Medical imaging

4.1 MEDICAL IMAGING BY ULTRASOUND

G. Kossoff
Ultrasonics Institute, Sydney, Australia

Abstract. Medical imaging by ultrasound is a relatively mature subject in that the method was introduced a little over thirty years ago and has had widespread clinical acceptance for over ten years. Ultrasound is an excellent form of energy to examine soft tissues in that a variety of interacting processes occur during the propagation process. Theoretically it is possible to measure six independent acoustic parameters of tissue namely acoustic impedance discontinuity, attenuation, velocity, echo scattering cross-section, impedance and Doppler frequency shift. Even partial attainment of these measurements would be of considerable assistance in the histopathological classification of the examined tissues reducing the need for surgical exploration.

Features unique to the method include the ability to examine tissues from many directions in the near field of focussed transducers. Unfortunately, soft tissue is a non-homogeneous, non isotropic, attenuating, dispersive, non linear medium in which the velocity of propagation can vary by up to 10%. The laws of propagation are complex and not fully understood. This in many instances has precluded the clinical attainment of objectives tested out in laboratory settings or in experiments on models.

SONOGRAPHIC IMAGING

Sonographic or pulse reflection imaging is by far the most clinically utilised form of imaging. In block diagram the method bears close resemblance to imaging of reflectance by radar and sonar but in waveform and signal processing it is more closely related to seismology. The image consists of a non linear display of acoustic impedance discontinuities in the examined tissues based on the assumption that the velocity of propagation is constant. The amplitude of the echo is dependent on several factors, one of the most important being the geometry of the interface. These range from interfaces such as the stromal organisation of tissue which are much smaller than the size of the ultrasonic beam to interfaces which occur at the boundaries of organs which are much larger than the beam.

A typical local amplitude display range spans between 30 and 40dB whilst the total range allowing for attenuation spans over 100dB. Relatively sophisticated signal processing is employed to display the clinically relevant information over the whole penetration on the limited grey scale display range of viewing devices. The axial resolution of the equipment is generally several times better than the lateral resolution and the resolution cell is an oblate ellipsoid. Its dimensions are a function of distance from the transducer and therefore difficult to compensate for by conventional deblurring techniques.

Attenuation is portrayed indirectly by appearances such as enhancement or shadowing behind regions in which local attenuation differs from the surrounding tissues. Although this may be considered to be crude representation, it is invoked frequently in clinical diagnosis.

Current emphasis is on equipment which acquire images in real time, the ultrasonic beam being swept and/or focussed either by mechanical or electronic means, Wells and Ziskin (1980). Applications include the examination of all soft tissue organs in the body, and in particular the fetus and the upper abdomen. A related application includes visual guidance of biopsy needles into areas of interest, pathological analysis being as yet the gold standard for determining the nature of the examined tissues.

Approximately 50% of the incident energy is reflected when ultrasound encounters a soft tissue/bone interface whilst over 99.9% is reflected at a soft tissue/air interface. Bone enclosed and air containing structures are therefore not amenable to examination by ultrasound. Most soft tissue organs such as the uterus, liver and kidney abut posteriorly against bone or air and can be examined in reflection mode only through the anterior abdominal well. Despite this limitation it is possible to examine most organs from many different directions using a technique known as compound scanning. Simple scanning which is the primary scanning mode on real time equipment only examines tissues either from one fixed direction as a sector scanning or from one image base as in linear scanning. Compound scanning in essence superimposes a number of simple scans obtained from different directions and bears close similarity to the back projection method of imaging commonly employed in computerised tomography. Advantages of compound scanning include improvement in lateral resolution and reduction in image speckle which represents the interference pattern obtained from very small structures. With compound scanning it is possible to employ a variety of post-processing reconstruction techniques such as peak, average and minimum detect, Robinson and Knight (1981). These emphasise the relative portrayal of specular and diffuse reflectors enhancing different features of clinical signs. The effect of these processing techniques on the echographic image of a pregnant uterus which was examined over an arc subtending at an angle of 120° around the patient are illustrated in Figure 1.

Figure 1(a). Simple scan of a pregnant uterus. The sonogram is a transverse section of a fetal abdomen. The placenta lies posteriorly. The fetal spine centrally, and the fetal ribs laterally, cast a deep shadow.
Figure 1(b). Compound scan at same level. Peak detecting reconstructing algorithm was employed. A better appreciation of the uterine content is obtained. The shadowing information has been removed.

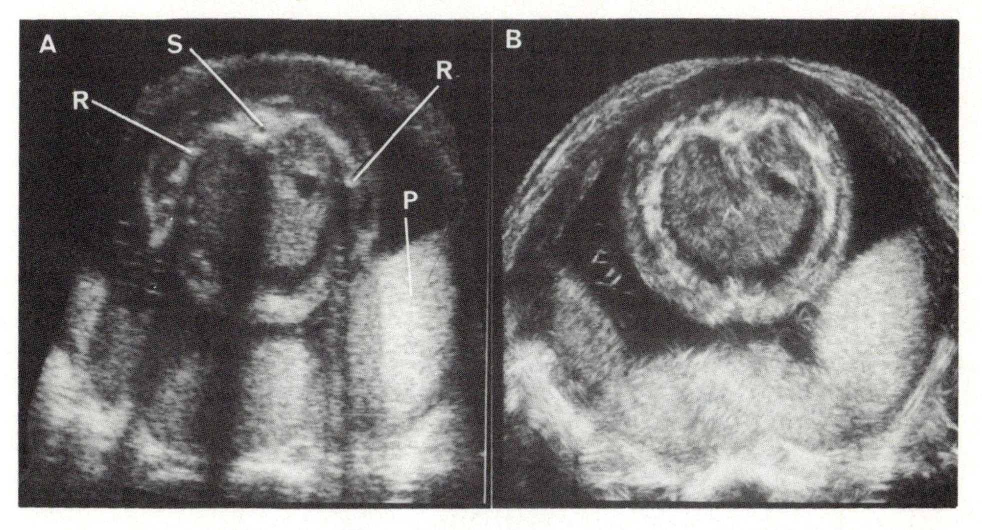

Figure 1(c). Compound scan using average reconstructing algorithm. The speckle in the image has been reduced and improved contrast resolution has been attained.
Figure 1(d). Minimum detect algorithm. This reconstruction scheme emphasises the display of small values. The shadowing from the spine from four transducers used to obtain the compound scan is obvious. Triangulation allows exact determination of position of the structure casting the shadowing.

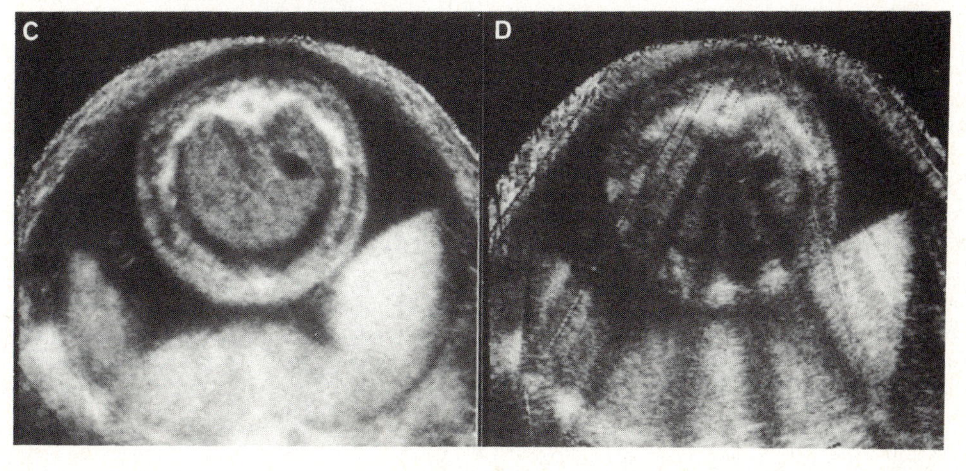

DOPPLER IMAGING

Red cells in flowing blood give rise to a Doppler shifted signal and equipment which maps out qualitatively the distribution of velocity of blood in vessels is in routine clinical use, White et al, (1980). A particular important application is the evaluation of blood flow in the carotid artery as a screening test to ascertain presence of stenosis to identify patients with a high risk for stroke. Colour coding is employed to demonstrate presence of increased velocities and an appreciation of the size of vessels and occlusions is made from the dimensions of the displayed information. A typical display of the carotid artery examined by this method is shown in Figure 2.

Figure 2. Doppler mapping of blood flow in the carotid artery. The red colour indicates normal flow velocity, yellow increased flow whilst blue is used to represent fast flow associated with obstruction. Increased flow is noted in the right external carotid artery whilst very fast flow is present in the common carotid in the left as indicated. This patient is at high risk for a stroke.

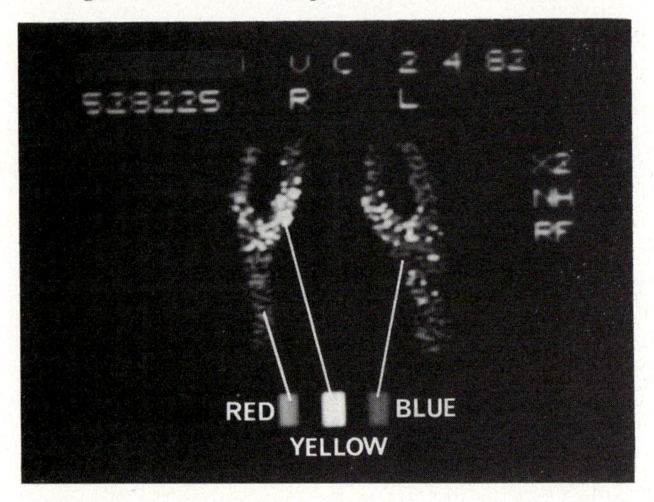

Appreciation of in vivo hemodynamics may be obtained by combining sonographic imaging with pulsed Doppler examination, Phillips, et al, (1983). Quantitative measurement of velocity requires knowledge of the angle subtended by the ultrasonic beam to the examined vessel which is obtained from sonographic imaging. Operation at high frequencies allows the attainment of small Doppler samples relative to the examined vessel. The Doppler sample is swept across the vessel providing information concering the distribution of blood flow as function of position of sample in the vessel and of cardiac cycle. Because of the need to operate at high frequencies the technique is suitable for the examination of relatively superficial vessels or on surgically exposed deep vessels.

Quantitative measurement of blood flow in litres per minute in superficial as well as deep vessels may be obtained by uniformly insonating the vessel of interest and measuring the average frequency corresponding to the average velocity of blood flow in the vessel, Gill, et al (1981). Again combined operation with sonographic imaging is employed to determine the angle subtended by the ultrasonic beam and to measure the area of the vessel. This information which may be obtained only by ultrasound is providing important new information on our understanding of the physiological development of the fetus and blood flow changes in organs affected by disease. The application of the method for the measurement of blood flow in the fetal umbilical vein is illustrated in Figure 3.

Figure 3. Quantitative measurement of blood flow. The image on the top left is a reduced scale representation of a transverse section of a pregnant interus. The image on the top right is an expanded view of the fetal abdomen. The doppler gate has been positioned over the umbilical vein. The curve at the bottom is the instantaneous value of flow in the umbilical vein in ml/min.

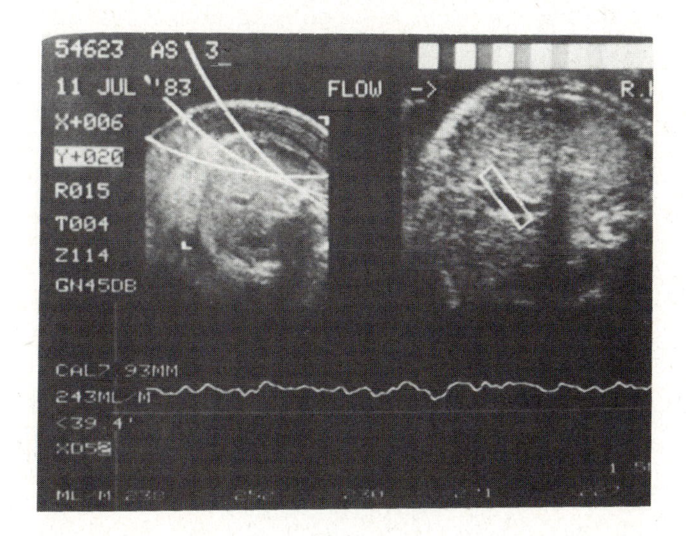

TRANSMISSION RECONSTRUCTION IMAGING

Transmission reconstruction imaging by ultrasound in a manner analagous to x-ray computerised tomography may be used to provide from amplitude and time delay of transmitted pulse quantitative images of attenuation and velocity in the scanned tissues. Unfortuantely, with the exception of the breast, few of the important organs in the body can be examined by transmission techniques. In this application the method suffers from aberrations in the propogation path by the varying velocities of the constituent tissues in the breast. Attempts to allow for refraction bending of rays have largely proved unsuccessful in the clinical setting and in practice the reconstructing process is based on the straight line propagation assumption, Greenleaf (1983). The velocity in the constituent tissues in the breast varies by over 10% and this introduces significant distortion in the geometry of the reconstructed structures and to the derived value of the imaged parameters.

In the breast the primary application of the technique is complementary to sonographic imaging which identifies areas of interest which are then quantified by transmission reconstruction imaging. One of the problems with sonographic imaging is related to the distinction between fatty lobules and malignancies which are both portrayed as areas of low level echoes on the sonogram. Clinical experience has demonstrated that the velocity of propogation of ultrasound in malignancies is significantly faster than in fat. Transmission velocity reconstruction imaging of areas of suspicion allows accurate differentiation between these tissues thus increasing the accuracy and reliability of the ultrasonic examination of the breast, Kossoff (1982). This combined application of the instrument developed in co-operation between the Queensland Institute of Technology and the Royal Brisbane Hospital is illustrated in Figure 4.

Figure 4(a). Sonogram of breast containing area of low level echoes within the glandular tissue which could represent either fatty infiltration or a malignancy.
Figure 4(b). Velocity reconstructed image at the indicated level on the sonogram. The velocity of propagation in the area seen on the sonogram is low confirming that the area represents fatty infiltration.

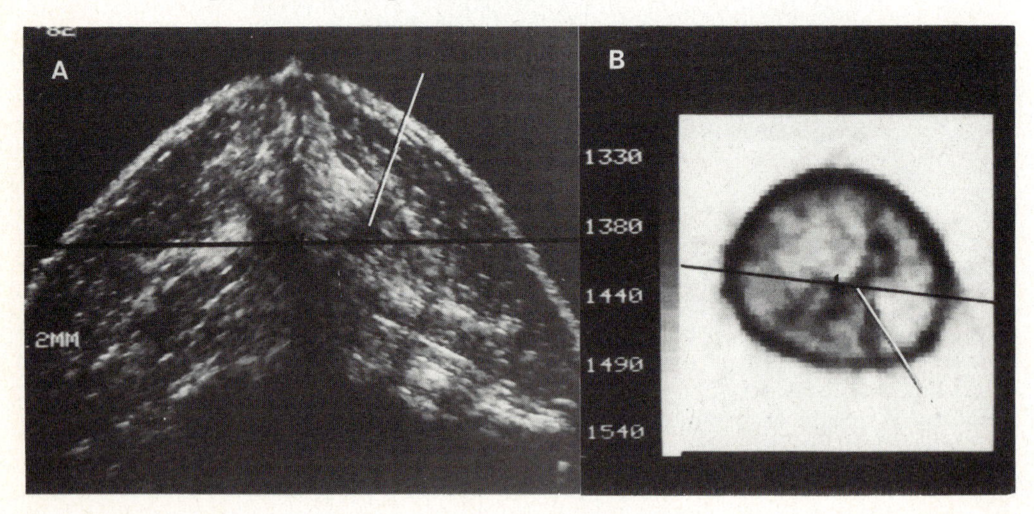

TISSUE CHARACTERISATION

Tissue characterisation relates to the quantitative measurement of acoustic properties of tissue and their frequency dependence. Although not directly related to imaging it represents the area of greatest potential development in ultrasound technology and is being based on information either identified by or related to imaging investigations.

Soft tissue on the average attenuates ultrasound at a rate of 0.5dB/cm MHz and the attenuation is linearly proportional to frequency. The ultrasonic pulse, being short, has a relatively wide frequency spectrum and the attenuation preferentially reduces the higher frequency content. If one can make the assumption that the size and distribution if interfaces in tissue is uniform, that multiple scattering does not take place and if appropriate corrections are made to allow for the effects of ultrasound the beam pattern, then attenuation in homogeneous tissue may be calculated from the change in the frequency content of echoes as function of penetration, Kuc and Taylor (1982). Sonographic imaging is employed to ensure that the data is selected from regions which satisfy the assumption of uniformity in scatterers. Figure 5(a) shows a sonographic image of a liver and the selected area over which the data was obtained whilst Figure 5(b) shows the value of attenuation on a normal patient as function of frequency. Although some theoretical problems still require resolution, the quantitative measurement of attenuation is likely to be the first successful clinical application of tissue characterisation.

Figure 5(a). Transverse section through the liver. Portion of liver containing uniform distribution of scatterers selected for the data analysis is indicated.
Figure 5(b). Attenuation in normal liver as function of data. Due to the spectrum of the insonating pulse the data spans a frequency range of 1.5 to 3.5 MHz. The attenuation is linearly proportional to frequency.

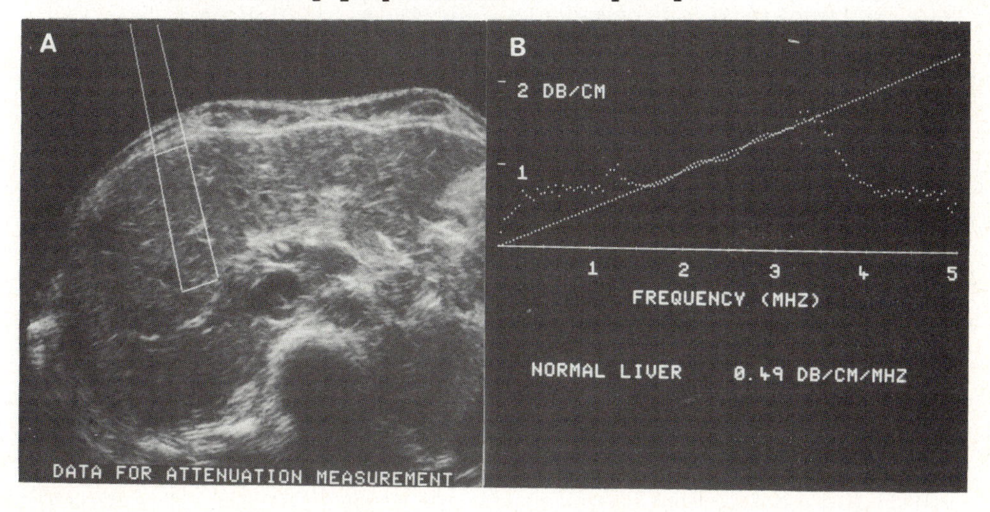

Sonographic imaging may be used to measure the velocity of propagation of ultrasound in tissue. The technique, as yet has only been developed in the analysis of a two velocity model and consists of examining the tissues by two simple scans. Refraction at the velocity change interface deviates the beam causing the echoes to be misplotted relative to their anatomical postion. The method is based on identifying the same structure on the two scans, measuring the degree of misregistration and from the contour of the velocity interface calculating the velocity in the second medium, Robinson et al, (1982). Despite the simplicity of the model, the in vivo and reproducibility of the method in the examination of the liver have been surprisingly good and the method promises to be the second clinical application of tissue characterisation.

Some reports have described measurement of echo scattering cross-section using methods somewhat analagous to those employed in crystalography analysis, Merton, et al (1982). Considerable work still remains to be done to determine the clinical validity of these measurements. Measurement of impedance has only been performed in laboratory settings and require extensive development in theoretical analysis to be applied clinically.

OTHER IMAGING TECHNIQUES

A variety of other imaging techniques have been attempted using ultrasound for medical purposes. Acoustic holography, Ash and Hill (1983), is the most extensively investigated of these techniques which, despite considerable effort, has failed to find clinical accpetance. The major limitation relates to the dispersion of the beam by the varying velocities in the examined tissues which gives rise to excessive interference and speckle to the holographically derived image.

Other studies include investigations into synthetic aperture focussing techniques, Ridder, et al (1981), and reflection reconstruction imaging, Friedrich, et al (1982). As yet these studies have been undertaken only on models or fixed tissues and it is uncertain how they will fare in a clinical setting.

REFERENCES

Acoustical Imaging (1983), Ed. E. Ash and C.R. Hill, Plenum Press, New York, Vol. 12.

M. Friedrich, E. Hundt and G. Maderlechner (1982). Computerised Ultrasound Echo Tomography of the Breast. Europ. J. Radiol. 2, 78-87.

R.W. Gill, B.J. Trudinger, W.J. Garrett, G. Kossoff and P.S. Warren (1981). Fetal Umbilical Venous Flow Measured In Utero by Pulsed Doppler and B-mode Ultrasound. In Normal Pregnancies. Am. J. Obstet. Gynecol., 139, 720-725.

J.F. Greenleaf. Computerised Tomography with Ultrasound (1983). Proc. IEEE, 7, 330-337.

G. Kossoff (1982). The Role of Computers in Ultrasound Cancer Diagnosis - Velocity Measurement Techniques, pp. 11-19 in Ultrasound and Cancer, Ed. S. Levi, Excerpta Medica, Amsterdam.

R. Kuc and K.J.W. Taylor (1982). Variation of Acoustic Attenuation Coefficient Slope Estimates for In Vivo Liver. Ultrasound in Med. & Biol. Vol. 8, 4, 403-412.

J. Merton, D. Nicholas, C.R. Hill, S. Grover, M. Queenan and D. Cosgrove (1982). Ultrasonic Diffraction Scanning of the Thyroid. Ultrasound in Medicine and Biology, 8, 145-154.

D.J. Phillips. F.M. Greene Jr., Y. Langlois, G.O. Roederer and D.E. Standness Jr (1983). Flow Velocity Patterns in the Carotid Bifurcations of Young, Presumably Normal Subjects. Ultrasound in Medicine and Biology, 9, 39-49.

J. Ridder, L.F. van der Wal and A.J. Berkhoot (1981). A New Synthetic Focussing Technique for Medical Ultrasound, pp. 56-60 in Recent Advances in Ultrasound Diagnosis 3. Ed. A. Kurjak and A. Kratochwil. Excerpta Medica, Amsterdam.

D.E. Robinson, F.C. Chen and L.S. Wilson (1982). Measurement of Velocity of Propagation from Ultrasonic Pulse Echo Data. Ultrasound in Medicine and Biology, 8, 413-420.

D.E. Robinson and P.C. Knight (1981). Computer Reconstruction Techniques in Compound Scan Pulse Echo Imaging. Ultrasonic Imaging, 3, 217-234.

P.N. Wells and M.C. Ziskin (1980). New Techniques and Instrumentation in Ultrasonography. Churchill Livingstone, New York.

D.N. White, C.E.J. Ketelaars and P.R. Cledgett (1980). Non Invasive Techniques for the Recording of Vertabral Artery Flow and their limitations. Ultrasound in Medicine and Biology, 6, 315-328.

4.2 REVIEW OF INSTRUMENTATION IN MEDICAL X-RAY AND GAMMA-RAY IMAGING

D. Rosenfeld
Imaging Science and Engineering Laboratory
School of Electrical Engineering
University of Sydney, Sydney, N.S.W., Australia 2006

Abstract: In this paper, recent advances in the following medical imaging modalities are reviewed: gamma-ray camera imaging; single photon and positron emission computed tomography; transmission computed tomography; digital radiography; and nuclear magnetic resonance.

I. GAMMA CAMERAS

The tradition Anger-type Gamma camera continues to be the main tool of the nuclear medicine practitioner. These cameras have improved steadily over the years. The most modern cameras consist of a large (460mm) NaI (Tc) crystal, about 9.5mm thick, optically coupled to 37 13mm photomultiplier tubes. Such cameras give fields of view up to about 400mm. The scintillating crystal is thick enough so that any gamma-ray is absorbed by it. When the gamma photon is absorbed, many light photons are emitted in all directions. These photons are detected in the many photomultiplier tubes. The outputs of the photomultiplier tubes are fed to a position and event computer which determines whether a detected gamma photon had energy within a particular range of values (to reject scattered photons) and then computes the x-y position of the event.

Anger-type detectors have distortions which need to be overcome, caused by three main processes:

- non-linear response of the position computing process, which causes spatial distortions leading to variations in count density over the field of view.

- non-uniformity in photopeak height over the field of view, causing count losses in off-center regions in a given energy window, thus resulting in false "cold" areas in the image.

- variations of sensitivity in the detector head and the collimator over the field of view.

The latest cameras coming on the market have corrections for all three types of distortions.

Linearity correction is to correct for the positioning errors of nuclear events. An orthogonal hole phantom is used to generate calibration data, which are later, during imaging, used to displace each pixel to its correct location. Different calibration data are used for images which were produced with different isotopes.

Energy Correction is to compensate for the fact that the NaI photopeak spectrum varies for different locations in the detecting crystal. Instead of one energy window being used, several position-dependent windows are used.

Sensitivity Correction: After energy and linearity correction, a uniform flood may still not produce a uniform image, due to the particular collimator used or other factors. In the latest cameras, a uniform flood calibration image is used to correct for this residual non-uniformity.

The gamma camera's utility as a diagnostic tool is becoming increasingly dependent on its ability to detect and process high count rates. The crucial parameter limiting count rate is the camera's dead time, due to intrinsic dead time in the crystal and photomultiplier tubes, and on the time needed to process the data. The latest cameras can process data at a rate of up to 500,000 discrete events per second.

When data is collected at high rates, it is possible that two photons interact in the crystal within a time interval shorter than the pulse-pair resolving time of the camera, simulating a single event with an energy higher than the two individual photon energies. This can create image artifacts. Reducing the systems dead time reduces such pile up artifacts.

All the above corrections are achieved by digitizing the pulse information as early as possible after the photomultiplier tubes, and by incorporating into the gamma cameras, fast bit-slice processors.

II. SINGLE PHOTO EMISSION COMPUTED TOMOGRAPHY (SPECT)

The clinical use of SPECT in nuclear medicine dates from the early 60's, when the first tomograph was produced by Kuhl and Edwards [1]. Since then, there has been a struggle for the survival of SPECT techniques. Enthusiasm has waxed and waned with the hopes and disappointments of innovations such as coded apertures and the adaptation of single and dual headed Anger-type cameras to rotating gantries. It now appears that the struggle for survival of SPECT may not have been in vain, as evidenced by some recent results on estimating regional cerebral blood flow using gamma rays [2,3,4]. Positron techniques have been successfully used for regional brain profusion studies for several years [5,6,7]. However, positron emission tomography (PET) studies are expensive (estimated at

[13] Zimmerman,
Emission Cc
Detectors",
and Other Se
Society of N

[14] Budinger,
"Emission C
from Proje
Vol.32, Topi
York, Spring

[15] Rosenfeld,
in Nuclear
Aided Diagno

[16] Rosenfeld,
for Tomogra
Sci., NS-24:

[17] Rosenfeld,
Applications
Opt. Soc. Am

[18] Rosenfeld,
Applications
J. Opt. Soc.

[19] Ter-Pogossia
Positron
Information"

[20] Yamamoto,
"Experimenta
Utilization
Emission Tom
MI-1:187, 1

[21] Garderet, P
Using Time-
Tomography S
IEEE ComSoc

[22] Nalciouglu,
View Recons
Nucl. Sc., N

[23] Wagner, W.,
Data: New Me
Nucl. Sci., N

$US1700 per study), limiting their use to only a few centers. The SPECT techniques, on the other hand, could have widespread clinical utility, limited only by the availability of imaging instrumentation. The development of N-isopropyl I-123 p-iodoamphetamine and other easy to distribute radiopharmaceuticals may lead to SPECT regional cerebral brain scanning becoming the procedure of choice for central nervous system lesions in clinical diagnostic medicine.

Imaging instruments for SPECT include:

- Single and dual headed gamma cameras attached to rotating gantries. Such systems are available from almost all gamma camera manufacturers.

- Four detector banks placed as closely as practical to the subject [9].

- A circular array of crystals with rotating and moving collimator fins [10,11].

- The Harvard Scanning Multidetector Brain System, which is modification of a device initially developed commercially by Union Carbide [12,13].

All the above techniques use some straightforward configuration to directly measure the projections. This is followed by reconstruction from projections, typically using the filtered back projections algorithm. The instruments suffer from poor resolution (around 10mm) and low signal-to-noise-ratio (SNR). At first, the instruments also suffered from inadequate correction for attenuation and scatter of the emitted photons. Reasonable, although not perfect, techniques for attenuation correction have been available for a while [14]. Today, all instruments use some form of such correction.

The low SNR problem still exists in available instruments. Source dependent imaging techniques, as developed by the author [15,16], show promise in improving noise performance in SPECT. These techniques recognize the inherent property of imaging systems that, as soon as one attempts to increase the photon collection efficiency beyond that of a simple pinhole or parallel hole collimator, either using coded aperture techniques, or by collecting more than one (overlapping) projection at a time for axial tomography (using time modulation), the SNR performance of the final image becomes dependent on the size and other properties of the emitting source. Since this cannot be overcome, source dependent imaging changes the collection efficiency of the imaging system according to what source is being imaged.

Two source dependent SPECT configurations have been studied. Figure 1 shows a cone-beam configuration for longitudinal tomography. Each pinhole view is a cone-beam projection of the source from a different direction. Figure 2 shows a fan-beam configuration for transverse section SPECT. Each pinhole view is a fan-beam projection of the 2-D

[24] Alvarez, R.E. and Macovsk , A., "Energy Selective Reconstructions in X-Ray Computerized Tomography", Phys. Med. Biol., 21:733, 1976.

[25] Robb, R.A., Ritman, E.L., et al, "The DSR: A High Speed 3-Dimensional X-Ray Computed Tomography System for Dynamic Spatial Reconstruction of the Heart and Circulation", IEEE Trans. Nucl. Sci., NS-26:2713, 1979.

[26] Robb, R.A., Hoffman, E.A., et al, "High Speed Three-Dimensional X-Ray Computed Tomography: the Dynamic Spatial Reconstructor", Proc. IEEE, 71:308, 1983.

[27] Boyd, D.P., Gould, R.G., et al, "A Proposed Cardiac 3-D Densitometer for Early Detection and Evaluation of Heart Disease", IEEE Trans. Nucl. Sci., NS-26:2724, 1979.

[28] Boyd, D.P., and Lipton, M.J., "Cardiac Computed Tomography", Proc. IEEE, 71:298, 1983.

[29] Kruger, R.A., Mistretta, C.A., et al, "A Digital Video Image Processor for Real Time X-Ray Subtraction Imaging", Optical Eng., 17:652, 1978.

[30] Ovitt, T., Capp, M.P., et al, "Development of a Digital Video Subtraction Scheme for Intravenous Fluoroscopy", SPIE, 206:73, 1979.

[31] Macovski, A., Alvarez, R.E. and Chan, J.L.H., "Selective Material X-Ray Imaging Using Spatial Frequency Multiplexing", Appl. Opt., 13:2202, 1974.

[32] Kelcz, F. and Mistretta, C., "Absorption Edge Fluoroscopy Using a 3-Spectrum Technique" Med. Phys., 3:159, 1976.

[33] Macovski, A. Brody, W., et al, "Future Trends in Projection Radiography:, SPIE, 206:19, 1979.

[34] Lehmann, L.A., Alvarez, R.E., et al, "Generalized Image Combinations in Dual kVp Digital Radiography", Med. Phys., 8:659, 1981.

[35] Stein, J., "X-Ray Imaging With a Scanning Beam", Radiol. 117:713, 1976.

[36] Bjorkholm, P., Annis, M., et al, "Digital Radiography: Contrast and Spatial Resolution", SPIE, 273:107, 1981.

[37] Lauterbur, P., "Image Formation by Induced Local Interactions: Examples Employing Nuclear Magnetic Resonance", Nature, 242:190, 1973.

[38] Demadian, R., "Tumor Detection by NMR", Science, 171:1151, 1971.

[39] Andrew, E.A., Bottomley, P.A., et al, Nature, 270:frunt cover, 1977.

[40] Bottomley, P.A. and Andrew, E.A., "RF Magnetic Field Penetration, Phase Shift and Power Dissipation in Biological Tissue: Implications for NMR Imaging", Phys. Med. Biol., 23:630, 1978.

DISCUSSION

R.H.T. BATES

Do you really mean to imply that a dual energy system can fully correct for beam hardening?

D. ROSENFELD

Dual energy beam hardening correction is not for any reconstruction of any combination of materials. For the materials found in the human body, however, near perfect correction can be achieved. This is because in the body, in the diagnostic energy range, to a high degree of accuracy, the energy dependent attenuation coefficient for X-rays can be decomposed into two basic functions representing the photoelectric and Compton scattering, respectively, and that the energy dependence of the photoelectric and Compton attenuation phenomena is the same for all materials of importance in the body. See, for example, J.P. Stonestrom, R.E. Alvarez and A. Macouski (1980). A framework for spectral artifact correction in X-ray CT. IEEE Trans. Biomed. Eng. BME-28, 128; and R.E. Alvarez and A. Macouski (1976). Energy-selective reconstructions in X-ray computerized tomography. Phys. Med. Biol. 21, 733.

5

Wide field radio mapping

5.1 AN INTRODUCTION TO WIDE FIELD MAPPING

T.W. COLE
School of Electrical Engineering, University of Sydney,
NSW 2006, Australia

The classical approach in interferometry is to describe it in terms of the van Cittert-Zernike theorem, relating as it does the object distribution and the resulting fringes from the interferometer as a Fourier transformation. This approach does, however, contain a number of assumptions which are extremely important when mapping a wide field of view and when the fractional bandwidth of the observation is large. A further complication in earth-rotation aperture synthesis is the general use of interferometer baselines which do not lie in a plane.

These problems are addressed in a number of papers of this symposium; all could be formulated as applications of coherence theory. It may be useful to briefly review that theory as it applies to aperture synthesis. In physics texts such as Born & Wolf (1975) interferometers are described as measuring the mutual coherence function $\Gamma(r_{12},\tau)$ which, for vector separation of interferometer elements r_{12} and added delay τ between the complex field amplitude V at elements 1 and 2, is given by

$$\Gamma(r_{12},\tau) = \langle V(r_1,t)V^*(r_2,t+\tau)\rangle \qquad (1).$$

Here $\langle\rangle$ represents time average. A more recent concept which relates closely to the techniques used in the radioastronomy case is the cross-spectral density $W(r_{12},\omega)$ measured as the crosscorrelation of the field amplitudes over a vanishingly small bandwidth centred on ω (Mandel & Wolf 1976).

The two measurements $\Gamma(r_{12},\tau)$ and $W(r_{12},\omega)$ are Fourier related and in the astronomy context (Cole 1979) one can write

$$W(r_{12},\omega) = \iint_{sky} \langle a(l,m;\omega)\, a^*(l,m;\omega)\rangle \exp[-ik(ul+vm+wn)]\, \frac{dl\, dm}{n} \qquad (2)$$

Here $k=\omega/c$ and the term in $\langle\rangle$ brackets is just the power per unit bandwidth at frequency ω from a solid angle element dl dm in direction (l,m). It is just $B_\omega(l,m)$ to be compared with the broadband brightness B(l,m). The exponential term is the geometrical delay introduced for the standard (u,v,w) and (l,m,n) co-ordinate systems. By using $W(r_{12},\omega)$ one avoids all of the assumptions about path lengths being small compared with the "coherence length" which one finds in the derivation of the van Cittert-Zernike theorem using $\Gamma(r_{12},\tau)$. Yet one still has power to address the imaging aberrations which occur in practice.

One such aberration is that of radial smearing due to a finite observing bandwidth. It can be shown [3] that this is due to a delay error for sources off-axis, an error which introduces quite naturally the autocorrelation of the radiation in the bandpass of the receiver. That is, although one corrects for the relative delay between signals $V(r_1,t)$ and $V(r_2,t)$ coming from sources at the point in the field where $l=m=0$ (and hence $V(r_1,t)=V(r_2,t)$), away from this direction one measures the mean product with a relative delay between the signals. This is just one point on the autocorrelation of the radiation in the passband of the receiver. The consequence is a weighting down of the value of the measurement compared to that at the field centre and is identical in effect to a taper on the spatial frequency plane. This taper is different for different source positions and the result is the important observation that for wide fields and wide bandwidths the instrumental response is position dependant. Global concepts such as Fourier transforms do not apply in these demanding cases and iterative, position-dependent processing is a natural consequence.

The chromatic aberrations can also be interpreted from the equations in another way. The exponential part of the equation is of form $[-ik(ul+vm+wn)]$ where once again $k=\omega/c$. But the exponent could take the SAME value for a range of values of u,v,w,l,m and n so long as the term k is suitably changed to suit. This means quite simply that observations at one frequency produce a map whose sidelobes and gratings are the same as from a DIFFERENT SIZED array observing at a different frequency. The equivalence concepts implied here have only partially been exploited in radio astronomy but an application in radar is presented at this symposium.

The problems of wide field mapping using aperture synthesis are explored in more detail in the next two papers. The problems are not insignificant and are not solved by moving to single dish mapping. One merely exchanges one set of problems for another. Nevertheless, the maps which are produced by modern radiotelescopes are adequate proof of the level of understanding of the problems which has already been achieved.

References.

Born, M. & Wolf, E. (1975). Principles of Optics. New York: Macmillan.

Cole, T.W. (1979). Mutual coherence function in radioastronomy. J.Opt.Soc.Am., 69, no.4, 554-7.

Mandel, L. & Wolf, E. (1976). Spectral coherence and the concept of cross-spectral purity. J.Opt.Soc.Am., 62, no.6, 529-35.

5.2 INVERSION OF NONPLANAR VISIBILITIES

R.N. Bracewell
Electrical Engineering Department, Stanford University,Stanford CA94305

INTRODUCTION

How to construct a radio astronomical brightness image $B(l,m)$ from complex visibilities measured in the (u,v)-plane is a well practised exercise that has led to images of extraordinary quality, unrivalled in angular resolution (10^{-8} rad) and dynamic range (60 db); but in newer applications where the interferometer vectors are rather long a more general theory has been needed. The problem is that the baseline vectors (u,v,w) may not lie in a plane, and certainly not in a plane perpendicular to the direction of the source. The classical Fourier relationship between brightness B and complex visibility $\mathcal{V}$ refers, however, to interferometric measurements in a plane.
In very-long-baseline interferometry where more than three stations are involved the baselines can be noncoplanar because the the stations need not lie in a plane. But a multielement interferometer at one observatory, even if the elements *are* coplanar, can also generate noncoplanar baselines as the earth rotates. Only baselines in planes perpendicular to the axis of rotation of the earth, mostly east-west baselines, can be coplanar with the new baselines generated as the earth rotates. Hitherto only approximate methods have been proposed for inverting nonplanar visibilities.

THE DISTRIBUTION OF COMPLEX VISIBILITY IN 3-SPACE

The fundamental inversion problem of interferometry is: given $\mathcal{V}(u,v,w)$ on the baseline space $\mathcal{B}$, find the brightness distribution $B(l,m)$, where (l,m,n) are direction cosines based on (u,v,w). As a tactic we may ask the opposite question, given $B(l,m)$ over a source $\mathcal{S}$ what is the complex visibility $\mathcal{V}_P$ at a point $\mathrm{P}(u_P,v_P,w_P)$. The answer is

$$\mathcal{V}_P(u_P,v_P,w_P)=S_0^{-1}\int\int_{\mathcal{S}} \mathbf{n}_0 n^{-1} B(l,m) e^{-i2\pi(lu_P+mv_P+nw_P)} dl dm, \tag{1}$$

where S_0 is the flux density of the source and $\mathbf{n}_0 \approx J_0(\arccos n)$ is related to the inclination factor of Huygens diffraction theory. Hitherto unproved, this relationship, which is exact (Appendix 1), has occasionally been surmised to be true and clearly reduces to the classical Fourier integral for small sources in the special case of $w=0$.

APPROXIMATE SOLUTION

If we were given values of $\mathcal{V}_P$ we could construct associated visibilities $\mathcal{V}_c(u,v)$ in the (u,v)-plane by assigning values $\mathcal{V}_P e^{i2\pi p_0}$ to points $\mathrm{P}_c(u,v)$ that are related to (u_P,v_P,w_P) in accordance with an association rule

$$u=u_P-(l_0/n_0)w_P \tag{2}$$

$$v=v_P-(m_0/n_0)w_P, \tag{3}$$

where (l_0, m_0, n_0) is the reference direction of the source and p_0, the distance from P to $P_c(u,v)$, is given by

$$p_0 = l_0(u_P - u) + m_0(v_P - v) + n_0 w_P. \tag{4}$$

Thus

$$\mathcal{V}_c(u,v) = S_0^{-1} e^{i2\pi p_0} \int\int_S n_0 n^{-1} B(l,m) e^{-i2\pi(lu_P + mv_P + nw_P)} dl dm. \tag{5}$$

If $\mathcal{V}_c(u,v)$ were corrected for area density and then subjected to Fourier transformation, an approximation to $B(l,m)$ would result. What relation would the result bear to the desired $B(l,m)$?

The answer to this question is not known. One way of expressing an answer would be to give the point spread function (PSF) describing the aberration introduced by the procedure. Since the PSF will depend on the spatial distribution of the observation points, which typically will be distributed discontinuously on more than one sheet in space, complications may be expected. Even in the case of a single cone corresponding to observations made with a continuous linear antenna running north-south, the boundary of the region on the (u,v)-plane over which associated visibilities can be assigned is unwieldy. The (u,v)-plane boundary may be bisectoral or it may be defined by a pair of overlapping ellipses. Within this reentrant boundary the density of associated visibilities will be strongly nonuniform even where the given visibilities are of uniform surface density on their cone.

A DISSECTION OF BASELINE SPACE

Thus it would be difficult to study the approximate solution analytically although a numerical study would be possible for a special case of particular interest. To avoid the geometrical complexity described above, suppose for the moment that visibilities are measured not on a cone but on an inclined *plane* through the origin, not perpendicular to the direction of the source.

If we apply the association rule (2)-(4) to assign visibilities to the perpendicular plane and then perform a two-dimensional Fourier inversion we obtain a brightness distribution with aberrations. We may also obtain a *rigorous* solution for comparison by first performing the Fourier inversion and then carrying out a shift in the directional coordinate system. In this way we can study a simple case of the aberration produced by the association rule free from certain complications of geometry.

Assume that the baseline components obey the condition

$$l_p u_P + m_p v_P + n_p w_P = 0 \tag{6}$$

and thus lie in a plane whose normal points in the direction (l_p, m_p, n_p). Then the offset p_0, from (4) and (6), is given by

$$p_0 = l_0(u_P - u) + m_0(v_P - v) - n_0(l_p u_P + m_p v_P)/n_p. \tag{7}$$

But it should be sufficient to consider a case where $l_p = 0$ (a baseline plane tilted towards the north). Likewise the position of the source should be unimportant, so we may take

$(l_0, m_0, n_0) = (0, 0, 1)$. Then

$$w_P = -(m_p/n_p)v_P = -\beta v_P \tag{8}$$

where $\beta = -m_p/n_p$ is the slope of the baseline plane toward the north and

$$p_0 = w_P. \tag{9}$$

Consider a unit source element ${}^2\delta(l - \lambda, m - \mu)$ within the region S. From (1)

$$\mathcal{V}_P(u_P, v_P, w_P) = \int\int_S \mathrm{n}_0\, {}^2\delta(l - \lambda, m - \mu) e^{-i2\pi(lu_P + mv_P + nw_P)} dl dm$$

$$= \mathrm{n}_0 e^{-i2\pi[\lambda u_P + \mu v_P + (1-\lambda^2-\mu^2)^{1/2} w_P]}.$$

The associated visibility $\mathcal{V}_c(u, v)$ constructed in the (u, v)-plane perpendicular to the source direction is, from (5),

$$\mathcal{V}_c(u, v) = e^{i2\pi w_P}\, \mathrm{n}_0\, e^{-i2\pi[\lambda u_P + \mu v_P + (1-\lambda^2-\mu^2)^{1/2} w_P]}$$

$$= \mathrm{n}_0 e^{-i2\pi\{\lambda u_P + \mu v_P + [(1-\lambda^2-\mu^2)^{1/2} - 1] w_P\}}$$

$$= \mathrm{n}_0 e^{-i2\pi\{\lambda u_P + \{\mu - \beta[(1-\lambda^2-\mu^2)^{1/2} - 1]\} v_P\}}. \tag{10}$$

Applying the Fourier transformation and making use of (2) and (3) to replace u_P by u and v_P by v (for the special case $l_0 = m_0 = 0, n_0 = 1$) we arrive, ignoring n_0, at the PSF

$${}^2\delta\{l - \lambda, m - \{\mu - \beta[(1 - \lambda^2 - \mu^2)^{1/2} - 1]\}\}. \tag{11}$$

Thus the source element remains a point element but it is shifted in the m-direction by an amount that is proportional to the slope of the baseline plane and to the square of the distance from the origin.
To see the general order of magnitude consider a zenithal observation with an east-west array at latitude 34° where $\beta = \cot 34^\circ$. Then we find

Distance from origin	(rad)	.001	.003	.01
	(arcmin)	$3'$	$9'$	$30'$
Shift	(arcsec)	$0.1''$	$1''$	$10''$

If the calculations are carried out for a general case, field of view centred at (l_0, m_0, n_0) and a field element ${}^2\delta(l - l_0 - \lambda, m - m_0 - \mu)$, we get instead of (10),

$$\mathcal{V}_c(u, v) \doteq \exp\{-i2\pi[(l_0 + \lambda)u + (m_0 + \mu)v]\} \exp[-i2\pi\, n_0(\lambda^2 + \mu^2) w_P / 2n_0]. \tag{12}$$

Thus in general there is an aberrational shift proportional to the square of the distance from the centre of the field of view, proportional to the slope of the baseline plane and in the direction of the slope. Fig. 1 illustrates the character of the distortion.

AN INVERSION ALGORITHM

The previous section shows us how the association rule will distort the desired brightness map but also teaches us how to avoid the distortion. Information is destroyed when the visibilities $\mathcal{V}_c(u,v)$, having been constructed in the (u,v)-plane, are combined without regard to their associated offsets. This is because different aberrations are associated with baseline planes of different slopes.

But the baseline vectors may first be assembled in groups that are coplanar or approximately so prior to application of the association rule. For example, a set of baselines defining a cone need not be projected directly onto the (u,v)-plane and there mingled with the projections of other cones; the cone could be regarded as a pyramid with plane faces and each inclined face could be treated separately by the association rule to generate a complex submap whose simple distortion could be corrected by (10) or (12) before that submap was combined with the submaps from other plane or approximately plane faces. Other ways of passing planes (6) through the occupied points of baseline space can also be imagined. Thus the powerful deduction that there exists a subset of baseline space for which the aberration of a point source reduces to simple distortion leads to an inversion algorithm.

To summarize: sort the visibilities $\mathcal{V}(u,v,w)$ into subsets on or near planes through the origin. With each subset associate visibilities $\mathcal{V}_c(u,v)$. Form a submap by Fourier inversion. Correct the submap for the distortion appropriate to the slope and orientation, and sum the submaps.

The procedure described is linear but space-variant; the PSF is different for different source elements. Given visibility samples sufficiently numerous and uniformly enough spaced to permit adequate correction for area density, one ought to arrive at the principal solution corresponding approximately to the (u,v)-plane boundary by following the above prescription.

Further work is needed to take this solution in principle and relax the strict requirements in order to arrive at a practical combination of reasonable computing time and acceptable

Fig. 1. Exaggerated impression of the quadratic distortion possessed by a complex submap.

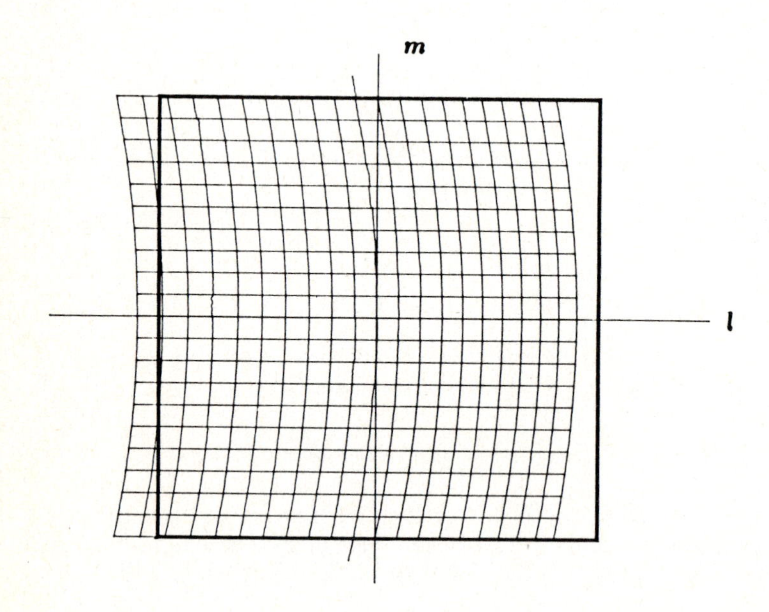

precision. An exact analytical solution to the inversion problem may be written down in terms of a sum over operations carried out point by point in the (u, v, w)-domain. What we need however is a way of grouping the data and taking advantage of approximations so that the computing time approaches proportionality to $N \log_2 N$ as with the Fast Fourier or the even faster Fast Hartley Transform.

APPENDIX 1

The purpose of this Appendix is to derive (1). Consider a rotated baseline coordinate system (u', v', w') such that

$$u' = l_1 u + m_1 v + n_1 w$$
$$v' = l_2 u + m_2 v + n_2 w$$
$$w' = l_3 u + m_3 v + n_3 w.$$

Then $u = l_1 u' + l_2 v' + l_3 w'$ and $v = m_1 u' + m_2 v' + m_3 w'$. A baseline (u_1, v_1) on the (u, v)-plane appears as (u_1', v_1', w_1') in (u', v', w')-space. Define the new quantity $\mathcal{V}'(u', v', w')$ by

$$\mathcal{V}'(u', v', w') = \mathcal{V}(u, v).$$

Then

$$\mathcal{V}'(u', v', w') = S_0^{-1} \int\int_S \mathrm{n}_0 n^{-1} B(l, m) e^{-i2\pi(L+M+N))} dl dm$$

where $L = ll_1 + mm_1$, $M = ll_2 + mm_2$, and $N = ll_3 + mm_3$. Let (λ, μ, ν) be direction cosines in the (u', v', w') system. Then $\lambda = ll_1 + mm_1 + nn_1$, $\mu = ll_2 + mm_2 + nn_2$, $\nu = ll_3 + mm_3 + nn_3$. Define $B'(\lambda, \mu)$ by the relation

$$\mathrm{n}_0' B'(\lambda, \mu) d\lambda d\mu = \mathrm{n}_0 n^{-1} B(l, m) dl dm. \tag{13}$$

Then

$$\mathcal{V}'(u', v', w') = S_0^{-1} \int\int_S \mathrm{n}_0' \nu^{-1} B'(\lambda, \mu) e^{-i2\pi\phi} d\lambda d\mu$$

where

$$\phi = (\lambda - nn_1)u' + (\mu - nn_2)v' + (\nu - nn_3)w'.$$

Consequently the basic brightness-visibility relation expressed in an arbitrary system (u', v', w') where baselines do not necessarily lie in the (u', v')-plane, becomes

$$\mathcal{V}'(u', v', w') = e^{-i2\pi n w_0} S_0^{-1} \int\int_S \mathrm{n}_0' \nu^{-1} B'(\lambda, \mu) e^{-i2\pi(\lambda u' + \mu v' + \nu w')} d\lambda d\mu.$$

The quantity $w_0 = n_1 u' + n_2 v' + n_3 w'$ is the projection of the baseline on the w-axis. Since the baseline lies in the (u, v)-plane and is therefore perpendicular to the w-axis it follows that $w_0 = 0$. Thus

$$\mathcal{V}'(u', v', w') = S_0^{-1} \int\int_S \mathrm{n}_0' \nu^{-1} B'(\lambda, \mu) e^{-i2\pi(\lambda u' + \mu v' + \nu w')} d\lambda d\mu.$$

Dropping the primes and inserting subscripts for any particular point P we obtain the result (1).

DISCUSSION

J.A. ROBERTS

In what sense is your procedure 'rigorously correct'? Do you mean that it yields a map which is the true sky convolved with an invariant point-spread function?

R.N. BRACEWELL

By setting aside the complications associated with the re-entrant shape of projected baseline space and the non-uniform density within that space one may concentrate on the simpler problem of the aberration caused by the projection procedure itself. This sub-problem can be discussed rigorously. If you use the algorithm resulting from this study you will still have to make your personal choice for handling area density compensation and the sharp cutoff at the edge of data space; so it is not possible to generalize about the map that would result.

R.H. FRATER

Isn't the result obtained by your procedure when taken to the limit in the number of planes used identical to that from the triple integral approach I described in my earlier papers?

R.N. BRACEWELL

There must be a connection although the two formulations do differ as to the presence of weighting coefficients. In the limit as the number of plane faces increases we get a string of collinear base-line vectors and the corresponding complex submap is a radial stripe. It does not seem necessary to further subdivide into point-by-point summation through three-dimensional baseline space because all the points of one infinitesimal face share the same distortion and may be lumped together. Additional grouping is inexact but alleviates the computing load.

F.R. SCHWAB

I don't believe that an explicit inversion formula or an especially tailored inversion technique is actually required. Instead, some iterative numerical inversion methods would require only that, given a hypothetical (or model) source, the visibility observation which the instrument would produce can be computed. Then you work with the visibility residuals. A disadvantage is that one has to return occasionally to the raw (ungridded) visibility data.

For example, with CLEAN one may iterate for a while ignoring the fact that the data are only approximately described by a convolution equation. Having got a model for strong features in the map, one may compute visibility residuals from the raw data, applying various instrumental corrections (the wz - term phase correction, attenuation of the amplitude due to finite integration time and finite bandwidth), compute a map from these residuals, and continue CLEANing. (And repeat the process.)

MEM methods could incorporate a similar modification. One can't make as effective use of the FFT algorithms though, as one might wish.

R.N. BRACEWELL

If you prefer iteration you may nonetheless wish to understand the mechanism of the aberration you plan to CLEAN. If an aberrated image suffered merely from distortion rather than from blurring CLEAN would not correct it. If submaps with different distortions were combined, blurring would result also; in such a case CLEAN would sharpen the blurred peak but would in general displace it from the true location. Now that we understand that the complex visibility data may be decomposed into subsets sharing a common distortion it would seem wise to incorporate that understanding into an iterative procedure. You could do this on your occasional returns to the three-dimensional complex visibility data.

J.E. BALDWIN

What are the relative advantages of (i) splitting up the (u,v,w) data into sets of points which lie on a finite number of facets of a pyramid and (ii) splitting up the map into a mosaic of a finite number of small maps each calculated by projecting the whole of the (u,v,w) data on to appropriate (u,v) planes?

R.N. BRACEWELL

By merging the projected data on planes $\ell_i u + m_i v + n_i w = 0$, where (ℓ_i, m_i, n_i) is the centre of the ith of N mosaic tiles, without keeping track of the facet $l_p u + m_p v + n_p w = 0$ from which each complex visibility datum originated, one accepts the aberration that my paper addresses. A procedure for making small maps is under development for the VLA. By accepting the computing load for making small maps one may reduce the aberration to a tolerable low level. There will be a misfit along the tile boundaries due to uncorrected aberration but it may sometimes be possible to place the boundaries in unimportant locations. There will also be boundary-matching irregularities associated with the means selected (a) for handling the changing outlines of the bisectoral or bielliptical projected baseline sub-spaces and (b) for compensating for the changing area density of data. Sorting the three-dimensional data by pyramidal facets, processing separately, and regridding each complex submap according to its own known distortion, seeks to eliminate aberration rather than to accept it.

N.H. FARHAT

I assume the accessed Fourier space cone is hollow. If so, would it be possible to fill up the cone or at least thicken its walls by using spectral diversity observations then apply the projection slice-theorem to retrieve projections of the 3-D object without distortions from slices of the Fourier manifold?

R.N. BRACEWELL

The conical space of baseline vectors for a north-south array is a surface. Broadening the spectrum changes the vectors but does not thicken the conical surface.

5.3 A SIMPLE EXPANSION METHOD FOR WIDE-FIELD MAPPING

D.J. McLean
Division of Radiophysics, CSIRO, Sydney, Australia

Abstract. Difficulties arise when mapping wide fields because the simple two-dimensional Fourier transform relation between brightness and visibility is not strictly valid. This paper shows that if the visibilities are transformed using the conventional two-dimensional transform, the resulting map is the true sky convolved with a point source response G. Although G is different in different parts of the map a simple expansion is presented,

$$G = G_0 + \nu G_1 + \nu^2 G_2 + \dots ,$$

where G_0, G_1, G_2, ... are independent of position in the map and ν is approximately half the square of the distance from the centre of the map. G_0 is the small field approximation to the beam, $G_0 + \nu G_1$ is comparable in accuracy to the methods described by Frater (1979), while three or more terms of the expansion permit accurate mapping of still larger fields.

Used in conjunction with CLEAN this expansion provides a straightforward method of coping with wide fields. Only minor changes to CLEAN would be necessary.

The expansion and its application are described; the simple case of planar arrays used for earth-rotation synthesis is then used in order to illustrate how the beam G is distorted towards the edge of the field.

INTRODUCTION

In modern radio telescopes the quantities recorded are the correlations between the radio noise signals arriving at pairs of aerials. Each correlation measured is a sample of the visibility $V(u,v,w)$, which is a function of the vector baseline, (u,v,w), separating the two aerials.

Given the sky brightness $B(\ell,m)$, the visibility $V(u,v,w)$ for a single observation is

$$V(u,v,w) = \iint_{\ell^2+m^2<1} B(\ell,m)\, e^{2\pi i(\ell u+mv+(n-1)w)} \frac{d\ell\, dm}{n} \, , \tag{1}$$

where (ℓ,m,n) are direction cosines measured from the coordinate axes (Ou,Ov,Ow), and therefore

$$n = +\sqrt{1-\ell^2-m^2} \, .$$

Here (u,v,w) is a coordinate system fixed to the sky with w measured toward the centre of the field being observed. Equation (1) includes a factor $e^{-2\pi iw}$, which represents the tracking phase introduced in the hardware.

Suppose that we have observations of V at a discrete set of points $\{(u_j,v_j,w_j)\}$. We wish to estimate $B(\ell,m)$ from the set of values $\{V(u_j,v_j,w_j)\}$. In terms of the sampling function,

$$S(u,v,w) = \sum_j \delta(u-u_j)\, \delta(v-v_j)\, \delta(w-w_j) \, , \tag{2}$$

two possible expressions for this estimate are:

$$M_0(\ell,m) = n\iiint S(u,v,w)\, V(u,v,w)\, e^{-2\pi i[\ell u+mv+(n-1)w]}\, du\, dv\, dw \tag{3}$$

and

$$M(\ell,m) = n\iiint S(u,v,w)\, V(u,v,w) dw\, e^{-2\pi i(\ell u+mv)}\, du\, dv \, . \tag{4}$$

Equation (3) has the advantage that for a point source anywhere in the field, say at (ℓ_0,m_0), the response in the direction of the source is

$$M_0(\ell_0,m_0) = \iiint S(u,v,w)\, du\, dv\, dw \, ;$$

i.e. the peak response to a point source is independent of position in the map. However, if the 'dirty' map so formed is to be 'cleaned' subsequently, this will be difficult and expensive because the sidelobes of the point source response vary across the field of view. Equation (4), which is a small-field approximation to (3), has the advantage that it is in the form of a simple, two-dimensional, Fourier transform.

Here we explore the implications of using equation (4) for a wide field. We shall find that the beam still varies across the field, but in a manner which can be approximated economically, yielding a simple method of cleaning.

Substituting equation (1) into equation (4) we find

$$M(\ell,m) = \iint\limits_{\ell_s^2+m_s^2<1} \frac{n}{n_s} B(\ell_s,m_s)\ G(\ell_s-\ell,m_s-m,n_s-1)\ d\ell_s\ dm_s \ , \qquad (5)$$

where $n_s = +\sqrt{1 - \ell_s^2 - m_s^2} \approx 1 - \frac{1}{2}(\ell_s^2+m_s^2)$,

and

$$G(\lambda,\mu,\nu) = \iiint S(u,v,w)\ e^{2\pi i(\lambda u+\mu v+w\nu)}\ du\ dv\ dw \ . \qquad (6)$$

Here G is the beam or point source response function, but it is seen to be a function of (n_s-1), half the square of the radial distance from the centre of the field.

For the image of a point source at (ℓ_s,m_s,n_s), $\lambda = \ell_s - \ell$ and $\mu = m_s - m$ measure the displacement in the map from the position of the source, and $\nu = n_s - 1$ depends only on the radial distance of the source from the centre of the field - i.e. it is independent of ℓ and m. Since (λ,μ,ν) are independent variables, the right-hand side of (6) is a three-dimensional Fourier transform.

EXPANSION OF G

It is straightforward to write down an expansion of G in powers of ν, either by expanding $e^{2\pi iw\nu}$ as a power series and integrating each term separately or by using a Taylor series and using the properties of the Fourier transform to evaluate the derivatives. Thus

$$G(\lambda,\mu,\nu) = G_0(\lambda,\mu) + \nu\ G_1(\lambda,\mu) + \frac{\nu^2}{2!}\ G\ (\lambda,\mu) + \ldots ,$$

where

$$G_j(\lambda,\mu) = \iiint (2\pi iw)^j\ S(u,w,v)\ e^{2\pi i(\lambda u+\mu v)}\ du\ dv\ dw \ .$$

In particular G_0 is the normal, small-field approximation. Approximating G by $G_0 + \nu G_1$ neglects terms of order $[\frac{1}{2}(\lambda^2+\mu^2)w]^2$ and higher, and so should be comparable in accuracy to the approximation employed by Frater (1979).

The advantages of this approach are: (a) it involves only simple two-dimensional Fourier transforms; (b) the expansion can easily be extended to any desired accuracy.

USE OF EXPANSION

This expansion lends itself to use with CLEAN in a very straightforward manner. Initially, instead of calculating just one beam, two or three must be calculated and stored. At each iteration, instead of subtracting a single beam, centred at the position of the brightest pixel, a linear combination of the two or three stored beams is used. The coefficients of this linear combination are determined by a single parameter, $(n_s-1) \approx -\frac{1}{2}(\ell_s^2+m_s^2)$.

INSTANTANEOUS OBSERVATIONS WITH A PLANAR ARRAY

The following uses the simple case of planar arrays to illustrate how the synthesized beam distorts as we look at different parts of the field. Consider an array for which all the elements are in a single plane, (u',v'), which is tilted by an angle θ relative to the u,v plane. For example, this is approximately true of the VLA used in snapshot mode.

Then we have:

$$v' = v,$$
$$u' = u\cos\theta + w\sin\theta,$$
$$w' = -u\sin\theta + w\cos\theta,$$
$$S(u,v,w) = \delta(w')\, S_p(u',v'),$$

where $S_p(u',v')$ is the two-dimensional sampling function, analogous to S.

Substitution in equation (6) yields for G

$$G(\lambda,\mu,\nu) = G_p(\lambda\cos\theta+\nu\sin\theta,\mu) \quad , \tag{7}$$

where

$$G_p(\lambda,\mu) = \iint S_p(u',v')\, e^{2\pi i(\lambda u'+\mu v')}\, du'\, dv'$$

is the beam for the case $\theta = 0$, i.e. when the field centre is along the vertical to the array. In this case the beam is independent of position in the map.

For forming instantaneous images we could use this invariance by working in the (u',v',w') coordinate system for the array geometry with the corresponding (ℓ',m',n') coordinates in the sky. However, in the next section we are interested in constructing a synthesized map, by using data collected over a period of hours, while the rotation of the Earth changes the orientation of the array relative to the sky. For this we must work in coordinate systems fixed relative to the sky, e.g. the (u,v,w) and (ℓ,m,n) systems. When we combine all the data from a 12-h observing period, it does not lie on a plane, but is spread through some volume of (u,v,w) space.

In the (ℓ,m,n) system we see that the instantaneous beam is spread by a factor $1/\cos\theta$ and shifted, in the negative u direction, by a distance $\nu\sin\theta$. These distortions are illustrated by Figure 1.

ROTATIONAL SYNTHESIS USING A PLANAR ARRAY

Over a 12-h observing period the direction and the amount of tilt of the array relative to the u,v,w coordinate system varies, and consequently so does the apparent shift of sources.

Figure 1. Beam distortions which result from using equation (4) for a tilted, planar array. In this example it is assumed that the array produces a beam which is circular for the direction perpendicular to the array. The crosses mark the true positions of point sources, and the ellipses the half-power contour of the response to those sources.

The vertical to the array is off to the right of the diagram, 1 rad away. The point source responses are elliptical because of the foreshortening of the array as seen from the centre of the field, and the ellipses are displaced from the true direction of the sources because the phasing of the array is exact only for the centre of the field. Both the stretching of the beam and the displacement are described by equation (7). Note that the displacement depends only on distance from the field centre, and is always in the same direction.

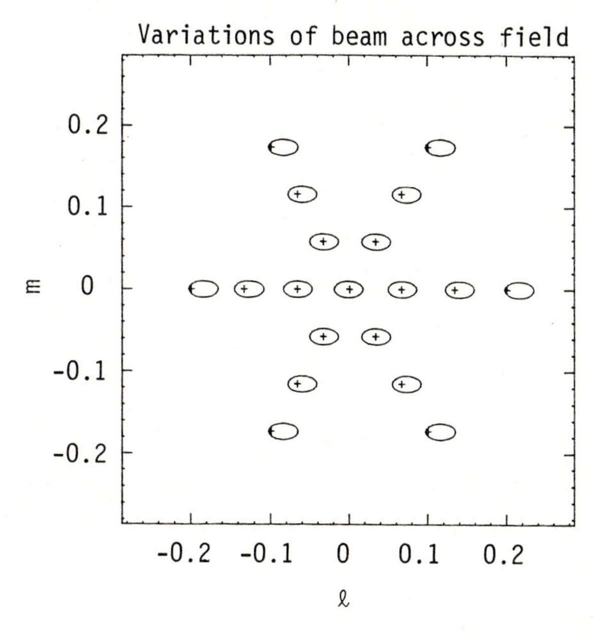

At hour angle equal to -6 h the apparent shift will be approximately to the east, and maximum. At transit the shift will be minimum, to the north for northern fields or to the south for southern fields, while at hour angle equal to +6 h the apparent shift will be approximately west and again maximum. The full synthesized beam, being the sum of these instantaneous beams, would therefore be approximately crescent-shaped for sources at a very large distance from the centre of the field. This is illustrated by Figure 2a. However, in most circumstances, even at the edge of the field of view, these shifts are small enough to produce only a small blurring of the beam, with a consequent minor loss of gain. This is illustrated by Figure 2b.

For example, consider an array with overall dimension D metres composed of dishes of diameter d metres operated at a wavelength of λ metres. The synthesized beam diameter is $\sim\lambda/D$ while the diameter of the field that it is useful to synthesize is $\sim\lambda/d$. Hence, in units of the beam

Figure 2. Half-power contours of some of the distorted and displaced instantaneous beams which, when summed, form the beam synthesized by a planar array. The changes of ellipticity, orientation, and displacement of these ellipses reflect the changing tilt of the array relative to coordinates fixed in the sky, as the earth rotates. It is assumed that for the direction normal to the plane of the array the instantaneous beam is circular. This figure represents only a small part of the total synthesized field.

Notice that the synthesized beam will have the same shape and orientation for all points at this distance from the field centre.

The axes are labelled in units of the synthesized beam radius.

(a) For a value of ν = 4.0 beamwidths, which for many applications is well outside the field of view. This figure illustrates how the synthesized beam which is the sum of the component beams is blurred into a crescent shape in these extreme conditions.

(b) For a value of ν = 1.0 beamwidth, which for a reasonable choice of parameters lies near the edge of the synthesized field.

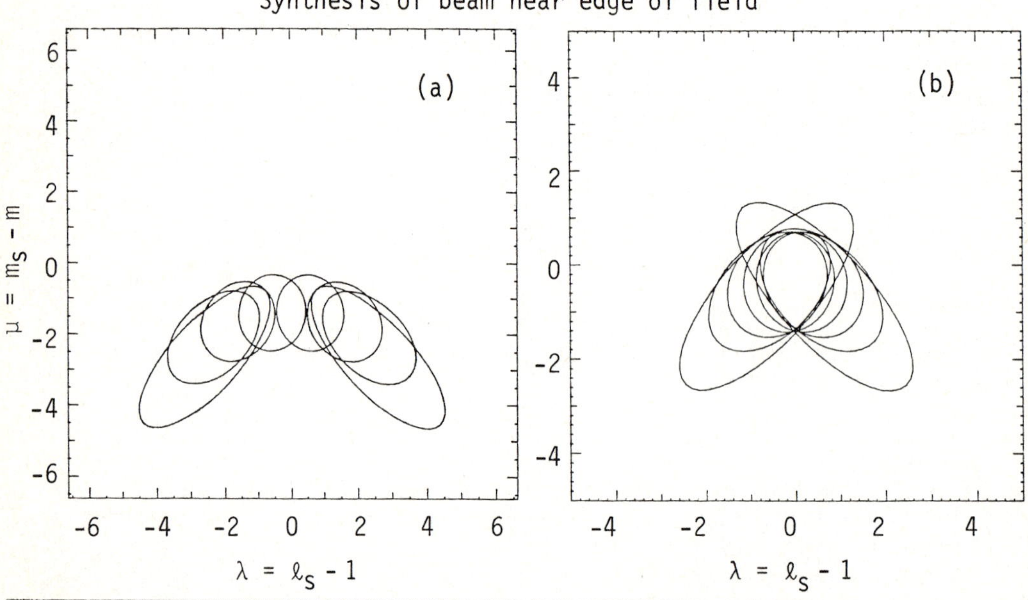

diameter, the maximum value of $\nu = (n_S-1) \approx \frac{1}{2}(\ell_S^2+m_S^2)$ is ν/beam = $\frac{1}{2}(\lambda/d)^2/(\lambda/D) = \frac{1}{2}(\lambda D/d^2)$. This is largest at long wavelengths, so consider the case (λ=0.22, d=22, D=6 × 10^3) ν/beam ≈ 1. Hence the shift is at most one beamwidth, and when the shift is large (i.e. $\theta \approx \pi/2$), so is the spread of the beam along the direction of the shift. Hence the beam will be spread somewhat rather than crescent-shaped.

Reference

Frater, R.H. (1979). The two dimensional representation of three dimensional interferometer measurements. In Image Formation from Coherence Functions in Astronomy, ed. C. van Schooneveld, pp. 19-26. Dordrecht: Reidel.

DISCUSSION

D.J. SKELLERN

Can you give an indication of the number of terms required to CLEAN a typical field which at present gives difficulty?

D.J. McLEAN

Unfortunately I have not yet made any estimate of this. My hope is that two or three terms will extend significantly the field which can be mapped accurately.

J.P. HAMAKER

In standard CLEAN you do both your component-finding and restoration at grid positions. I would think that your beam distortions imply that you have to do either one or the other at intermediate positions and therefore do some sort of interpolation.

D.J. McLEAN

Yes, you are correct. Normal CLEAN uses a convolution with the 'CLEAN-beam' as a final step. I think that the most straightforward adaptation for wide fields might be to combine the convolution step with the restoration on to a grid, so that the CLEAN beam becomes the interpolating function. An alternative approach is to restore on to a distorted grid. Either of these techniques would add a little, but not too much, to the computing cost.

U.J. SCHWARZ

How looks the distorted beam for sidelobes and grating rings? Is the procedure also correct?

D.J. McLEAN

Yes, the procedure is correct for the sidelobes and grating rings, as well as the main beam, but I do not have a simple picture of what they look like.

5.4 A SCANNING TECHNIQUE FOR SYNTHESIZING LARGE FIELDS OF VIEW - APPLICATION TO THE OOTY SYNTHESIS RADIO TELESCOPE

A. Pramesh Rao and T. Velusamy
Radio Astronomy Centre
Tata Institute of Fundamental Research
P.O. Box 8, Ootacamund 643 001, India

1. INTRODUCTION

In aperture synthesis we are often confronted by the conflicting requirements between sensitivity and field of view. While large size antennas provide greater sensitivity, they restrict the size of the sources that can be mapped to the primary beams of the individual elements. In principle one can construct a map of a large field by combining several small maps (mosaic mapping), but in practice with incomplete u-v coverage, it is very difficult to "clean" the maps. The problem is particularly severe while mapping very complex extended regions much larger than the primary beam (eg. sources close to the galactic plane). In this paper, we describe a technique to map and "clean" fields larger than the primary beam. This technique has been successfully used to map large sources using the Ooty Synthesis Radio Telescope.

2. OOTY SYNTHESIS RADIO TELESCOPE

The Ooty Synthesis Radio Telescope (OSRT) which operates at 326.5 MHz consists of the Ooty Radio Telescope (ORT) and several small antennas distributed over 4 km. The ORT is an equatorially mounted parabolic cylinder which is 530m long in the north-south direction and 30m in the east-west (Swarup et al. 1971). An array of 1056 dipoles is mounted along the focal line of the cylinder and pointing in declination is done by phasing this dipole array using diode switches. As the diode switches are under computer control, it is possible to change the declination of the ORT within a few milliseconds. The other elements of the OSRT are also parabolic cylinders but are much smaller than the ORT, having a size of 25m x 9m. While the small antennas have a large primary beam $\sim$ 500' EW x 170' NS (voltage radiation pattern) the ORT with its large north-south extent has a narrow primary beam $\sim$ 170' x 8'. The existing digital delay line and correlator system allows us to correlate the signal from any one antenna with those from five other antennas. Since the ORT has the largest collecting area ($\sim$8000m^2) we keep the ORT as the common antenna and correlate it with all the smaller antennas. Thus the field of view of the OSRT is determined by the voltage radiation pattern of the ORT (170' x 8'). Such a narrow north-south field of view is quite inadequate for mapping large extended sources. To overcome this, it is planned to divide the dipole array of the ORT into several smaller elements (each with field of view > 30' in the north-south) and have a 128 channel digital correlator system in which every element of the OSRT can be correlated with every other element. Since this hardware is not available at present we have evolved a scheme in which by using only the existing 5 baseline digital correlator system, we can map a large field of view (170' x 170') in a single observing session.

3. THE SCANNING TECHNIQUE

The effect of the large north-south extent of the ORT can be visualised in one of two ways as shown in Fig. 1 where we have shown the ORT correlated with one of the small antenna (MKVT). In the

Fig.1 (a) Schematic of one pair of antennas of the OSRT. (b) North-south voltage radiation pattern of these antennas. (c) u-v coverage for this baseline for $\delta=0°$. Horizontal solid line shows the u-v track for the centres of the two antennas. Cross represents data at a given HA. The vertical solid line denotes the north-south extent of ORT and the dashes along this line indicate values of v at which the true visibility has been determined. The hatched area corresponds to the full u-v coverage possible with this baseline.

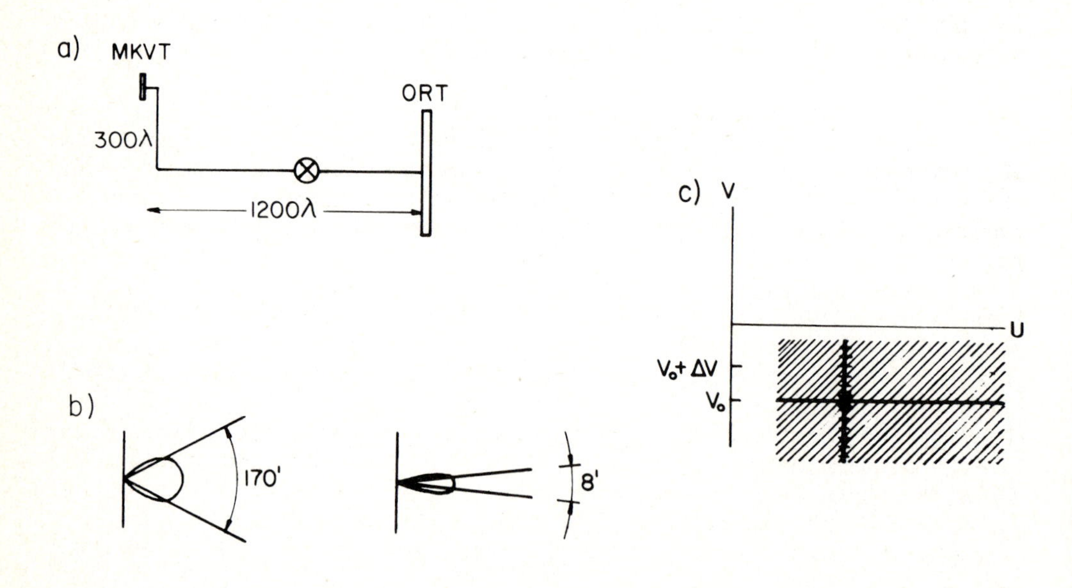

sky plane (Fig. 1b) the field of view of the interferometer pair is restricted by the smaller north-south primary beam of the ORT. In the u-v plane the effect is to convolve the true visibility of the source by the north-south aperture of the ORT giving an average visibility at v_o. The solid line in Fig. 1c shows the u-v track for the baseline. The visibility measured by the antenna pair at any point on the u-v track (shown by the cross) is the true visibility averaged in the north-south over the shaded area which is the physical extent of the ORT. Thus, each measurement contains information on the true visibility over a range of v but to make a map of the entire source, it is necessary to remove the effect of this convolution and determine the true visibility at each v (shown by dashes) within shaded area in the u-v plane (Fig. 1c). We describe below a method for determining the true visibility at different points within the shaded area.

If B(x,y) is the brightness distribution in a field, the visibility function is related to it by

$$V(u,v) = \iint_{-\infty}^{\infty} B(x,y)\, e^{-2\pi i(ux+vy)}\, dx\, dy \qquad (1)$$

The apparent brightness distribution seen by the interferometer pair in Fig. 1a is $B(x,y)\, P(y-y_o)$ where $P(y-y_o)$ is the north-south voltage radiation pattern of the ORT when it is pointed at a declination y_o. For simplicity we have assumed that the voltage radiation pattern of the small antenna, and the east-west voltage radiation pattern of the ORT are much broader than the field being mapped and so can be set equal to a constant. Thus the visibility measured by this interferometer pair is

$$V_{obs}(u_o,v_o,y_o) = \iint_{-\infty}^{\infty} B(x,y)P(y-y_o)e^{-2\pi i(u_o x+v_o y)}\, dx\, dy. \qquad (2)$$

If, for a given value of u_o and v_o, we measure the visibility with the ORT pointed at different declinations y_o and, if we perform a one dimensional Fourier transform of the observed visibilities over y_o, we get

$$\begin{aligned} V'(u_o,v_o,\Delta v) &= \int_{-\infty}^{\infty} V_{obs}(u_o,v_o,y_o)e^{-2\pi i\, \Delta v y_o}\, dy_o \\ &= \iint_{-\infty}^{\infty} B(x,y)e^{-2\pi i(u_o x+v_o y)}\Big[\int_{-\infty}^{\infty} P(y-y_o)e^{-2\pi i\, \Delta v y_o}\, dy_o\Big]dx\, dy \\ &= \iint_{-\infty}^{\infty} B(x,y)e^{-2\pi i(u_o x+v_o y)}\big[P'(\Delta v)e^{-2\pi i\, \Delta v y}\big]dx\, dy \end{aligned} \qquad (3)$$

$[P'(\Delta v)e^{-2\pi i\, \Delta v y}]$ is the Fourier transform of the voltage radiation pattern which is the complex north-south aperture illumination of the ORT when it is phased for declination y, and $P'(\Delta v)$ is aperture illumination corresponding to zero declination.

$$\begin{aligned} V'(u_o,v_o,\Delta v) &= P'(\Delta v) \iint_{-\infty}^{\infty} B(x,y)e^{-2\pi i\{u_o x+(v_o+\Delta v)y\}}dx\, dy \\ &= P'(\Delta v)\, V(u_o,v_o+\Delta v). \end{aligned} \qquad (4)$$

Thus we see that the Fourier transform of the visibility measured as a function of declination gives us the true visibility function of the source at different points in the shaded area in Fig. 1c weighted by the complex aperture illumination pattern of the ORT. After correcting for the aperture illumination pattern of the ORT the estimated visibility function can be gridded and Fourier transformed to get the dirty map of the full field. The weighting by the illumination pattern of the ORT in equation 4 ensures that we do not try to estimate visibilities for values of v outside the shaded area in Fig. 1c for which no information exists in the measured data. By observing a point source for which $V(u,v)$ is constant, we can determine $P'(\Delta v)$, as seen from equation 4. Ideally $P'(\Delta v)$ should be unity but in practice it is complex reflecting gain and phase errors along the dipole array. While solving for the true visibilities using equation 4 we flag the visibilities at values of Δv for which $P'(\Delta v)$ has abnormally low gain. In practice this leads to only a few holes in the v coverage whose effects on the map can be taken out by "clean".

4. APPLICATION TO THE OSRT

The practical implementation of the scheme described above became possible only after the recent installation of the diode phase shifters on the ORT, which enables us to change its declination pointing within a few milliseconds. The basic sampling interval for acquiring data from all the baselines is kept at 200 ms because there is no hardware for stopping the fringes. When mapping large fields we not only sample the visibility from all the baselines but we also update the declination pointing of the ORT every 200 ms so that the declination of the ORT cycles between specified limits. Thus, even when mapping a 180' field in the north-south, (61 settings separated by 3'), the entire cycle of declination scanning is completed every 12 seconds, during which period there is negligible change of u and v for the field centre. The offline programs sort and average the visibility at each declination over a specified period (1 to 5 minutes), calibrate the data and perform the Fourier transform over declination. The transformed visibility after correction for the aperture illumination pattern for the ORT forms the visibility data base corresponding to the full field to be mapped. For all subsequent analysis, namely gridding, mapping and cleaning we follow standard procedures.

For mapping extended sources the visibility at short spacings is very important. To measure short spacings we separate a section of the ORT 25 wavelengths long in the north-south and correlate it with the rest of ORT as if it is an independent element of the OSRT. With declination scanning, this north-south baseline gives us the visibility at u=0 over a range of v from about 25 to 500 wavelengths. While the output of this baseline does not change with hour angle, this baseline is continuously recorded since it is a good monitor for interference and scintillation.

In Fig. 2 is shown the supernova IC443 mapped with the OSRT using the declination scanning technique. The map shows remarkable agreement with those published in the literature (Duin & van der Laan (1975)).

5. CONCLUSIONS

The declination scanning technique has been shown to be a powerful method for mapping extended sources using the presently available hardware for the OSRT. While primary beam scanning has been discussed for similar applications (eg. Ekers & Rots (1979)), to the best of our knowledge, this is the first time such a scheme has been implemented successfully. The mapping of the wide field has been achieved at the expense of sensitivity since due to scanning, only a fraction of the total synthesis time is spend at each declination. However, when mapping large fields, the declination scanning technique enables us to calibrate on point sources within the field in a very natural way, especially when the point sources are separated in declination from the main source. The reconstructed visibility corresponding to the extended field suffers from all the problems of wide field mapping like curvature effects and we are investigating practical schemes for minimising these effects.

6. ACKNOWLEDGEMENT

We thank P.S. Janardhanan for his assistance in writing the offline programs.

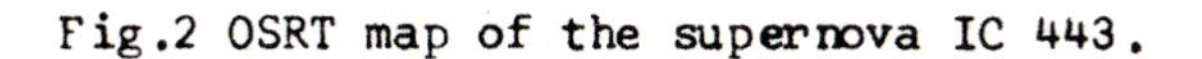
Fig.2 OSRT map of the supernova IC 443.

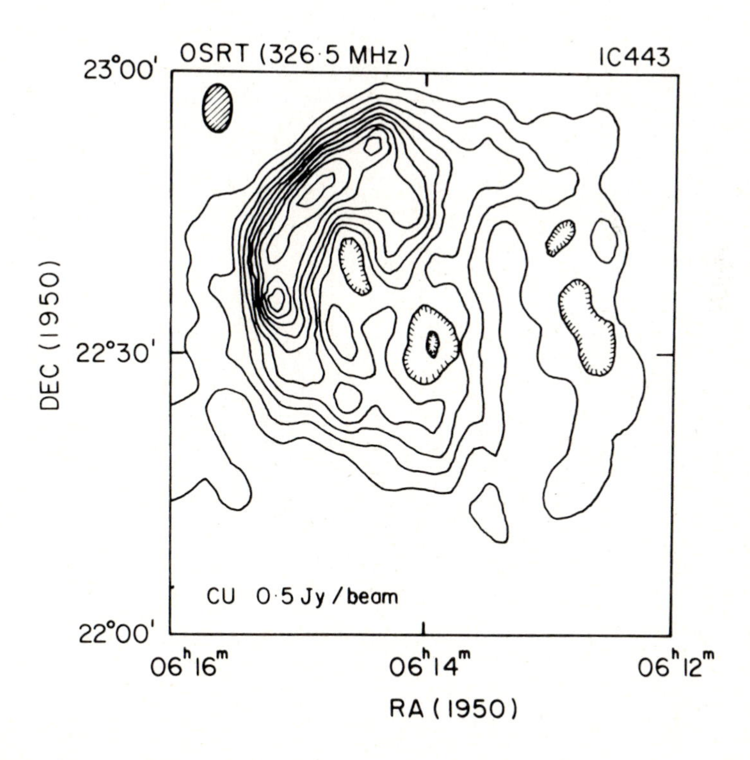

References
Duin, R.M. & van der Laan, H. (1975). Astron. Astrophys. 40, p. 111.
Ekers, R.D. & Rots, A.H. (1979). Image Formation from Coherence Functions in Astronomy, D. Reidel Publishing Co., p. 61.
Swarup, G. et al. (1971). Nature Phys. Sci. 230, 185.

DISCUSSION

R.D. EKERS

Is your method the same as dividing the big antenna into smaller sections and using them sequentially?

A.P. RAO

Yes. Both the methods are equivalent.

J.D. O'SULLIVAN

I would like to point out that to get some of the benefits of self-cal methods, one does not actually need to "close" the phase closure loops.

A.P. RAO

But for self-cal to work each antenna must be correlated with a representative section of the other antennas. In our case, because the Ooty Radio Telescope is so much larger than the other antennas, the only useful correlations for all but the very strong sources are those of small antennas with ORT or sections of ORT. Under such conditions self-cal will not work very well.

5.5 MAPPING OF LARGE FIELDS WITH SINGLE-DISH TELESCOPES

R. Wielebinski
Max-Planck-Institut für Radioastronomie, Bonn, FRG

Abstract. The single-dish radio telescope is the ultimate wide-field instrument, allowing in fact the mapping of the whole sky. However, to make such maps numerous methods had to be developed, which like in synthesis telescopes, base mainly on data processing techniques. In this contribution various techniques developed at the MPIfR for mapping of radio continuum at centimetre wavelengths are described.

1 INTRODUCTION

Many astronomical objects subtend a large angle in the sky. For studies of our own Galaxy the necessary coverage is 4π. For the Andromeda Nebula we need to study a $3° \times 1°$ field. Numerous supernova remnants are more than 1° in diameter. Also the weak lobes of some radio galaxies subtend more than 1° in the sky. For all these objects large fields must be mapped, requiring special observing techniques.

In the classical case the single-dish radio telescope is used with a single, sensitive receiver. Mapping is achieved by scanning the sky, in a TV fashion, arranging individual scans to cover a desired area. From these scans, after regridding, a two-dimensional array is generated from which a map can be made. Sensitivity can be improved by increasing the integration time, which in effect means repeating the maps and adding. Data processing techniques not only increase the through-put but also by correct data handling increase the sensitivity close to the theoretical noise limit.

Numerous facilities must be included in a data reduction system to obtain optimal maps. First of all the interference spikes must be rejected at an early stage. This is possible either in electronic hardware or in early software processing. The data must be treated by the sin x/x function to match to the illumination function of the telescope. For each scan baseline fitting is important to remove effects of varying ground radiation, and receiver zero level drifts. To increase the dynamic range the sidelobes surrounding the main beam must be cleaned near strong sources. Finally, at the highest frequencies (10 GHz and above) weather effects, posing the most serious problem, may be rejected by multibeam switching techniques.

These techniques, developed for the continuum data reduction at the

MPIfR, will be described in the present paper. (The descriptions of the programmes are found either in the NOD2 Manual or the Hitchhiker's Guide, both MPIfR internal reports; see also Haslam (1974) and Reich (1983)).

2 BASELINES

In a scanning mode, when the signal is weak, the determination of the baseline is not a straight-forward matter. The simplest method is to fit a linear baseline to the ends of a scan. An improvement can be made by fitting a polynomial baseline. Further methods call for the subtraction of the mean scan from each scan in a map. This method is particularly necessary when scans are made which include considerable ground radiation. Another method applied by Sofue & Reich (1979) was to smooth a map by a rather large beam and thus derive a 'mean map'. This mean map is then subtracted from the full resolution map and the smoothing/subtraction process repeated. Using some four iterations, a satisfactory result is usually obtained.

Maps can be made by driving the telescope in orthogonal directions (λ,β-scans). When such pairs of maps are added, considerable improvement in baselines results. A further step of baseline optimisation is to fit the scans at the cross-over points in an iterative manner. This procedure, termed 'Basket Weaving', was developed by C.J. Salter for λ-β-scans. The procedure starts with the determination of temperature differences at the cross-over points. The λ and β scans are then corrected by means of a fit of these differences in an iterative manner. Usually after the third iteration sufficient correction is reached. Further variants of this general idea of baseline correction were the nodding scans used by Haslam et al. (1974). The procedure PLAIT (Hitchhiker's

Figure 1. An example of 'Basket Weaving'. The SNR CTA 1 observed in orthogonal directions and the final result on the right (W. Sieber, MPIfR).

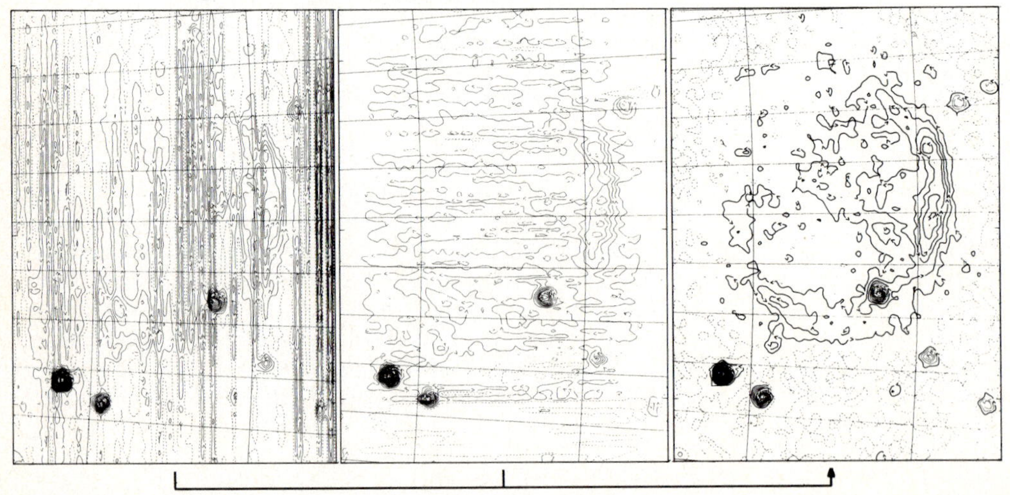

Guide) optimises cross-over points for maps made at arbitrary scanning angles and is of great importance in producing deep maps at high radio frequencies.

3 DYNAMIC RANGE

It is a well-known fact that the dynamic range near strong sources is limited by the sidelobe structure of the telescope beam. The sidelobe patterns can be determined down to -30 dB using naturally occurring radio sources. Various procedures have been adopted for increase of dynamic range. The simplest is to subtract the strong coma lobe only by replacing it with an appropriate negative source at the correct position relative to the main beam. A subtraction of a map of an area containing a point source from one of a source with halo has been used by Andernach *et al*. (1979). A more sophisticated programme, REBEAM, replaces a dirty beam of the map with any chosen beam by a method similar to the Fourier cleaning of synthesis maps. Fourier filtering of the high-frequency components also helps in improving the dynamic range of observations.

A complete procedure for increasing the dynamic range was developed by Reich *et al*. (1978). Observations of a source are made over a wide range of hour angle in the altitude-azimuth system of the telescope. The rotating (astronomy) components are separated from the non-rotating ones (sidelobes) in an iterative procedure. A dynamic range of over 40 dB has been achieved with this technique.

4 MULTIBEAM OBSERVATIONS

Beam switching was introduced rather early into radio astronomy to overcome the weather variations. Originally beam-switched systems were used for the measurement of point sources. This method of observation was analysed for extended sources by Emerson *et al*. (1979). It was shown that the only information lost was due to the discrete spacings of the horns, and thus could be fully restored. A practical reduction programme for a 3-beam (horns) 2-channel (amplifiers) system at λ2.8 cm was developed and has led to many maps of weak extended sources (e.g. Klein & Emerson 1981; Klein *et al*. 1982). Since the individual maps are scanned in azimuth, the addition of individual coverages by the PLAIT programme improves the baselines as well. Maps with r.m.s. noise of $\sim$ 1 mK have been made at 10.7 GHz with these methods. The only limitation of the method is the field of view; sources greater than 5 times the largest beam separation cannot be mapped satisfactorily. However, in a secondary focus beam throws of up to 20 x HPBW are possible. The use of beam switching at millimeter wavelengths, possibly in conjunction with nutating subreflector, should also allow mapping of radio sources in this newly opened frequency range. The use of many receiver channels also increases the overall system sensitivity.

Figure 2. Steps in multibeam observations. (a) total power map using one horn, (b) beam-switched difference map, (c) restored map. (Courtesy of H. Morsi, MPIfR)

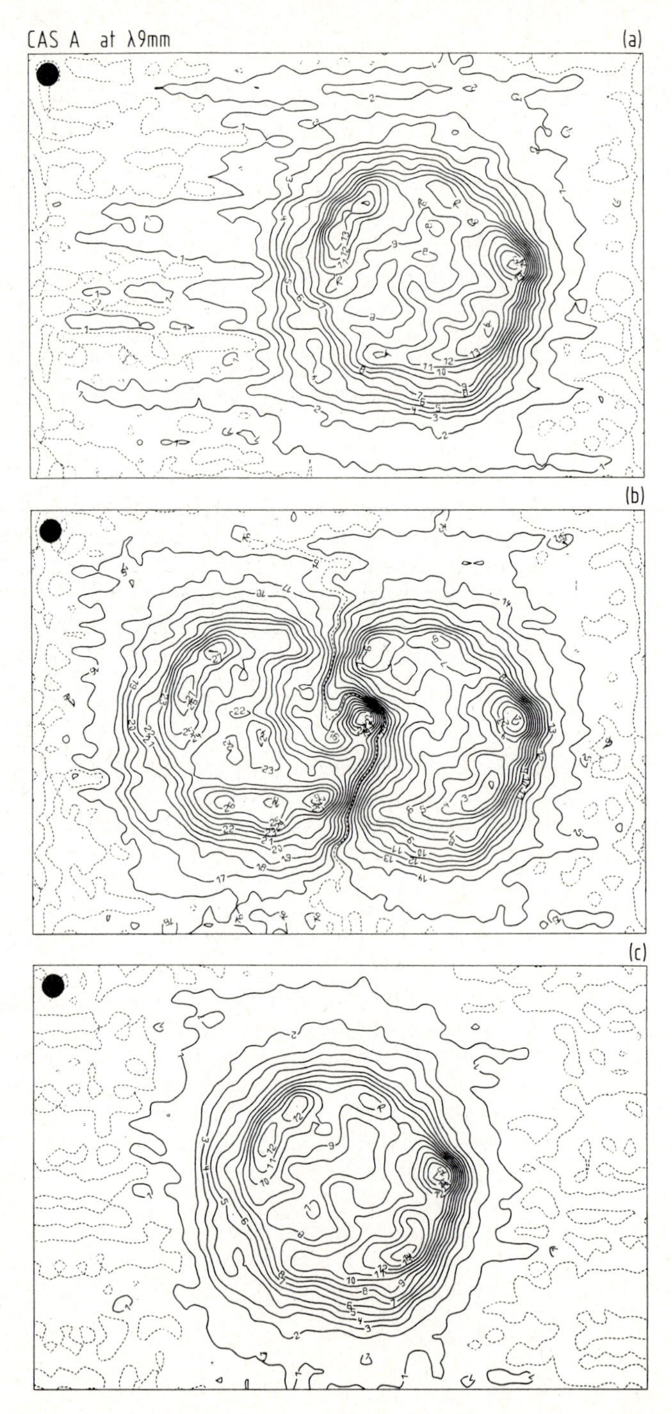

5 SOFTWARE BEAM SWITCHING

In a normal beam-switched receiver two feed horns are connected through waveguide runs to a switch followed by an amplifier. It is usual to switch the same polarisations so that in a three-horn system two beam pairs are possible (see Emerson *et al*. 1979). It is difficult to find a geometrical arrangement in which the same polarisations at many different beam spacings could be switched. A modular construction, where each horn is followed by two detectors (one for each polarisation) and a multiplier (to determine the polarisation), allows great flexibility. The beam switching which is needed for the reduction of weather effects can then be done in software. The stringent requirement, on the total power stability, which is a prerequisite of software switching, is met by many types of receivers (e.g. FET amplifiers, mixers) in use today.

To test this observational method a two-beam, four-channel receiver was constructed for λ9 mm (see Figure 3). In this receiver each module detects the two senses of circular polarisation and provides Q and U signals from a multiplier. All the data processing including the channel subtraction is done in the on-line computer. This receiver system was successfully used to map extended radio sources and appears to hold great prospects for the future. The two modules could be moved to different separations with great ease. The construction of further modules will improve the system sensitivity and allow larger beam separations; hence wide-field mapping capabilities.

Figure 3. A block diagram of a module of a 9 mm test receiver. Two such units are placed in the secondary focus at present. The beam switching is done in software.

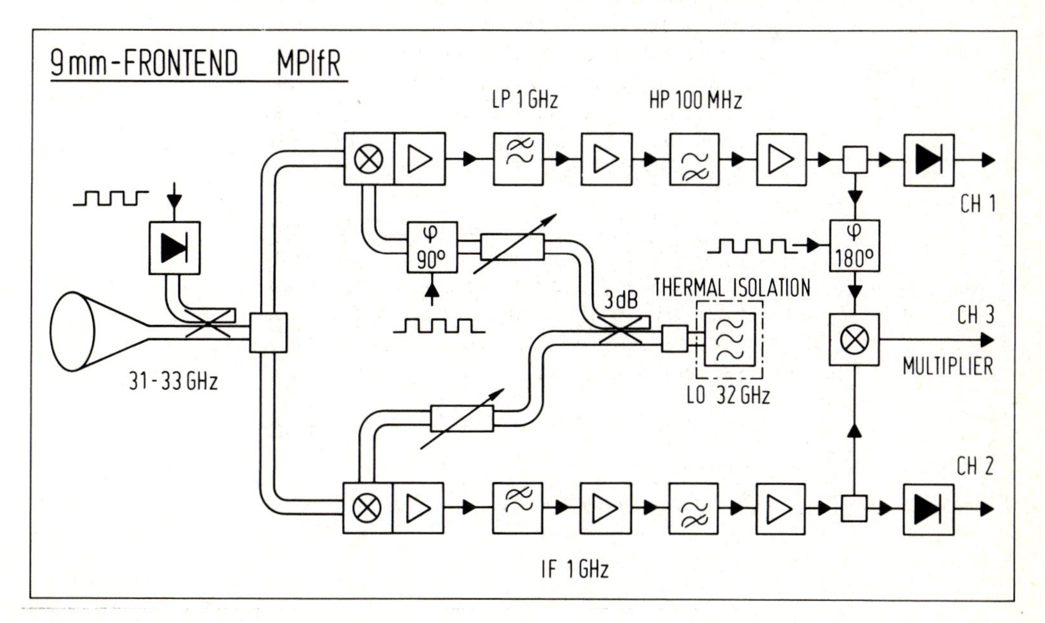

A 4-feed 8-channel receiver system using cooled FET amplifiers at λ2.8 cm is being constructed for the 100-m telescope. Fields of some 2° x 2° will be mapped with this system. Future plans for the 100-m telescope include a new high surface accuracy subreflector and a sensitive multi-beam (possibly 8 beams) system at λ9 mm.

6 CONCLUSION

Single-dish wide-field mapping techniques have been developed in recent years at high frequencies which make a large single-dish an ideal complement to the lower frequency capabilities of the synthesis arrays. Westerbork at 610 MHz has the same angular resolution as Effelsberg at 10.6 GHz. The VLA at 1.4 GHz matches the 100-m telescope beam at 32 GHz. The use of the 327 MHz band at the synthesis telescopes will allow the study of larger astronomical objects at a very wide frequency base.

ACKNOWLEDGEMENT

I wish to acknowledge discussions with G. Haslam, U. Klein and W. Reich who developed a number of the techniques described in this paper.

REFERENCES

Andernach, H., Baker, J.R., von Kap-herr, A. & Wielebinski, R. (1979). The radio continuum halo of M87. Astron. Astrophys. 74, 93-9.

Emerson, D.T., Klein, U. & Haslam, C.G.T. (1979). A multiple beam technique for overcoming atmospheric limitations to single-dish observations of extended radio sources. Astron. Astrophys. 76, 92-105.

Haslam, C.G.T. (1974). NOD2 A general system of analysis for radioastronomy. Astron. Astrophys. Suppl. 15, 333-8.

Haslam, C.G.T., Wilson, W.E., Graham, D.A. & Hunt, G.C. (1974). A further 408 MHz survey of the northern sky. Astron. Astrophys. Suppl. 13, 359-394.

Klein, U. & Emerson, D.T. (1981). A survey of the distributions of 2.8 cm radio continuum in nearby galaxies I. Observations of 16 spirals. Astron. Astrophys. 94, 29-44.

Klein, U., Beck, R., Buczilowski, U.R. & Wielebinski, R. (1982). A survey of the distribution of λ2.8 cm radio continuum in nearby galaxies II. NGC 6946. Astron. Astrophys. 108, 176-187.

Reich, W., Kalberla, P., Reif, K. & Neidhöfer, J. (1978). High dynamic range observations with the Effelsberg 100 m telescope. Astron. Astrophys. 69, 165-170.

Reich, W. (1983). Radioastronomische Meß- und Reduktionsverfahren zur Erfassung ausgedehnter Kontinuumsemission. Kleinheubacher Berichte 26, 183-8.

Sofue, Y. & Reich, W. (1979). Radio continuum observations of the North Polar Spur at 1420 MHz. Astron. Astrophys. Suppl. 38, 251-263.

6

Calibration and errors

6.1 SELFCALIBRATION

T.J. Cornwell,
National Radio Astronomy Observatory, Very Large Array,
P.O. Box 0, Socorro, New Mexico 87801.

P.N. Wilkinson,
Nuffield Radio Astronomy Laboratories, Jodrell Bank,
Macclesfield, Cheshire, United Kingdom.

1 INTRODUCTION

The puritanical idea that one cannot get something for nothing might lead us to reject many of the data processing methods being discussed at this conference. Thus, we would lose CLEAN, MEM, speckle imaging and selfcalibration: some of the most important advances in astronomical imaging in the last decade. Selfcalibration, or as it has been called elsewhere, hybrid mapping, appears, at first sight, to be impossible. How could we possibly determine both the source structure and the instrumental calibration necessary to view the source through our instrument ? In this paper we intend to provide an answer to this question and to outline the history of selfcalibration, it's present uses and possible future developments.

2 THE CLOSURE PRINCIPLE AND APPLICATIONS

In common with other radio interferometrists in the fifties Jennison (1958) was faced with the problem of inferring source structures from visibility amplitude alone since it was then difficult to maintain the phase of an interferometer. For some sources a major ambiguity remained, for example, the distinction between double and triple structure. Clearly phase information was required : Jennison's inspiration was to notice that in a system of three interferometers formed from three antennae one good phase observable, the closure phase, is available. One forms the closure phase Ψ by summing the visibility phase ψ around the closed loop of baselines :

$$\Psi_{123}(t) = \psi_{12}(t) + \psi_{23}(t) + \psi_{31}(t) \qquad (1)$$

This observable is "good" in the sense that any phase errors ϕ which can be ascribed to individual antennae cancel out. The observed phase on baseline 12, for example, is related to the true phase by :

$$\psi_{12,obs}(t) = \psi_{12,true}(t) + \phi_1(t) - \phi_2(t) + \varepsilon_{12}(t) \qquad (2)$$

where ε is a noise term due to receiver noise, sky background, etc. Substituting equations such as this into the above equation we find that, apart from terms due to noise, the observed and true closure phases are identical. A similar observable, the closure amplitude, was discovered by Twiss, Carter & Little (1962) :

$$\Gamma_{1234}(t) = (A_{12}(t).A_{34}(t))/(A_{23}(t).A_{14}(t)) \quad (3)$$

where the amplitude on baseline 12 is given by :

$$A_{12,obs}(t) = A_{12,true}(t).a_1(t).a_2(t) + \xi_{12}(t) \quad (4)$$

and ξ is a noise term. Again, apart from a noise term, the observed and true closure amplitudes are identical. (Smith (1952) used a quantity analogous to the closure amplitude). These two observables can also be derived from the equation relating the true and observed vector visibility :

$$V_{ij,obs}(t) = g_i(t).g_j^*(t).V_{ij,true}(t) + \eta_{ij}(t) \quad (5)$$

where η is a complex noise term and g_i is the complex gain of the i'th antenna. One curiosity should be noted : since the absolute phase at an antenna is meaningless we can refer all phases to one reference antenna and thus a closure phase exists for only three antennae whereas the closure amplitude requires at least four (see Cornwell and Wilkinson (1981) for further elucidation of this point).

Closure quantities are completely insensitive to all properties of the visibility function which can be factorised by antennae. In particular, the closure phases are insensitive to changes in atmospheric and antenna instrumental delays while closure amplitudes are not affected by variable antenna gains, atmospheric attenuation or pointing (on small sources only; see section 4). Furthermore, position and strength information is lost. To prove the first point note that the visibility of a source contains a factor which depends upon the vector $\underline{s}$ pointing to the source centroid :

$$\exp\,(2\pi j.(\underline{b}_i - \underline{b}_j).\,\underline{s}) \quad (6)$$

where $\underline{b}_i$ is the position vector of the i'th antenna. These two terms can be absorbed into the complex antenna gains and therefore have no effect upon the closure phase. Similarly, absolute amplitude information factors out of the closure amplitudes. Since closure quantities are completely insensitive to strength and position a certain slackness in pointing and positioning the antennae in an array is permissible. (The upper limits in slackness being set by the antenna primary beam widths and the receiver bandwidths.)

Jennison's concept of closure phase was described in his Ph.D thesis over thirty years ago but only within the last decade has significant use of it been made. The reason for this time lapse between conception and utilisation is mainly due to the computational problems involved. By their very nature, direct use of the closure quantities in the Fourier inversion of the visibility function is impossible. However, much ingenuity has been expended upon their indirect use; see e.g. Rhodes & Goodman (1973),Rogers *et al.* (1974),Fort & Yee (1976). In modern times Rogers *et al.* (1974) were the first to utilise the closure phase; they used it in the most urgently needed application, VLBI, where phase information is usually corrupted by clock errors. In a more popular algorithm Readhead & Wilkinson (1978) used the CLEAN method (Hogbom 1974, Schwarz 1978) to generate a representation of the brightness distribution while imposing positivity and confinement constraints. The algorithm can be summarised :

a. Find an initial model of the brightness distribution by e.g. model fitting to the amplitudes alone.
b. Transform the current model to obtain predicted visibility and force consistency with the amplitudes and closure phases.
c. Use the CLEAN algorithm to deconvolve and to reject regions of brightness which violate positivity and confinement.
d. Goto step b. unless convergence has been obtained.

Further refinements and extensions were made by Cotton (1979) and Readhead *et al.* (1980). As a result of this work, at the end of the seventies radio astronomers could image radio sources using very poorly calibrated data such as is obtained in VLBI or with MERLIN. It then became attractive to adapt the same techniques to arrays such as the VLA and WSRT which are limited in dynamic range by calibration errors.

3 SELFCALIBRATION

The concept of closure quantities is particularly useful for arrays containing small numbers of antennae but for instruments such as the VLA (Thompson *et al.* 1980) and modern VLBI-arrays the sheer number becomes overwhelming : for N (27) antennae there are N(N-1)(N-2)/6 (2925) closure phases, of which only N(N-1)/2 - (N-1) (325) are independent. In such circumstances it becomes attractive to find some way of obeying the closure relations without ever explicitly calculating them. For this purpose it is important to notice that the closure quantities are invariant with respect to the complex antennae gains. Consequently, if, in trying to correct the observed visibilities, one only allows changes to complex antenna gains then the closure quantities will be conserved. This observation lead Schwab (1980) and Cornwell & Wilkinson (1981) independently to suggest that one should choose complex antenna gains by minimising a weighted sum of squared errors of the data from the model :

$$L(t) = \Sigma_{ij,i<j}\, w_{ij} . | V_{ij,obs}(t) - g_i(t).g_j^*(t).V_{ij,true}(t)|^2 \quad (7)$$

Cornwell and Wilkinson added a term to incorporate prior information about the relative stabilities of the gains. Schwab & Cotton (1983) further generalised this formula to allow estimation of antenna delays and fringe rates in VLBI.

This method of choosing complex antenna gains is embedded in the same iterative algorithm as before :

a. Choose an initial model, usually a point source will suffice. For partially phase-stable instruments such as the VLA some CLEAN components from the initial image can be used.
b. Calculate the complex gain corrections $g_i(t)$ which minimise the difference L(t) between observed and predicted visibilities. Apply gain corrections to the observed data.
c. Apply image plane constraints via CLEAN or any other method.
d. Goto b unless convergence is attained.

One great advantage of this approach over the closure phase formalism is that noise is treated correctly as an additive complex quantity. The weighting terms w_{ij} add flexibility : they can be chosen to be the inverse of the variance of η_{ij} in which case the solution signal to noise is optimised or if the model is particularly poor for some baselines then the weight can be reduced to de-emphasis their effect in the solution.

4 SOME PRACTICAL POINTS

The success of selfcalibration relies upon its sensible application: a number of choices of control parameters are required. Here we present a short discussion of these choices: for fuller treatment see Cornwell (1982) and Wilkinson (1982).

The most obvious question about selfcalibration is "Why does it work ?". To answer this we need to consider the effect of calibration errors. If the visibility data is miscalibrated then the true image is seen convolved with an error beam which violates two reasonable constraints upon the image : positivity and confinement. Enforcement of positivity and confinement can be performed using various methods; first, however, it is necessary to remove the sidelobes due to incomplete u,v coverage which also violate these constraints. Both CLEAN (Hogbom 1974) and MEM (Wernecke & D'Addario 1976, Gull & Daniell 1978) are well suited to this role although the former may be preferable for small images of high dynamic range. CLEAN windows are ideally suited to rejecting components from some regions of the brightness distribution and to force positivity one can stop CLEANing just before the first negative component. In MEM positivity comes free and confinement could be enforced by the use of windows. Unfortunately, positivity and confinement are sometimes not strong enough constraints : for arrays with well behaved beams, such as linear arrays, calibration errors can be converted into believable structure such as jets so one must be careful to anticipate such ambiguities.

Lax use of the known image plane constraints is perhaps the major cause of failure of selfcalibration; for example, a large number of CLEAN components will model some of the calibration errors and thus convergence to an acceptable image will be delayed or prevented. One useful rule, widely used at the VLA, is to only include CLEAN components which occur before the first negative component. Another ad hoc rule has been suggested by R. R. Clark (private communication) : one can calculate the effective noise level expected on a selfcalibrated image from the r.m.s. misfit of the data from the model after adjusting the complex antenna gains (see equation 7). For example, if at some stage in the process the misfit is M times the average noise per visibility point then it is likely that there will be systematic errors present in the trial image at approximately M times the thermal noise limit. It is therefore sensible to halt the cleaning process well before this likely systematic error level has been reached so that spurious clean components will not be passed around the selfcalibration loop. The depth of cleaning will automatically increase as the process converges and the misfit decreases. Figs. 1a,b show the practical effects of applying these rules. Fig. 1a shows an image made from MERLIN data (see Cornwell and Wilkinson 1981) where clean components close to the expected systematic error level were passed around the iterative loop, 8 iterations in all were made. Fig. 1b was made in exactly the same way but with a cutoff at approximately 3 times the expected error level. This image is not only superior but in the selfcalibration 5 times fewer clean components were passed around the loop, thus saving considerably in CPU time. If cleaning is instead stopped at the first negative component - in this case a somewhat more severe limitation - then a marginally worse image than Fig. 1b is obtained. We cannot generalise about the best cutoff rule from this one case but the overall message is very clear - use only a minimum of clean components especially in the early stages of selfcalibration.

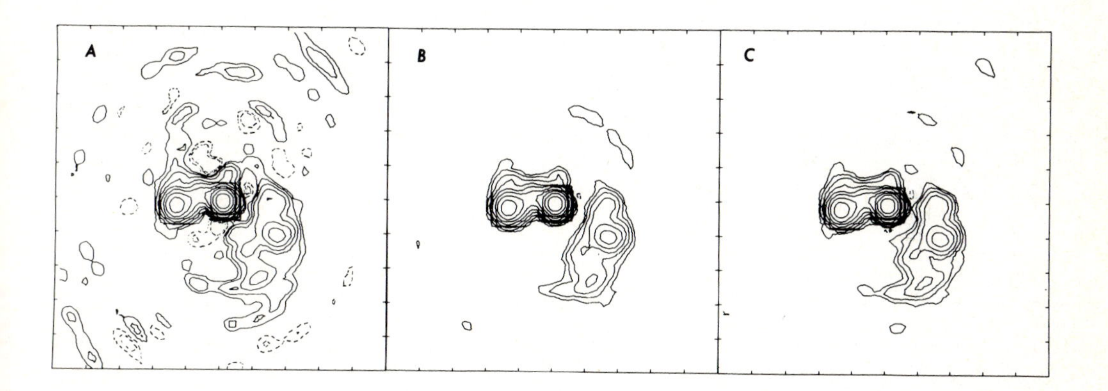

Figure 1. Tests of selfcalibration strategies. See text. Contour levels 0.1, 0.2, 0.4 .... 51.2 % of peak brightness.

A further saving in CPU time can be acheived if one adopts a difference imaging philosophy. One should clean only the data formed from the difference between the previous and current corrected visibility data. The number of clean components subtracted must be restricted by a version of R. Clark's rule and negative components must be included to allow the method to correct itself. The relatively few clean components subtracted from this difference image are then transformed and added to the transform from the previous iterations thus "updating it". Complex gain corrections are then calculated in the usual way and the whole process is iterated. We have tested this approach and have shown that it converges to produce an acceptable image. Fig. 1c shows an image made from the same data as Figs. 1a,b but with about 4 times fewer clean components in all. There are minor differences at low levels but it seems clear that this is a promising approach which can both save CPU time and operate in a "hands-off" manner.

If the visibility is not much greater than the noise on a sufficiently large number of baselines then selfcalibration can yield completely spurious results. For example, with VLA data, merely by adjusting complex gains one can pull a point source of strength ~ 5 sigma from a noise field! Cornwell (1981) has considered the noise behaviour of selfcalibration : only for a point source is a simple solution possible. Suppose that the atmosphere is stable for a time τ and that the resultant noise in an image made from the data collected in that time is σ_{image} then for reliable amplitude and phase selfcalibration the flux of an "almost unresolved" source must obey :

$$F >> \sigma_{image}.(N_{ant}-3)^{-1/2}. \qquad (8)$$

If this holds then the noise in the resulting image is increased over the thermal limit by a factor $((N-1)/(N-3))^{1/2}$. For extended sources equation (8) seems to work reasonably well if F is replaced by the median visibility amplitude. By comparison, the limit for detection from amplitudes alone is :

$$F >> \sigma_{image}.(\tau/T)^{1/4}. \qquad (9)$$

and, of course, for coherent integration :

$$F >> \sigma_{image}.(\tau/T)^{1/2} \qquad (10)$$

where T is the total integration time.

Some common types of error in visibility data do not obey the antenna related error model (equation 5). Instead, an additional term $G_{ij}(t)$ must be factored into the first term of the left hand side of equation 5. $G_{ij}(t)$ can be due either explicitly to correlator problems or, more commonly in arrays with digital correlators, to antenna based problems which do not factorise. Some examples of the latter are (a) Confusion on shorter baselines. (b) Non-matching passbands in the antennae amplifiers. We then have an equation such as equation (5) for each element of the passband. Integrating over non-identical passbands leads to terms like $G_{ij}(t)$ (Thompson & D'Addario 1982). To the extent that the passbands are constant in time the $G_{ij}(t)$ can be removed by performing a simple baseline related calibration (see e.g. Cornwell and Wilkinson 1981). (c) Pointing errors can yield differential gain errors over a source large in comparison to the primary beam-width of the antennae. (d) Non-isoplanicity of the atmosphere over the source prevents simple factorisation of the gains. (e) Bandwidth and time-averaging smearing also produce effects which cannot be factorised. (f) Second-order changes in the gains during the integration time.

Clark (1981) has considered the order of magnitude of some of these effects. In most high dynamic range VLA images time variable effects like (b) seems to be a limit; as yet there seems no way to overcome these problems. However, in an array which is very stable instrumentally then considerable progress can be made : in an impressive extension of antenna based selfcalibration Noordam & de Bruyn (1982) have very successfully used the redundancy of WSRT to remove these effects thereby attaining the highest dynamic range (>40dB) yet acheived in radio interferometry.

5 SELFCALIBRATION AND ADAPTIVE OPTICS

A very close analogue to selfcalibration has arisen in optical astronomy, namely adaptive optics (see e.g. Code 1973, Buffington et al. 1978, Woolf 1982, Hardy 1978). We now digress to consider this interesting convergence.

The resolution of an optical telescope of size greater than about 10-20 cm. is nearly always limited by short term phase errors introduced by turbulent cells in the first few kilometers of atmosphere above the pupil (see e.g. Muller & Buffington 1974). A clever method of real time phase correction has been described by Muller & Buffington (1974). In this scheme a phase correction device (a "rubber mirror") is placed in the pupil plane and controlled so as to maximise some measure of image sharpness. Detection and correction must take place within the characteristic time scale of the atmosphere, about 10-20 milliseconds, and hence a strong source lying within one isoplanatic patch is required. Typically two trial values of the phase for a particular part of the pupil are inserted and the resultant sharpness measured ; the optimum phase is then interpolated.

Various forms of the sharpness measure have been suggested; Muller and Buffington define a sharpness measure to be a function of the perturbing phases which is maximised for the true image. Some typical measures are: the energy of an image (i.e the sum of squares of pixel values), the moment of inertia, a central reference intensity, various moments of the pixel values, and various norms expressing a distance from an ideal image. Of these, the energy or sum of squares of the pixel values is most often used and is sometimes called the sharpness S. Hamaker et al. (1977) have demonstrated that by maximising S we are simply requiring that redundant samples of the Fourier transform of the object have the same phase, a constraint which can be trivially enforced in redundant spacing interferometry.

In radio interferometry we could follow the same scheme as adaptive optics : changing antenna phases within the characteristic time of the atmosphere, few minutes to several hours, to optimise some measure of the image quality. However, this approach requires that a good image, free from other defects such as poor sampling, be formed in that time scale, which for most arrays is simply not possible. Indeed, for sparsely filled arrays direct optimisation of the sharpness of the dirty image seems to be prevented by the presence of these sidelobes which overwhelm the sidelobes due to calibration errors (Steer, Ito & Dewdney 1983). Even for arrays with relatively good instantaneous u,v coverage such as the VLA this scheme is expensive in terms of computer time since the visibilities affected by each perturbation of an antenna phase must be individually transformed to yield the resultant change in an image. For this reason we prefer to deal with the entire data set using the iterative approach outlined in the previous section.

It is interesting to consider why selfcalibration has revolutionised radio interferometry while adaptive optics have so far been of peripheral interest in optical astronomy. Several factors are important : first, rubber mirrors are difficult to build and operate while the computer based manipulations required for selfcalibration are simple. Secondly, in optical astronomy the number of atmospheric coherence times per typical observation is of order 1,000-100,000 compared to, at most, ~ 100 in radio interferometry. As a consequence, many radio sources can be selfcalibrated whereas adaptive optical systems are effective for relatively few optical sources. Optical speckle imaging has succeeded in the imaging of relatively weak objects by relying purely upon the amplitude information, integrated incoherently. Radio speckle imaging is also possible simply by averaging the visibility amplitudes. Note, however, a major advantage of radio over optical speckle imaging : in the former we have access to the individual visibilities before coherent addition (but see Brown 1978) and, consequently, the noise level is not increased by the decorrelation of the same complex visibility measured between different parts of the array. In the terminology of speckle imaging, the modulation transfer function is unity everywhere in the aperture.

The signal to noise limitations mean that selfcalibration/adaptive optics will probably be of marginal importance for ground based optical arrays. Instead, speckle imaging seems much more promising.

6 TWO STATE-OF-THE-ART SELFCALIBRATED IMAGES

To demonstrate the power of selfcalibration we here show two state-of-the-art selfcalibrated images.

The first, shown in Figure 2, kindly supplied by Craig Walker, John Benson, George Seielstad and Steve Unwin shows an image of 3C120 made from data collected in a fourteen station MkII VLBI observation at λ18cm. (An image of comparable quality is shown by Simon *et al.* 1983). In figure 3 we show the second image, which was kindly supplied by Rick Perley and John Cowan, a VLA image of Cygnus A at λ20cm with data from the A,B and D arrays. The dynamic range (peak/rms) for this image is about 3000.

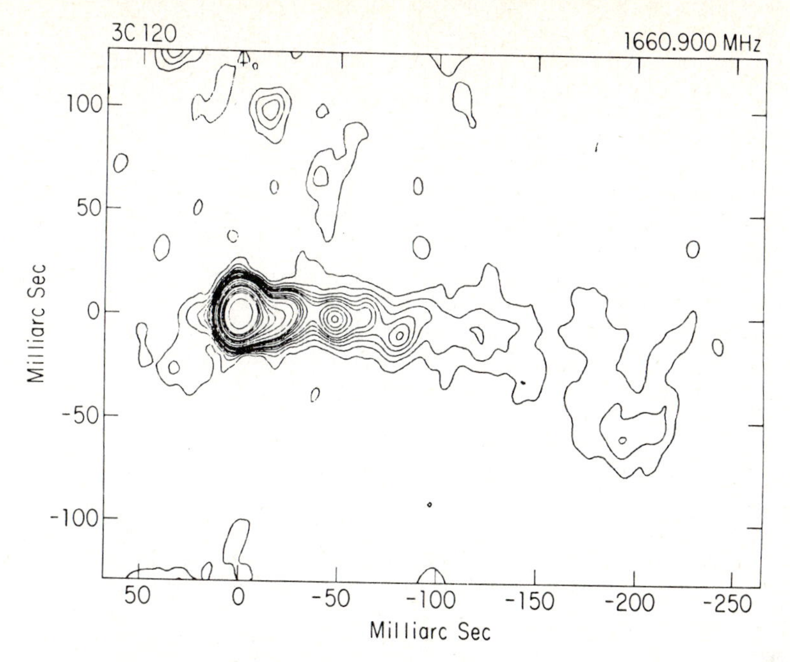

Figure 2. VLBI image of 3C120 at λ18cm. See text. The contour levels are 4, 8, 12, 16, 20, 24, 32, 40, 48, 56, 80, 120, 160, 200, 240, 280, 320, 360, 400, 1000 mJy per beam and the peak is 2080 mJy per beam.

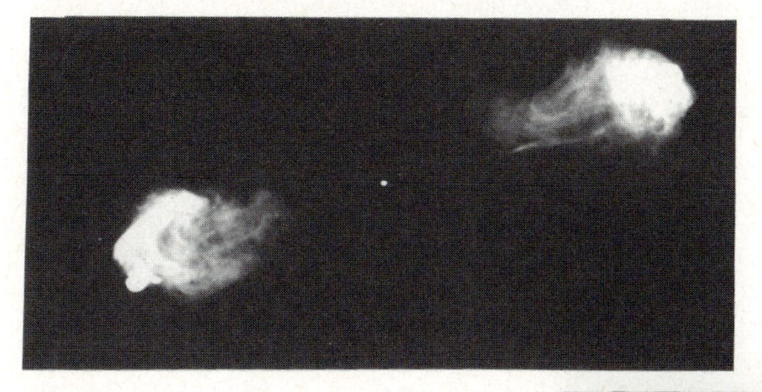

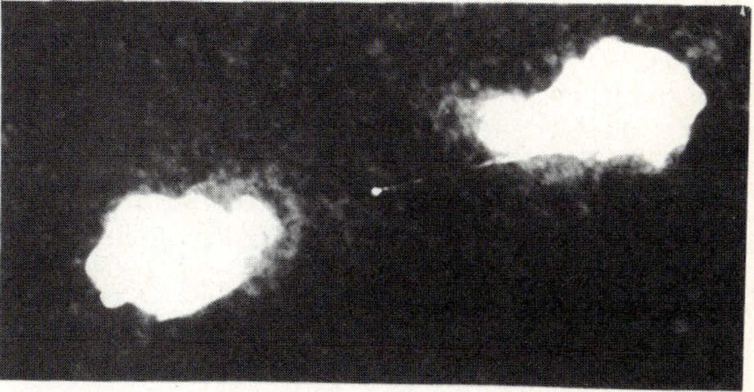

Figure 3. VLA image of Cygnus A at λ6cm. See text. The lower radiograph saturates two magnitudes below the upper radiograph.

7 THE FUTURE

We will now discuss a number of technical challenges in the future of selfcalibration.

First, the signal to noise limit may be lowered by using models for the atmosphere which have fewer degrees of freedom. For example, in a small linear array it may be advantageous to couple the phase parts of the gains by a linear dependence on distance, thus reducing the number of phase degrees of freedom from N-1 to 1. In a similar vein Basart et al. (1983) are adapting the Kalman filter formalism to the estimation of gain errors. In an attempt to define the ultimate signal to noise limits for selfcalibration we note that if the atmosphere can be tracked across the array then the effective time between solutions for the gains is the crossing time for the whole array rather than a single antenna. This increase in intervals between solutions yields a concomittant decrease in the signal to noise limit. However, even with all this cleverness the coherence time may be so short that any selfcalibration is not possible. We may then have to resort the radio speckle imaging (see section 5). Since reconstruction from amplitudes alone seems possible for non-pathological objects in two or more dimensions (see e.g. Fienup 1983) the only real drawback to this approach is the telescope time needed.

Non-isoplanicity seems to present no real conceptual problems but the practical problems include a substantial increase in computational complexity and a moderate increase in the number of degrees of freedom. One can envisage one higher level of complexity over selfcalibration : the sky would be split into a mosiac of isoplanatic patchs for each of which one set of antenna phases would be optimised. The VLA is expected to be mildly non-isoplanatic at an operating frequency of 327MHz.

One cause of failure of selfcalibration is the presence of error terms in the measurement equation which cannot be attributed to antennae (see section 4). Baseline redundancy (Noordam & de Bruyn 1982) can be used to remove correlator-based effects but one pays a large price in efficiency of use of the antennae which becomes apparent when imaging large fields at low dynamic range. Furthermore, we note that strict redundancy is not always required. To understand this consider the effective redundancy which arises when imaging a field of size θ. The sampling theorem then tells us to sample at least every $1/(2\theta)$ wavelengths in the u,v plane; samples closer than this are then effectively redundant and are *implicitly* used in selfcalibration to eliminate other degrees of freedom such as antenna and correlator gains. The use of CLEAN windows currently enforces such effective redundancy.

8 ACKNOWLEDGEMENTS

The National Radio Astronomy Observatory is operated by Associated Universities Inc., under contract with the National Science Foundation. We thank the authors cited in section 6 for providing the illustrations.

9 REFERENCES

Basart,J.P., Mitkees,A.A.,Mansy,F.M.M., and Zheng,Y. (1983). Scientificreport, Dept. of Elec. Eng., Iowa State Univ.
Brown,T.M., (1978). J.Opt.Soc.Am., 68,883-889.
Buffington,A.,Crawford,F.S.,Pollaine,S.M.,Orth,C.D.,and Muller,R.A., (1978). Science, 200,489-494.
Clark,B.G., (1981). VLA scientific memorandum 137.
Code,A.D., (1973). Ann. Rev. Astron. and Astrophys., 11,239-268.
Cornwell,T.J. and Wilkinson,P.N., (1981). M.N.R.A.S. 196,1067-1086.
Cornwell,T.J., (1981). VLA scientific memorandum 135.
Cornwell,T.J., (1982) lecture 13,NRAO summer school on synthesis mapping,Socorro N.M. 1982, ed. A.R.Thompson.
Cotton,W.D., (1979). A. J., 84,1122-1128.
Fienup,J.R., (1983). paper presented at OSA meeting on "Signal recovery and synthesis with incomplete information and partial restraints", Lake Tahoe.
Fort,D.N. and Yee,H.K.C., (1976). Astron. Astrophys., 50,19-22.
Greenaway,A.H., (1982) Optics Comm., 42,157-161.
Gull,S.F. and Daniell,G.J., (1978). Nature, 272,686-690.
Hamaker,J.P.,O'Sullivan,J.D. and Noordam,J.E., (1977). J.Opt.Soc.Am., 67,1122.
Hardy,J.W., (1978). Proc. IEEE, 66,651-697.
Hogbom,J., (1974). Astron. Astrophys. Suppl., 15,417.
Jennison,R.C., (1958). M.N.R.A.S., 118,276.
Muller,R.A. and Buffington,A., (1974). J.Opt.Soc.Am., 64,1200-1210.
Noordam,J.E., and de Bruyn,A.G., (1982). Nature, 299, 597-600.
Readhead A.C.S. and Wilkinson,P.N., (1978). Ap. J., 223,25-36.
Readhead,A.C.S.,Walker,R.C.,Pearson,T.J. and Cohen,M.H., (1980). Nature, 285,137-140.
Rhodes,W.T. and Goodman,J.W., (1973). J.Opt.Soc.Am., 63,647-657.
Rogers,A.E.E. et al, (1974). Ap.J., 193,293-301.
Schwab,F.R., (1980). SPIE, 231,18-24.
Schwab,F.R. and Cotton,W.D., (1983). 88, 688-694.
Schwarz,U.J., (1978). Astron. Astrophys., 65,345-356.
Simon, R.S. et al. (1983). Paper to be published in the proceedings of IAU symposium 110, Bologna, Italy, 1983.
Smith,F.G., (1952). Proc. Phys. Soc B, 65 971.
Steer,D.G., Ito,M.G. and Dewdney,P.E., (1983). paper presented at OSA meeting on "Signal recovery and synthesis with incomplete information and partial restraints", Lake Tahoe.
Thompson,A.R. and D'Addario,L.R., (1982). Radio Science, 17,357-369.
Thompson,A.R.,Clark,B.G.,Wade,C.M. and Napier,P.J., (1980), Ap. J. Supplement, 44,151-167.
Twiss,R.Q.,Carter,A.W.L. and Little,A.G., (1962). Aust.J.Phys., 15,378.
Wernecke,S.J. and D'Addario,L.R., (1976). IEEE C-26,351-364.
Wilkinson,P.N., (1982). paper presented at CNES conference on VLBI techniques, Toulouse,France,1982.
Woolf,N.J., (1982). Ann. Rev. Astron.and Astrophys., 20,367-398.

DISCUSSION

J. NOORDAM

An automatic process is dangerous because a range of possible maps may be consistent with the data (especially if there are few telescopes and consequently few constraints).

T.J. CORNWELL

There will always be brightness distributions which are too complex to recover from any particular data set, be it from 4 telescopes or 27. We agree that we do not know as yet how to make the computer decide whether SELFCAL is likely to work on a particular data set. Given that the astronomer has made this decision, all we are saying is that it is possible to codify some of the rules which an experienced astronomer would apply, by hand, each iteration.

R.H.T. BATES

The trouble with comparing radio self-calibration with optical wavefront correction is that the latter is hopeless for faint objects, and it gets worse the larger the pupil is compared to the diameter of the average seeing cell.

T.J. CORNWELL

The two drawbacks that you mention also apply to self-calibration but the dividing lines are such that self-calibration is worth while for a relatively large fraction of radio sources.

P.E. DEWDNEY

You tended in your talk to gloss over the enforcement of the positivity constraint by simply 'throwing away' the parts of the interim map which you 'don't like'. However, this is the only non-linear process in the loop and is very important. Have you thought about devising an 'optimum' non-linear process which might produce convergence in minimum time.

Although some persons took this question to mean that I was suggesting maximum 'entropy' as such a scheme, I didn't really intend this. It would, however, fall into the class of such non-linear operations. We need some theoretical backing for deciding how many (say) CLEAN components to keep at each iteration to arrive at the conclusion in minimum time.

T.J. CORNWELL

We do not know of any such scheme. The suggestion of Ron Clark is intuitively appealing and seems to be effective.

J.G. ABLES

It would be of interest to compute the image 'entropy' at each stage of the iteration to see if the entropy measure changes monotonically as the SELFCAL algorithm proceeds.

T.J. CORNWELL

I agree but I am confident that the entropy will decrease since the dynamic range increases during iteration. It should be noted that in an accompanying talk Steer et al. found that self-calibration and maximum entropy were not compatible. Instead they introduced a mechanism to artificially decrease the entropy of an image prior to gain calculation. Thus minimum entropy reconstruction would seem to be preferable for self-calibration perhaps with mollification of the final image.

M.M. KOMESAROFF

(1) Since positivity is one strong constraint which makes SELFCAL work, does this mean that it cannot be used to map all Stokes parameters?

(2) Isn't it impossible to determine by SELFCAL the phases of a set of RH polarized feeds relative to a set of LH feeds?

T.J. CORNWELL

(1) For Stokes parameters the constraint analogous to positivity is that the fractional polarization should be less than unity. The power of this constraint is not known to us.

(2) Yes, one cannot determine the phase difference between the RH and LH feeds by self-calibrations. At high frequencies ($>$1 GHz) the difference is usually constant and can be calibrated by reference to a source of known polarization. At low frequencies time-variable ionospheric effects must be calibrated by other methods.

6.2 A REAL-TIME CALIBRATION SYSTEM FOR THE CULGOORA RADIOHELIOGRAPH USING SELF-CALIBRATION TECHNIQUES

D.J. McLean and N.R. Labrum
Division of Radiophysics, CSIRO, Sydney, Australia

In the 22 years since the original design of the Culgoora radioheliograph (Wild 1967; Sheridan et al. 1973; Labrum et al. 1975) was begun, the advances in digital electronic components have been remarkable; if a similar instrument were designed today it would be built as a correlation instrument and would have many advantages in terms of improved image speed and quality over the present system (see McLean 1979; McLean et al. 1979). In fact, to avoid the expense of converting the existing radioheliograph system to modern technology we are now well advanced with a less expensive scheme to improve the image quality. This consists of a four-point approach involving:

(1) Measurement and correction of gain errors in real time (solar observations only).

(2) Measurement and correction of phase errors in real time, using self-calibration techniques (solar observations only).

(3) Detection of and compensation for missing aerial channels, caused by mechanical failure of the aerial drives, or failure of the associated electronics, transmission line, etc.

(4) Determination of the nature and importance of various sources of error in order to improve the quality of non-solar observations also.

AMPLITUDE CALIBRATION AND COMPENSATION FOR MISSING AERIALS

Figure 1 illustrates how the new equipment relates to the existing system. At the beginning of each second the relative gains are determined from measurements of the total power in each channel due to the Sun compared with the total power when the aerials are replaced by cold resistive loads (to determine the contribution due to receiver noise), and these gains are used to set gain-equalization attenuators for that second. As part of this process faulty channels (very low or very high gain or noisy) are recognized and disconnected. To compensate for this the gain of the diametrically opposite aerial is increased by 6 dB (see Fig. 2). It is obvious that this would be correct if the image were formed from separately recorded correlations between aerial pairs. Since the actual image-forming hardware operates linearly on these correlations, it is also correct in this case. This part of the

Figure 1. Block diagram showing the relationship between the existing radioheliograph (shaded blocks) and the new calibration scheme. Apart from the new gain equalization attenuators, and the transfer of phase information from the calibration computer to the existing phase computer the pre-existing radioheliograph system is essentially unmodified.

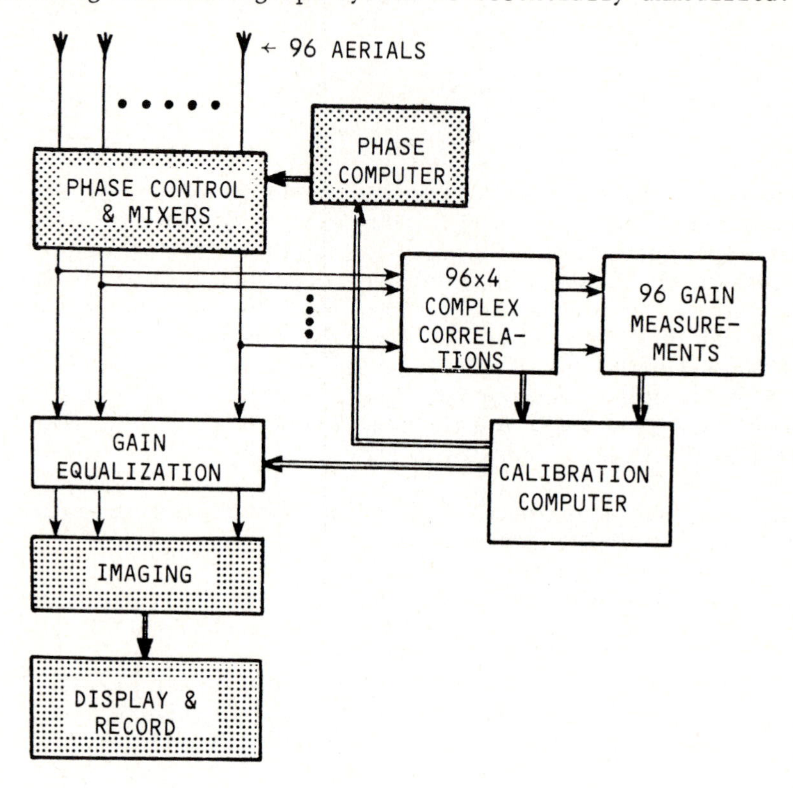

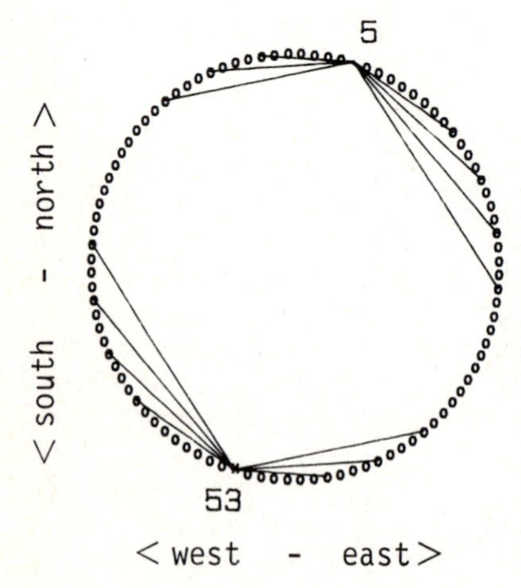

Figure 2. Illustration of how the redundancy of the Culgoora array will be exploited to compensate for missing aerials. Supposing aerial 5 is missing, the figure shows some of the missing spacings normally formed with this aerial, but the same spacings are also formed with aerial 53, diametrically opposite. Hence if the gain of aerial 53 is boosted by 6 dB this will compensate for the fact that aerial 5 is unserviceable.

new calibration scheme is straightforward, and need not be discussed further.

PHASE CALIBRATION

The phase calibration functions as follows.

Correlation products are formed for all the four sets of closest spacings of the array. These correspond to the four inner rings of the u-v diagram shown in Figure 3. For a compact source somewhere in the field of view which is not fully resolved by the short baselines, the phases of the correlations should vary sinusoidally around each of the inner circles of Figure 3. The difference between a sinusoidal fit and the measured phases is taken as the phase difference between the corresponding pair of aerials. Working around the circle from each of three reference aerials, the phase differences are used to assign phase errors to individual aerial channels. The philosophy of this process is close to that of the self-calibration techniques described by Schwab (1980) and Cornwell & Wilkinson (1981), although the algorithm is much simpler and faster. The phase corrections so determined are then fed back to the computer, which controls the phasing of the array for incorporation in subsequent images.

Figure 3. U-V coverage at the zenith for the Culgoora array, which consists of 96 aerials, uniformly spaced around a circle 3 km in diameter. The crowding of points around the inner circles indicates that the shortest baselines are highly redundant.

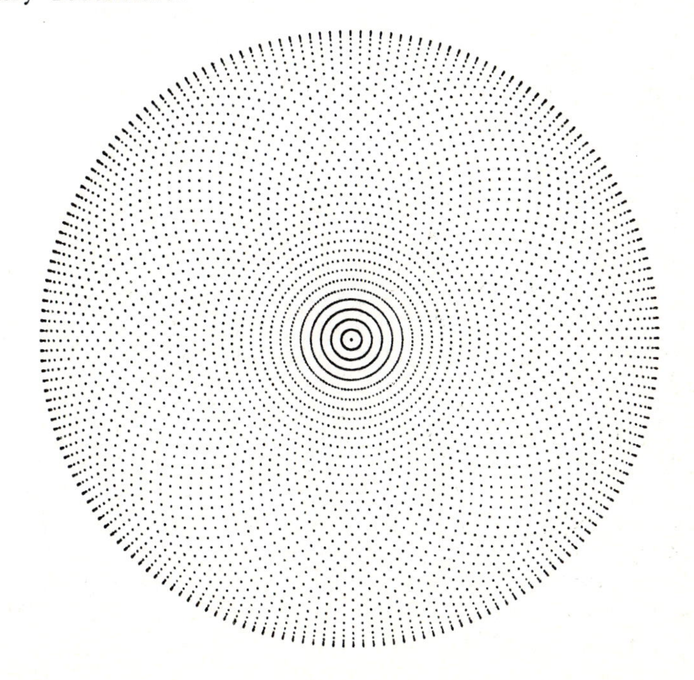

In deriving individual phase errors from phase differences between aerial pairs, we can easily calculate several independent estimates by working clockwise or anti-clockwise around the array from one of the reference aerials, and by moving in steps of 1, 2, 3 or 4 unit spacings. These eight estimates are combined with weight reflecting their estimated accuracy, and the errors in the result are also estimated. If these errors are high the corrections will not be made. This will protect us against errors which would otherwise be introduced when the source distribution is more complex than we have assumed.

COMPUTER SIMULATIONS

The new system is not yet completed. In this paper we present the results of a computer simulation used to verify that the system will function well despite a number of sources of error which cannot be removed by any simple calibration system.

Figure 4 shows the simulated beam when phase errors of 30° r.m.s., but no other errors, have been introduced, and before phase calibration has been carried out. As expected, the sidelobe level is quite high and the gain is correspondingly reduced. Figure 5 summarizes the variation of gain and sidelobe level with varying levels of phase errors. From comparison of these results with sidelobe levels currently obtained with the radioheliograph we now believe that the level of phase errors in the radioheliograph with the cold calibration may have been as high as 50° r.m.s., with a consequent large loss of gain. This serves to emphasize the need for the new calibration scheme. When the phase calibration technique is applied to these simulations, the sidelobe level decreases by about a factor of one-half for each iteration of the self-cal procedure. After a few iterations the result is essentially error-free.

Clark (1981) listed the types of errors which cannot be removed by calibration. Of these, those which concern us most are variations of the phase and gain errors across the band-pass of our receivers. Such problems arise as a result, for instance, of reflections between discontinuities on transmission lines, and are different from one channel to another. Figure 6 shows an example of a simulation which included such errors. The passband was divided into 10 parts. For each channel random phase errors (r.m.s. 30°) and gain errors (r.m.s. 6 dB) were assigned independently to each sub-band, together with a further phase error component (also 30° r.m.s.) which was applied uniformly across the whole pass-band for that channel. In addition to these errors five aerials were treated as unserviceable. The phase calibration procedure, the gain calibration procedure and the compensation for missing aerials were all applied, with the results shown in the figure. There are residual, 2% sidelobes both positive and negative close to the main beam, but nothing as high as 1% further out. The regular pattern revealed by the zero level contour points to a limitation of the missing aerial compensation scheme. As illustrated by Figure 7, when the compensation scheme is applied, there are some spacings (10 in our example) which should get double weight but which in fact get quadruple weight

Figure 4. Example of a simulated beam (0°.3 off axis) with r.m.s. phase errors of 30°, no other errors, with the phase calibration not done. The contour interval is 1% from -6% to +6% and 10% from 10% to 90%. Negative contours are dashed, the zero contour is a full heavy line, and positive contours are full lines.

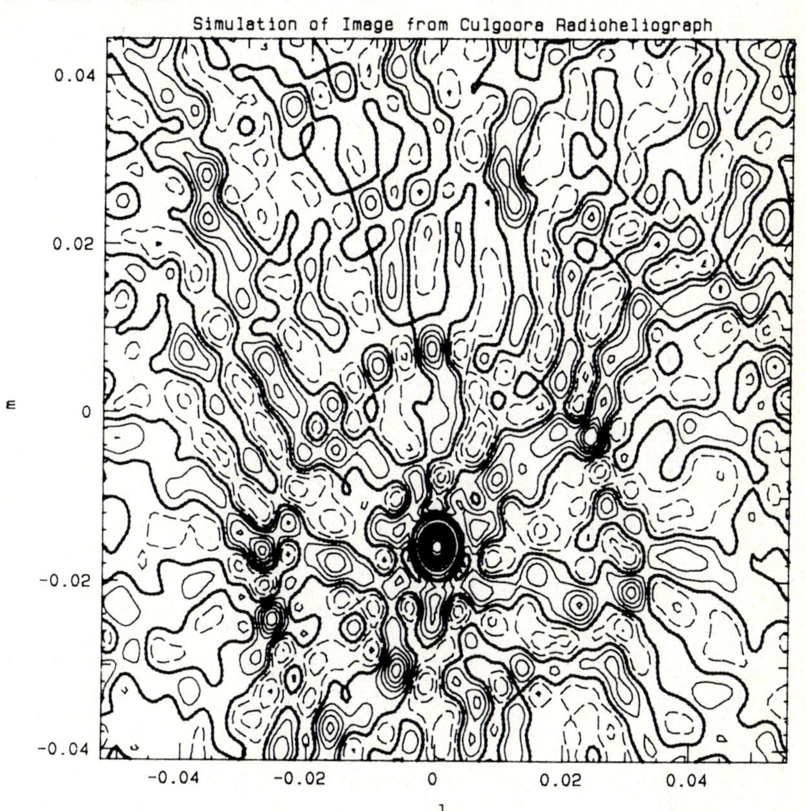

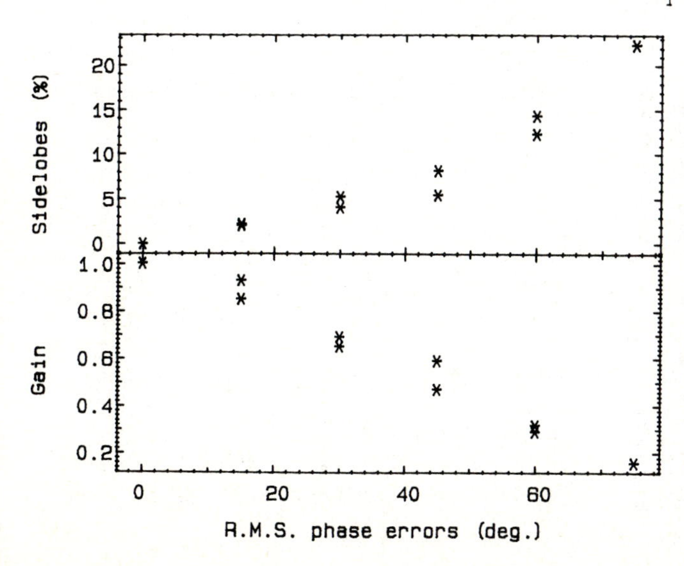

Figure 5. (a) Sidelobe levels for different levels of phase errors from a number of simulations similar to Figure 4; and (b) gain errors for different levels of phase errors. The r.m.s. sidelobe level is calculated for all pixels outside a box enclosing the main beam. We use 3 r.m.s. as an estimate of the peak sidelobe level. For most cases two different sets of phase errors, with the same r.m.s., were tried.

Figure 6. Example of a simulation of the beam incorporating 30° r.m.s. phase errors, plus other errors as described in the text. Aerials 5, 17, 48, 60, 90 were missing, and 53, 65, 0, 12, 42 were boosted to compensate. The phase calibration was done (seven iterations). Over most of the field the errors are <1% but close to the beam there are 2% sidelobes. Contour levels are as specified for Figure 4.

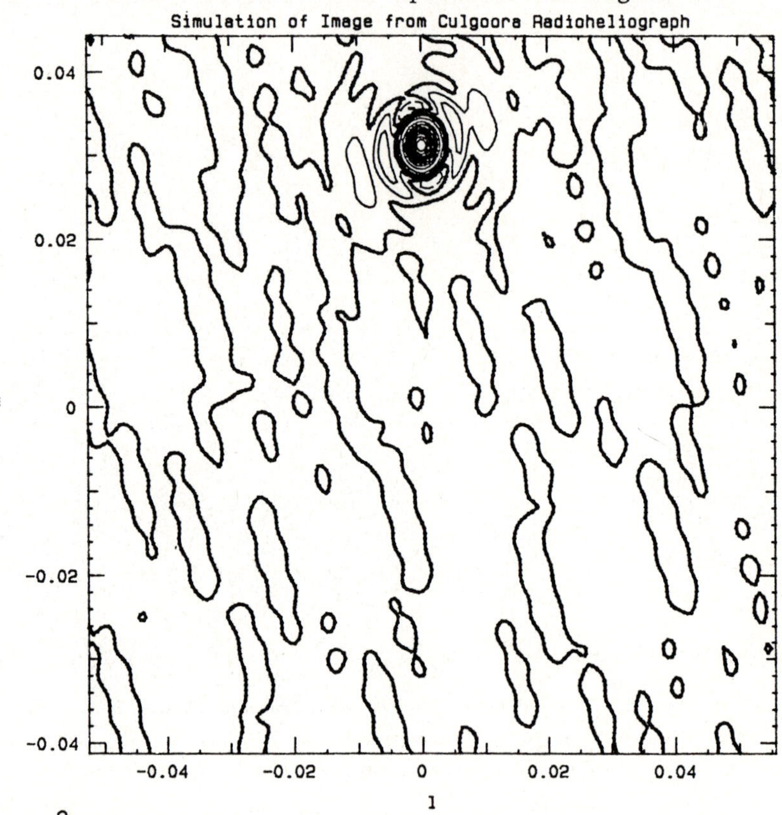

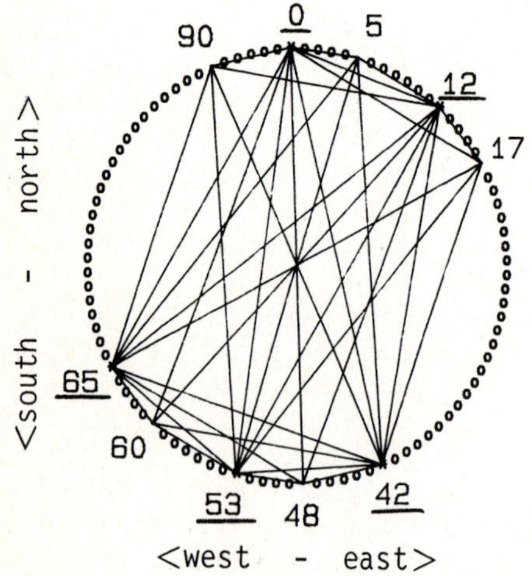

Figure 7. Figure illustrating the origin of the regular pattern in Figure 6. Aerials 5, 17, 48, 60, 90 are missing, and aerials 53, 65, 0, 12, 42 are boosted 6 dB to compensate. However, some spacings are still missing: for example 42-48 spacing is missing because 48 is out and the same spacing on the other side of the array, 0-90, is missing because 90 is out. Furthermore, some spacings are boosted twice: for example, aerial 12 and aerial 42 are both boosted, so the spacing 42-12 gets quadruple weight instead of just double weight.

because the gains of both aerials have been boosted. There are other spacings (25 in our example) which are lost completely. The combined effect of these few incorrectly weighted spacings is the ripple pattern visible in Figure 6.

From Figure 6 and a number of other similar tests, we expect to improve the dynamic range of the Culgoora radioheliograph by about one order of magnitude by the steps that we are taking.

APPLICATION TO NON-SOLAR OBSERVATIONS

The information gained during solar observations will be used to improve the quality of non-solar observations. Our plan is to log the variations in phase errors through the day, and analyse these at the end of the run in order to separate at least three main sources of error: ionospheric (which we cannot apply to the night-time observations), fixed errors due to difficulties with the old calibration scheme, and matching errors in the switched elements of the stepped delay lines, which will have to be tabulated and corrected for as appropriate. At this stage we cannot predict how much improvement this will yield for night-time observations, but it should be quite significant.

References

Clark, B.G. (1981). Orders of magnitude of some instrumental effects. VLA Scientific Memo No. 137.

Cornwell, T.J. & Wilkinson, P.N. (1981). A new method for making maps with unstable interferometers. Mon. Not. R. Astron. Soc., 196, 1067-86.

Labrum, N.R., McLean, D.J. & Wild, J.P. (1975). Radio heliography. Methods Comput. Phys., 14, 1-55.

McLean, D.J. (1979). A proposed correlator back-end for the Culgoora radioheliograph. In Image Formation from Coherence Functions in Astronomy, ed. C. von Schooneveld, pp. 159-64. Dordrecht: Reidel.

McLean, D.J., Beard, M. & Bos, A. (1979). A proposed correlation back-end for the Culgoora radioheliograph. Proc. Astron. Soc. Aust., 3, 371-5.

Schwab, F. (1980). Adaptive calibration of radio interferometer data. Proc. S.P.I.E., 231, 18.

Sheridan, K.V., Labrum, N.R. & Payten, W.A. (1973). Three-frequency operation of the Culgoora radioheliograph. Proc. IEEE, 6, 1312-7.

Wild, J.P. (1967). The Culgoora radioheliograph. I - Specification and general design. Proc. IREE (Aust.), 28, 279-91.

6.3 INTERDEPENDENCE OF INSTRUMENTAL ERRORS

J E Noordam
Netherlands Foundation for Radio Astronomy
Dwingeloo, The Netherlands

Abstract. Calibration techniques make use of the interdependence of instrumental errors from sample to sample. The "classical" assumption of slowly varying errors is invalid for long baselines. It has to be replaced by the "modern" assumption of telescope-based errors, necessary for a highly successful technique called SELFCAL. The reliability of SELFCAL, especially in the case of extended sources, can be improved by exploiting redundant baselines (if available). This is called the "luxury" relation because of its extra power, but also because of the popular belief that it should be expensive in telescopes.

INTRODUCTION

Indirect images, the subject of this symposium, are usually obtained by measuring the Fourier Transform (or Visibility Function) of a brightness distribution, and transforming it back. The reconstructed image will be convolved with some point-spread function (PSF) determined by the measuring instrument. Since the PSF sidelobes may be quite large, it is often desirable to remove them from the image by deconvolution. To this end, we must accurately know the actual PSF, which may be distorted by instrumental errors.

Deconvolution with the theoretical, rather than the actual PSF will leave an unremovable error-pattern which is the Fourier Transform of the unknown errors. If the errors per measured visibility sample are independent (noise), its RMS level over the image will be proportional to the RMS error, and to the RMS sidelobe level of the theoretical PSF, (Clark 1981).

If the errors are correlated from sample to sample, their contributions will add up in the image. Often this produces a characteristic pattern that may help to localise the problem, but it may also be mistaken for real structure. In any case it limits the attainable RMS dynamic range, which is defined as the ratio between the highest peak in the map and the weakest believable structure. In general this ratio will be smaller in the vicinity of strong sources.

Independent errors per sample cannot be calibrated, so the best possible image will be "noise-limited". Calibration is the art of recognising and exploiting relations between sample errors, that cause images to be "dynamic-range-limited". Three of them will be treated here for the example of aperture synthesis in radio astronomy. They are termed, somewhat facetiously, the "classical", the "modern" and the "luxury" relation. Some of the concepts should also be useful with other indirect imaging instruments.

RADIO APERTURE SYNTHESIS

The (complex) visibility function is sampled by correlating the signals from a pair of radio telescopes. Such an interferometer is characterised by the (projected) length and orientation of its baseline, and further by its frequency and polarisation response. The baseline changes continuously with the rotation of the Earth, causing the interferometer to follow an elliptical path in the Fourier plane. Consequently, integration times per sample can only be short, ranging from a few seconds to a few minutes, depending on field size and baseline length.

In order to save time, arrays of telescopes are normally used, in which each combination of two telescopes constitutes a separate interferometer. A set of samples from all available interferometers in a given integration interval will be called a "scan". The number of scans per observation varies from a few for a "snapshot" to several hundred for a full 12 hour synthesis.

Each sample in a scan is a complex number with an amplitude, a phase, and, of course, a gain- and a phase-error. These instrumental errors may be caused by drifting receivers, cables or mechanical parts, but also by troposphere and ionosphere. Since the latter two are least subject to our control, they tend to dominate in modern synthesis instruments.

The RMS sidelobe level of the PSF referred to above is usually a few percent, although it may be much higher than that for short observations.

A general error-model for the phase F'_{ij} measured by interferometer ij at any time is:

$$F'_{ij} = F_{ij} + f_i - f_j + f_{ij} \tag{1}$$

Here f_i and f_j represent phase-errors that can be attributed to telescopes i and j respectively, while f_{ij} is the remaining interferometer-based phase error (including random noise). F_{ij} is the true (celestial) phase due to the source.

A similar expression for the <u>logarithm</u> of the amplitude is:

$$G'_{ij} = G_{ij} + g_i + g_j + g_{ij} \tag{2}$$

All phase and gain errors are functions of time.

Examples of telescope arrays are the Very Large Array (VLA) in New Mexico, MERLIN at Jodrell Bank (England) and the Westerbork Synthesis Radio Telescope (WSRT) in the Netherlands. (See also table I). Out of these, only the WSRT has redundant baselines. Very long baseline interferometry (VLBI) with baselines up to thousands of kilometers, is also practised regularly with telescopes all over the globe.

THE CLASSICAL INTERDEPENDENCE: Slowly varying errors.

The number of independent errors reduces drastically, if one may assume that they vary only slowly with time. They can then be fully determined per interferometer by referring to a nearby celestial calibrator source at regular time intervals during the observation. Care should be taken, that no new errors are introduced by the actual movement of the telescopes. The calibration frequency is limited to about once every 10 minutes by the time it takes to slew the telescopes back and forth, and by the minimum acceptable duty cycle. In the case of amplitude errors, this is sufficient: The electronics can be made stable to better than 1 percent over many hours, and short-term atmospheric extinction variations do not usually exceed this value. Very importantly, the amplitude stability is independent of baseline length.

The situation is worse for short-term phase error variations, caused predominantly in the ionosphere below about 1.5 GHz, and in the trophosphere at higher observing frequencies. They increase roughly linear with baseline and frequency (or baseline measured in wavelengths), with RMS values ranging between 1 and 10 degr/km/GHz.

Consequently, at the usual observing frequencies of a few GHz and with baselines longer than about 30 km, the assumption of slowly varying errors is no longer valid for the phase. Without other means at our disposal, images would have to be reconstructed from the amplitudes alone, which cannot be done in general. But even for shorter arrays and low frequencies, the phase fluctuations limit the possible dynamic range to about 1 percent at best, which for many sources is at least a factor of 100 above the noise.

It is fortunate therefore, that there are other relations between instrumental errors to exploit, especially the assumption of telescope-based errors (see below). For absolute positions, however, we need either optical identifications, or calibrator sources and slowly varying phase errors. For this reason, the position accuracy in radio astronomy is limited to a few tenths of an arcsecond, while the resolution may be a thousand times better.

It is not always necessary for good results to know the instrumental errors explicitly. For instance, very good spectral dynamic range can be obtained by comparing the outputs of simultaneously measured frequency channels in a suitable way. Because the channels are close together (relative to the observing frequency), and share the same signal path at the same time, most of the errors are eliminated in this way. A similar procedure may be followed for three out of the four Stokes polarisation parameters, provided they are measured simultaneously.

Although these are also examples of relations between instrumental errors, they are not directly related to imaging and will not be pursued here.

THE MODERN INTERDEPENDENCE: Telescope-based errors only.

A totally different way of reducing the number of independent instrumental errors is to assume, that they can all be traced back to individual telescopes. This is equivalent to stating that the interferometer based errors f_{ij} and g_{ij} in equations (1) and (2), which include the receiver noise (!), are negligibly small.

With an array consisting of N telescopes, a maximum number of N(N-1)/2 different interferometers can be formed. The "reduction factor" which is the ratio between measured samples and independent errors, becomes

$$N(N-1)/2N = (N-1)/2 \qquad (3)$$

A high reduction factor is crucial for a very successful technique called SELFCAL (or SELF CALibration), about which much more will be said by others at this Symposium. The method uses the differences between the data and an approximate model of the brightness distribution, to solve for telescope-based errors in a least-squares sense. All scans are treated separately. A new (and hopefully better) model is then derived from the corrected data, and the process continues iteratively until the residues are smaller than some specified criterion. Note, that the only stability-requirement now is, that the errors are constant during the integration-interval.

It is clear, that the data's resistance to the wrong model increases with a larger reduction factor, although it is difficult to quantify this effect because it also depends on the source structure: Extended sources have a small "correlation length" in the Fourier-plane, and many different brightness distributions will be consistent with the data if the visibility function is undersampled. If given half a chance, the method will try to manipulate the telescope phase and gain errors in such a way, that low level extended structure is suppressed in favour of strong compact sources, that make their presence felt in all baselines because of their long correlation length. The problem is not one of reconstruction uniqueness, but of ending up with the wrong data, and thus diminishing the chance of ever reconstructing the right image.

Several things may be done to lessen the hazards of converging to the wrong model. The first is, obviously, to have the largest possible reduction factor, which requires a lot of telescopes or the presence of redundant interferometers in the array (see below). It is also a good idea to have all telescopes participate in long as well as short baselines, and to assign more weight to the baselines for which the model is assumed to be most accurate (usually the longer ones). A much recommended precaution is finally to allow only the phases to vary in the first few iterations, and meanwhile to rely on the (calibrated) gain stability of the telescopes.

The great success of SELFCAL must be attributed to the fortunate circumstance that the sky at even the highest resolution seems to be dominated by strong compact sources and is otherwise quite empty (if not, we shall never know). It is also the only way to obtain useful results with long baseline interferometry arrays like MERLIN and VLBI. But even with smaller arrays like the VLA and the WSRT, it is now used extensively to get a very much better dynamic range.

It is not surprising, therefore, that the main effort in understanding instrumental errors nowadays concentrates upon the simple question whether the underlying assumption of telescope-based errors is true. Possible problem areas are the correlator, any transmission line shared by a number of telescopes, and the way the telescopes filter the received signal, which should be identical for all of them.

A well-designed digital correlator produces only very small errors of its own. Experience with the WSRT line backend suggest RMS values of 0.02 percent (0.01 degr) per GHz observing frequency, constant over many days. A notable exception seems to be connected with very strong sources, for which the correlated fraction of the signal is more than 10 percent (Cas A, Crab). Another problem is the "ghost" mechanism treated elsewhere in this volume by A. Bos.

Sometimes it is necessary (or practical) for two or more telescopes to share the same waveguide for signal transmission to the correlator. This could conceivably be a source of interferometer-based errors.

Many problems are connected to not observing exactly the same brightness distribution with all telescopes, because of pointing errors or telescope shape, but also because of different bandshapes or polarisation response. Although the causes are telescope-based, the effects on phase and gain will depend on source characteristics and thus on baseline length and orientation.

The above three problems can be minimised by careful instrument design, but the next two are more fundamental and will ultimately limit the performance.

The first is decorrelation due to rapid phase error fluctuations within an integration interval. The effect is proportional to both baseline length and observing frequency, and whereas it is merely annoying at radio frequencies, it might well prove to be a major limitation in optical and infrared interferometry.

The second and most unavoidable problem of all is random receiver noise, which is clearly an interferometer-based effect. This is the most serious limitation that stands in the way of general use of SELFCAL. Even when the source structure is simple enough, there often is not sufficient signal to produce well-defined data, especially at the longer baselines. All telescopes for which an error solution is required should participate in at least one baseline with enough signal-to-noise ratio (> 2), and in order to limit noise-propagation through the least-squares error-solution, the input data should be weighed with the square root of their amplitude.

The presence of interferometer-based errors may be detected by measuring the so called "closure phase" for a set of three telescopes. For this reason they are often referred to as "closure errors". Using equation (1) we get:

$$F'_{ij} + F'_{jk} - F'_{ik} = F_{ij} + F_{jk} - F_{ik} + (f_{ij} + f_{jk} - f_{ik}) \quad (4)$$

Since all telescope-based errors have dropped from this equation, the interferometer based-errors can be isolated if we know what phases F_{ij} to expect from the (calibrator) source. A similar expression can be written for the "closure amplitude" of four telescopes.

A more accurate way of detecting interferometer errors, because it does not depend on a calibrator-model, is by comparing redundant baselines.

Interferometer-based errors can be as large as a few percent (degrees). They set a limit to the effectiveness of SELFCAL because they cannot, in general, be distinguished from incompleteness of the utilised model. Fortunately, however, they seem to be quite constant over long periods of time, allowing them to be corrected with values obtained from calibration measurements. They may even be recognised in the SELFCAL residues, because model incompleteness cannot cause constant phase-residues over the whole observation. It remains important to find out <u>why</u> they should be constant in this way.

THE "LUXURY" INTERDEPENDENCE: Redundant baselines

The number of independent errors per scan may be reduced even further by comparing (subtracting) the outputs of redundant interferometers in the array. Since these values should be identical, any difference can be interpreted in terms of instrumental errors only, independent of source structure, (Noordam et al 1982)

Using equations (1) and (2) for the redundant baselines ij and kl, we get:

$$F'_{ij} - F'_{kl} = f_i - f_j - (f_k - f_l) \quad (5)$$

$$G'_{ij} - G'_{kl} = g_i + g_j - (g_k + g_l) \qquad (6)$$

A necessary condition is, that all errors are telescope-based, just like with SELFCAL. Therefore, the interferometer-based errors have been ignored here. Again, all scans are treated separately.

Absolute errors cannot be detected by a comparison technique, although there may be enough redundant baselines to solve for all relative errors between the telescopes in an array. For each scan, we will be left with (at least) one absolute gain error and an unknown phase gradient over the array. This may be easily seen by polynomial expansion of f_i and g_i in terms of telescope positions, and substituting them in equations (5) and (6).

It should be mentioned here, that an image-sharpening technique for optical instruments ("rubber mirrors") implicitly makes use of redundant spacings. The aperture is subdivided in many small elements, each of which is assumed to have a single phase-error. The RMS intensity of the image is maximized by adjusting the elements, and it can be shown that this is equivalent to making the outputs of redundant interferometers in the aperture equal (Hamaker et al 1977). Unfortunately this technique cannot be used for determining the remaining absolute errors per redundandized scan, since there are no redundant interferometers connecting them.

In the extreme case of "full redundancy", the reduction factor (needed for reliable convergence of SELFCAL) is equal to the number of different interferometers per scan. One WSRT scan, for instance, has 38 different interferometers out of a total of 91. Since there are 14 telescopes, the number of independent errors is reduced to that number in normal SELFCAL, yielding a reduction factor of 2.71. But after redundandizing, one only has to determine one gain factor and one linear phase gradient from 38 (complex) differences between data and model: A reduction factor of 38(!) If there are not enough redundant spacings to connect all telescopes in the array, the reduction factor will be smaller but still worth the effort: Any reduction in the number of independent errors helps.

	Max baseline	No of telescopes	No of interferometers	reduction factor
MERLIN	127 km	6	15	2.5
VLA	30	27	351	13.0
WSRT	3	14	38	2.7 (38)

Table I: Maximum reduction factors for some telescope arrays. The highest value for the WSRT is obtained by the use of redundant baselines.

There are many advantages to internally "perfect" scans. Not only do they make SELFCAL much more reliable because of the high reduction factor, but the way is opened to other, more model-independent techniques.

A good example is the so called centroid method. If all telescopes are in a linear array, each scan represents a one-dimensioned projection of the celestial brightness distribution. Since total flux and source centroid position are invariant under projection, these quantities may be used to "align" the different scans in a completely model-independent way. Especially the total flux per scan may be determined with great accuracy by first a one-dimensional CLEAN (deconvolution) followed by a polynomial fit to pick up low-level extended structure that would be very hard to model for SELFCAL.

Another possibility is the study of variability by comparing internally perfect scans observed before and after the change. Complex extended structure is used as an accurate reference here, instead of confusing the issue.

Regularly spaced arrays have not been built for their redundant baselines, but for the low sidelobe level of the resulting beam (PSF). Since the invention of CLEAN, this is no longer so important for compact sources. In this sense, redundant baselines must be considered a luxury, and arrays like the VLA have been designed with no redundancy at all, in order to maximise the return per (expensive) telescope.

In the case of extended sources, however, deconvolution by CLEAN becomes an uncertain process, and low sidelobes are still highly desirable. Also, extended structure is very difficult to model accurately, requiring the highest possible reduction factor for reliable results with SELFCAL.

Both these considerations can be met by a linear array with telescopes at coordinates 0,1,2,3,4,8,13,18,23 ... and so on.It provides regular and complete coverage of the central part of the Fourier plane with a minimum number of telescopes, while allowing a full redundancy solution.

If resolution is more important than complete coverage, one might double the longest baseline with each new telescope in the following fashion: 0,1,2,3,4,8,16,32,64 ... The basic group of 5 equidistant telescopes ensures a gain- as well as a phase-solution. For a small number of telescopes (say 6), a more optimal configuration possibly exists, but may not be easily extendable.

CONCLUSION

High dynamic range images in radio aperature synthesis can only be produced with SELFCAL, although this technique has its limitations: All instrumental errors should be telescope-based (implying a high signal-to-noise ratio) and the observed field should be dominated by a few strong, compact sources.

The reliability of SELFCAL can be improved by reducing the relative number of independent errors as much as possible, either by using more telescopes simultaneously or by exploiting redundant baselines in the array. The reduction factor can be very high in the latter case.

The use of redundant baselines also opens the way to less model-dependent techniques, like the centroid method. In view of this, the question, while designing new arrays, should not be whether one can afford to have redundent baselines, but whether one can afford not to have them. Especially since they can be much less expensive in telescopes than is popularly believed.

References

B. G. Clark, (1981) "Orders of magnitude of some instrumental effects", VLA Scientific Memorandum No. 137.

J. P. Hamaker, J. D. O'Sullivan, J. E. Noordam, (1977) "Image sharpness, Fourier optics and redundant-spacing interferometry", J. Opt. Soc. Am., 67, 1122-1123.

J. E. Noordam, A. G. de Bruyn, (1982) "High dynamic range mapping of strong radio sources, with application to 3C84". Nature, 299, 597-600.

DISCUSSION

R.N. BRACEWELL

What is the relative importance of redundancies resulting from crossings in the u-v plane?

J.E. NOORDAM

Probably small, but all available model-independent information should be used.

T.J. CORNWELL

On any source crossing-points can be used to derive a self consistent set of correlator errors for all crossed baselines. An attempt to exploit this form of redundancy at the VLA was thwarted by time-variability of the correlator errors.

J. HAMAKER

You do not list the effect of bandwidth decorrelation (or the delay beam, if you wish) amongst the interferometer-based errors. It limits the effective field seen by the longer baselines, so different interferometers see different source distributions. Have you seen the effect in the extended sources you have worked on?

J.E. NOORDAM

No, because in the fields we have reduced until now most of the flux was close to the centre.

R.D. EKERS

Your method of using the 1D redundancy constrained self-calibration to search for variability will still be limited by the pointing errors of the elements of the array.

J.E. NOORDAM

Yes, I agree.

6.4 ON GHOST SOURCE MECHANISMS IN SPECTRAL LINE SYNTHESIS OBSERVATIONS WITH DIGITAL SPECTROMETERS

A. Bos
Netherlands Foundation for Radio Astronomy
Dwingeloo, the Netherlands.

1) Introduction.

The Westerbork Synthesis Radio Telescope (WSRT) is an earth rotational synthesis instrument consisting of 10 fixed and 4 movable antennas on a maximum East-West baseline of 3 km.
Each antenna has a receiver system connected to a digital crosscorrelation spectrometer (cf. Bos et al., 1981). This spectrometer measures the real part of a complex spatial and temporal correlation function $C(\underline{x},\tau)$ for N positive and N negative values of the lag variable τ ($\underline{x}$ is the spatial variable). The temporal Fourier transform of this correlation function produces N-1 physically meaningful equidistant points (frequency channels) of the spectrum of the complex visibility $R(\underline{x},\omega)$.
A subsequent spatial Fourier transform yields N-1 maps of the sky brightness $B_s(\underline{1},\omega)$ at N-1 equally spaced frequency intervals B/N (B is the receiver bandwidth). The effective frequency resolution of the spectrometer is determined by the weighting function in the temporal Fourier transform. Maps made using a uniform weighting function in particular contain a spurious response (a 'ghost') diametrically with respect to the phase centre from each source component (Fig. 1).
These spurious responses influence in particular observations with a high spectral dynamic range (e.g. in the study of faint recombination lines) and also the high dynamic range mapping (>40 dB) of continuum sources which became possible through the use of redundant baselines (Noordam and De Bruyn, 1982).

2) The basic mechanism.

The basic mechanism is introduced by the real to complex Fourier transform of $C(\underline{x},\tau)$ which is sampled and truncated in both the spatial and the temporal domain.
The complex visibility $R(\underline{x},\omega)$ consists of an even (real) cosine term and an odd (imaginary) sine term. When the source has a flat spectrum the sine term has a discontinuity at $\omega=0$ and due to the periodicity $R(\underline{x},\omega)$, one at $\omega=2\pi B$ (see Fig. 2). Sampling and truncation in the temporal domain produces a spectrum that is periodic and smoothed by the Fourier transform $W(\omega)$ of the weighting function $w(\tau)$.

Fig. 1. An observation of the calibrator source 3C147 at 610 MHz about 4' out of the phase centre of the map using a uniform weighting function. (The phase centre is the centre of the map.) The map of frequency channel number 2 (out of 15) shown here shows a ghost image with a positive peak intensity of 4.3% of the intensity of the original source component. The data were calibrated using the redundancy scheme described by Noordam and De Bruyn (1982).

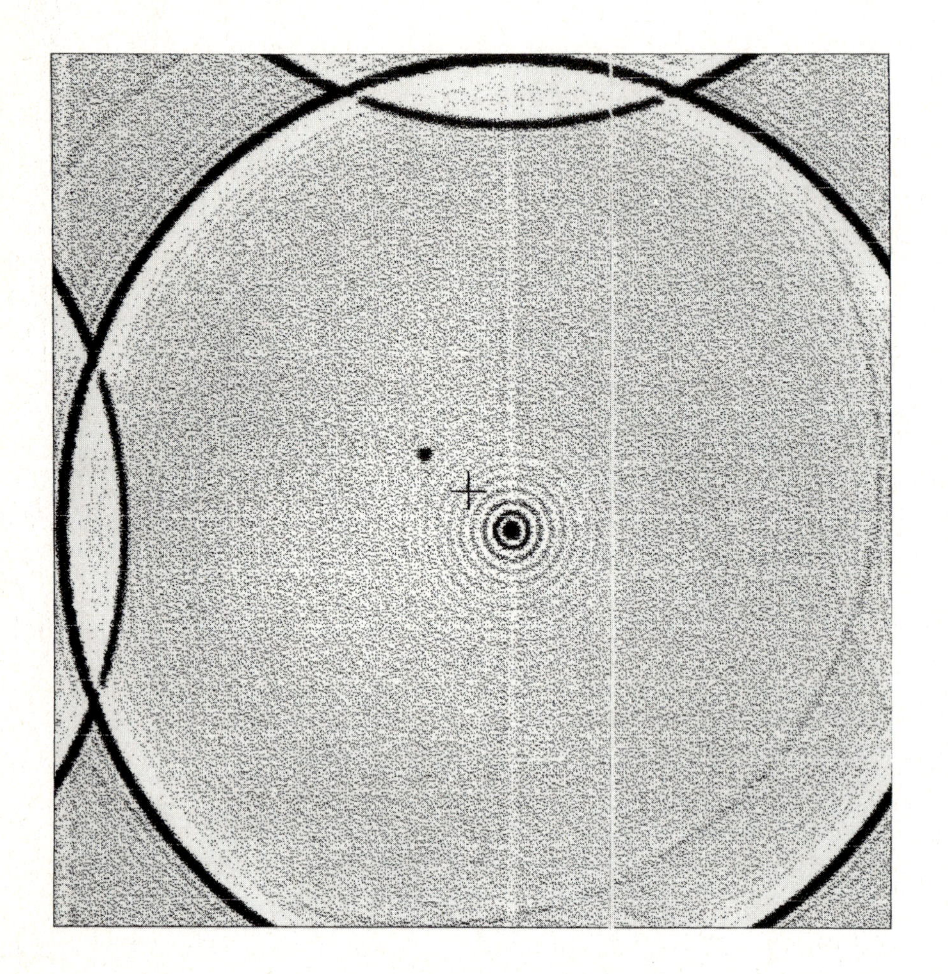

The convolution of a flat spectrum with $W(\omega)$ produces ripples on the sine spectrum due to the sidelobes of $W(\omega)$ and, in the vicinity of $\omega=0$ and $\omega=2\pi B$, the finite width of $W(\omega)$. (In Fourier theory this effect is known as the Gibbs phenomenon. (Papoulis, 1962, Bracewell, 1978)). For a uniform weighting function the magnitude of the ripple pattern has a peak value of 18% (9% of the stepsize of the discontinuity). As the ripple pattern is absent on the cosine term, the complex visibility has an imaginary error term $\Delta R(\underline{x},\omega)$ (the ripple pattern) whose shape is

determined by:
a) the shape of the receiver passband
b) the shape of the smoothing function $W(\omega)$ or the weighting function $w(\tau)$
c) the shape of the source spectrum.

In a subsequent (2-dimensional complex to real) spatial Fourier transform the imaginary error term yields a double image with odd symmetry from each source component. One image (the 'hidden' component) coincides with the error free image; the other (the 'ghost' component) appears with opposite sign diametrically with respect to the phase centre of the map. The two components can be identified as the contribution of the sidelobes of $W(\omega)$ at positive and negative frequencies respectively.

Fig. 2. The cosine and sine spectrum for a square passband and a flat source spectrum smoothed by a sinx/x smoothing function. (The Fourier transform of a uniform weighting function).

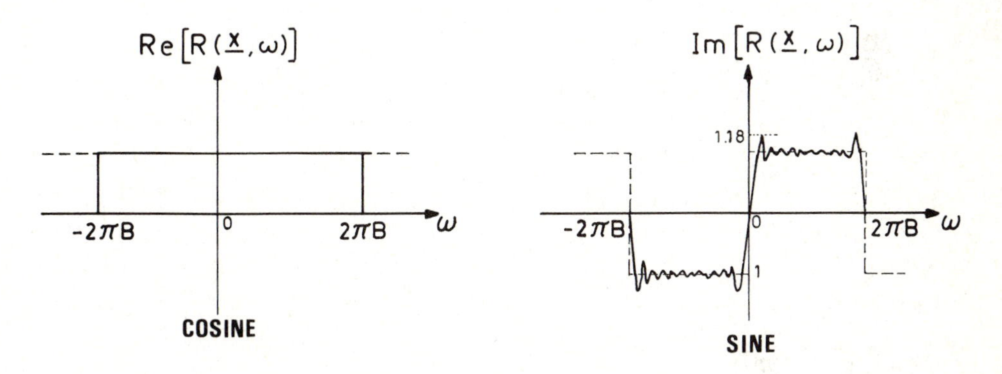

Apart from the presence of the ghost component the consequences are:
1) In the phase centre of the map the ghost and hidden components cancel each other. This implies that the effective receiver passband is position dependent.
2) The dependence of $\Delta R(\underline{x},\omega)$ on the shape of the source spectrum implies that the effective receiver passband also depends on the source spectrum. Continuum and line spectra will therefore have different effective receiver passbands. (In particular at the edges.)
3) For non-identical receivers the effective shape of the passband in the phase centre depends on fixed phase offsets present in the receiver system during observation. These phase offsets should be correctly compensated during the observations as correction afterwards does not remove the ripple $\Delta R(\underline{x},\omega)$.

These consequences have implications in particular for the method used to calibrate the receiver passband and the subtraction of a continuum background. In the following we will concentrate on the tools available to alleviate the related calibration problems rather than dealing with the calibration procedures themselves.

3) Suppression techniques.

The intensity of the error components in a map is in addition to the shape of the source spectrum determined by the receiver passband and the weighting function used in the temporal Fourier transform. The passband of the WSRT spectrometer has a steep edge at 3 kHz and a more gradual slope at $\omega=2\pi B$. The error pattern is therefore most dominant in the first frequency channels.

As the shape of the receiver passband is fixed the suppression of the ghost images is controlled by the choice of the smoothing function $W(\omega)$ (See Harris (1978) for a summary of various types of smoothing functions) as the passband ripple is determined mainly by the sidelobe pattern of $W(\omega)$. However low sidelobe levels broaden the mainlobe of $W(\omega)$ and decrease spectral resolution. The net effect is that the errors are concentrated in the first and last frequency channels which should be deleted. For the remaining channels the error pattern at Westerbork is typically smaller than 0.15% when a Hanning (raised cosine) function is used for $W(\omega)$ and a total of 3 channels is deleted. The ghost intensities can be made as small as 0.01% by using for example a Dolph Chebychev weighting function, at the eexpense of a few more unusable edge channels.
For observations with a small number of frequency points (15 or 31) the loss in spectral resolution combined with the loss of the channels at the bandpass edges is quite substantial, although it may be acceptable when a sufficiently large number of channels is available (>128).

Another possibility is the use of a 90-degrees phase switching technique. When $\pi/2$ is added to the local oscillator signal phase with subsequent multiplication by $-i$ (in software after the temporal Fourier transform) the ripple pattern is moved to the cosine (real) part of the complex visibility. A spatial Fourier transform now yields spurious components with even symmetry. The average of two successive samples with and without a $\pi/2$ phase offset removes the ghost image and yields a position independent shape of the receiver passband which is also independent of the fixed phase offsets during observation. (Note that the receiver operates as a time shared complex cross correlation receiver). The suppression of the ghost component is in practice dependent on the radial distance to the phase centre. For the Westerbork array which works with 10 second samples, the suppression is equal to about a factor of 60 at the 3 dB points of the primary antenna beam and a factor 300 at 0.1 of that distance.
The hidden component is still present which has still consequences for the calibration of the passband because of the dependence of $\Delta R(\underline{x},\omega)$ on the source spectrum.

For continuum observations the (N-1) spectral line channels are averaged after the temporal Fourier transform. As the ripple pattern oscillates with alternating sign in successive channels an extra suppression for both the ghost and the hidden components of about a factor 10 can be obtained when the channels are added with proper weights.

4) Conclusion.

The spurious responses are introduced when the real part of a truncated complex crosscorrelation function is measured. The level of the effects can be adequately controlled by the use of a proper weighting function during the temporal Fourier transform. Use of 90-degrees phase switching removes the ghost component and it allows a more efficient use of the receiver hardware when the number of frequency channels is small.

Passband calibration should take into account the fact that spurious responses may be different for source and calibrator spectra. Otherwise baseline ripples and slopes may show up in the calibrated spectra at in particular the passband edges.

It should finally be noted that other mechanisms for the production of ghost sources exist. One is related to the temporal Fourier transform of an asymmetrical correlation function, the other one due to a finite suppression of the unwanted sideband in single sideband mixers. As these effects are relatively small, this paper has been restricted to the most fundamental and dominant mechanism.

References.

Bos, A., Raimond, E., van Someren Greve, H.W., 1981, Astron. Astrophys. 98, 251-259.

Bracewell, R.N., 1978, The Fourier transform and its applications, New York, McGraw Hill.

Harris, F.J., 1978, Proc. I.E.E.E., 66, 1, 51-83.

Noordam, J.E., de Bruyn, A.G., 1982, Nature, 299, 597-600.

Papoulis, A., 1962, The Fourier integral and its applications, New York, McGraw Hill.

7

Processing – clean

7.1 CLEAN AS A PATTERN RECOGNITION PROCEDURE

J.A. Högbom
Stockholm Observatory, S-13300 Saltsjöbaden, Sweden

INTRODUCTION

A decade ago when the first CLEANed maps were produced, radio sources looked very different from the way they look today. The resoving power of synthesis telescopes, though very impressive for the time, was such that most of the intense extragalactic emission regions remained unresolved or just barely resolved. Very few sources were observed to be so complex that they could not be adequately described as a set of one to three unresolved or barely resolved components. In practice, the radio sky away from the plane of the Galaxy looked very much like a field of randomly scattered point sources. Synthesis maps were produced with arrays on accurately aligned E-W or N-S axes. It was known that only with regular spacings would the main sidelobes of the synthesized pattern be concentrated to specific well defined locations where they could be easily recognized.

With less regular arrays the situation was not so simple. The sensitivity often permitted intense sources to be registered with an adequate signal/noise in a few hours' integration time. It then seemed a waste of time to track such a source for days on end just to cover the u,v plane so completely that a direct Fourier transform inversion would deliver a satisfactory map. In 1968 I observed a number of 3C sources with the Green Bank interferometer, time-sharing between the sources as they passed a set of three to six pre-set hour angles. Each source was observed at a total of 50-100 spacings somewhat irregularly distributed over the u,v plane, and I went home with the hope that problems caused by the irregular coverage would be solvable some way or other. The result, after exploring a number of blind alleys, was CLEAN.

Most people will agree that the procedure works very well with fields containing only sources that are small compared with the synthesized beam, and this was indeed a good approximation to the reality of yesterday. Today, sources are examined with much higher angular resolution, emission can be measured over hundreds of beam areas and much of the structure is far from pointlike. The basic CLEAN algoritm (Högbom 1974) is not very efficient at handling such extended regions of emission. Thousands of iterations may be used and still not give a fully satisfactory result. The high resolution achieved with relatively small antennas implies a very large number of pixels per field and brute force cleaning becomes exessively expensive in terms of computer time. A common problem is that of regular ripples over extended emission regions. The ripples correspond to Fourier components not covered by the measurements and are a consequence of an unsuccessful interpolation of the visibility between the measured spacings. In its basic form, CLEAN cannot be applied to amplitude only measurements such as are produced with certain long baseline interferometers. Finally, apart from the pioneering work by Schwarz (1978,1979), very little seems to have been done about the very important problem of reliability: exactly under what circumstances and to what extent can specific details on a cleaned map be relied upon?

DEVELOPMENTS OF THE BASIC ALGORITM

In view of all these difficulties it seems most remarkable that CLEAN is still being used! There are, however, some good reasons for this surprising fact. Compared with potentially more powerful procedures such as model fitting or MEM, the algoritm is very simple indeed; when it runs one has, possibly mistakenly, a feeling of understanding in detail what is going on. Consequently, when things go wrong, one has a fair chance of inventing a simple modification that will help solve the problem. One example of such a modification is the setting of windows restricting CLEAN to work in areas to which you believe that all significant emission is concentrated (Schwarz 1978). This of course is a way of introducing a priori information about source structure such as may be available e.g. from earlier low resolution maps of the same region. In addition, window cleaning is faster since only part of the complete field is processed. A variation of this theme is to begin cleaning at a low resolution, use the resulting cleaned map to set windows for a higher resolution clean and so on. The tendency to add ripples over extended components can be attacked in a number of ways: to make complementary clean runs at a reduced resolution, to subtract a rough model from the measurements in advance and let CLEAN work on the differences only (for a discussion of these aspects, see Cornwell 1982 and Clark 1982). It can also be discouraged more directly by introducing a bias towards smooth distributions in the algoritm itself (Cornwell 1983).

An important modification was introduced by Clark (1980). Most of the transform map from a multi-antenna array such as the VLA will be sufficiently close to zero to be unimportant during a substantial number of consecutive subtraction cycles. For the purpose of defining the corresponding clean components, the subtraction step need only consider the high points of the map and a small central region of the synthesized beam pattern, the so called 'beam patch'. A large number of clean components can be defined before the effects of these simplifications will influence the selection of further components in any significant way. At this point the whole set of components is Fourier transformed and subtracted in the spacing domain. After subtraction, a return transform gives a map from which all traces of these components have been removed and on which a further series of incomplete subtractions is performed and so on. The gain in time comes because a double Fourier transform with the FFT algoritm needs less computer time than the many complete beam pattern subtractions which it replaces. The Clark algoritm has made it feasible to run thousands of iteration cycles with very large maps, an operation which would often be unrealistic with the basic algoritm.

The CLEAN algoritm has also found use as an integral part of more sophisticated procedures including some that are designed to invert measurements with bad or absent phase information (Readhead and Wilkinson 1978, Readhead et al 1979) and in speckle imaging (Bates et al 1982).

A good simple solution to the problems of errors and reliability still does not exist. However, it seems that in practice a simple set of rules and tests can give sufficiently good error estimates for any particular cleaned map. These include checking that you have not in fact solved for more unknowns than there are independent measurements and to make sure that the transforms of vital features are not confined to gaps in the u,v coverage. One can repeat the operations with a different loop gain, after omitting a substantial fraction of the measurements and/or after degrading the measurements by adding extra noise; reliable features should survive such treatement albeit with a lower dynamic range.

A PATTERN RECOGNITION VIEW OF THE CLEAN ALGORITM

The construction of maps from measurements which only give a diluted coverage of the spacing domain is recognized to involve a great deal of a priori information. CLEAN makes use of some such information, but does encounter difficulties, in particular with weak extended distributions, because of its inherent insistence that everything has to be described in terms of point sources. CLEAN works in the map domain and the procedure must clearly be influenced by the way in which the ('dirty') map is derived from the measured visibilities. Quite a variety of synthesized patterns ('dirty beams') can be produced by choosing different sets of weights w_k for the measured visibilities V_k in the Fourier transform ('dirty map') calculation:

$$DM(l) = \mathrm{Re} \sum w_k \cdot V_k \cdot \exp(-i \cdot 2\pi \cdot u_k \cdot l) \qquad (1)$$

u_k (k=1,N) are the spacings at which the visibilities V_k have been measured.

A desirable pattern will be characterized by a distinct main lobe and a low sidelobe level. The first requirement determines the overall envelope weighting (taper) over the spacing domain while the second has to do with details of individual weights. It is well known that equal weights, w_k=const, will give the lowest possible global rms sidelobe level. However, we are in practice only concerned with the sidelobe pattern out to a certain distance from the main lobe, a region comparable in size with the effective field of view as determined e.g. by the antenna envelope beam or by the known overall size of a dominating source. This leads to lower optimal weights for neighbouring measurements; a consequence of the fact that such measurements will not be independent and so should not be used with full weight. When such map is cleaned it is clear that this a priori knowledge about the effective size of the field will have influenced the final cleaned map. There are various standard procedures in use for downweighting close measurements in this way, and the step is often combined with gridding the measurements in preparation for the FFT.

A possible way of introducing more general a priori information about what sources may look like would be to subject the data to a pattern analysis searching for, and extracting from the data, features of different shapes according to some preference ranking derived from the a priori information.

A specific type of feature may be looked for by convolving the map with a function designed to be especially sensitive to that shape. A significant maximum on the convolved map then signals the presence of a feature whose shape is strongly correlated with that previously defined. A convolution in the map domain is equivalent to a multiplication (weighting) of its Fourier transform by the transform of the convolving function; we see that a CLEAN subtraction, at a maximum of the dirty map, is equivalent to a pattern recognition and subtraction operation, the pattern looked for being that defined by the weights given to the measurements in the spacing domain. The pattern appearing on the map, of course, is that of the 'dirty beam' but we let CLEAN interpret such patterns in terms of discrete small diameter 'point' features in the sky. The 'point' feature is, according to our a priori information, the most likely member of a large family of brightness distributions, each of which would produce a 'dirty beam' pattern on the map.

Most fields observed with the high resolution telescopes of today, however, contain a great deal of resolved structure. From a pattern analysis point of view, such structures will be ill matched to the standard cleaning procedure.

In order to keep things simple, let us assume that we want to find out whether there is in the direction $l=0$ a feature whose shape (brightness distribution) is described by a normalized symmetric function $F(l)$. This could be for instance a gaussian of a certain width or, in the two dimensional case, a gaussian with specified axes and orientation. Such a feature will contribute an amount $S \cdot f(u_k)$ to the measured visibility V_k. Here S is the integrated flux density of the feature and $f(u)$ the suitably normalized transform of $F(l)$. When the map is calculated as the discrete transform of the measured visibilities, then this feature will make a contribution

$$S \cdot \sum w_k \cdot f(u_k) \tag{2}$$

to the deflection at the corresponding point on the map ($l=0$). However, the actually measured visibility V_k differs from the value $S \cdot f(u_k)$ by an amount ΔV_k which has two components: a) errors of measurement ('noise') and b) contributions to V_k from all other sources in the field ('confusion'). These components produce an error in DM(0) of

$$\Delta DM = \mathrm{Re} \sum w_k \cdot \Delta V_k \tag{3}$$

What set of normalized ($\Sigma w_k=1$) weights should we choose if we want to detect a source having the specific shape (brightness distribution) F(l) with the smallest possible error? We shall be primarily interested in the case when the error is dominated by the confusion contribution.

If the errors ΔV_k could be regarded as independent and normally distributed with a standard deviation σ_k, then the standard error in DM would be

$$\sigma_{DM} = \left(\frac{1}{2} \sum w_k^2 \, \sigma_k^2 \right)^{1/2} \tag{4}$$

The confusion errors will never in practice be independent and normally distributed. This would be a reasonable approximation only if the field contains a large number of randomly distributed point sources and the measurements are located at well separated but irregularly distributed points in the spacing domain. This idealized case will be used only as a norm relative to which real life situations can be discussed in a qualitative way. In order to detect the feature discussed above at an optimum signal/error, the weights should be

$$w_k \sim f(u_k) \; / \; \sigma_i^2 \tag{5}$$

Thus, the optimum choice of weights will depend upon what kind of a feature we are looking for, and no one weighting can be ideal for all shapes. The most commonly used overall weighting function (taper) is a gaussian down to $\simeq 0.25$ of its central value at the largest spacing. Such a taper implies weights that are suitable for a field of slightly resolved sources if the measurements are of equal quality (σ_k=const) or, alternatively, a field of point sources if the errors happen to be proportional to $w_k^{-1/2}$. This should not be too far from the real situation if the observed field consists mainly of randomly distributed small diameter sources and this is indeed the situation in which CLEAN performs best.

High pass cleaning

Extended emission regions contribute mainly to the short spacing measurements. Thus, when cleaning a complex brightness distribution, the short spacing visibilities must be regarded as having large and strongly correlated confusion errors. For the purpose of cleaning sharp features in such a field, the short spacing measurements should according to (5) be used with very low weights.

The same conclusion follows from a simple consideration of how one does in practice analyze point sources on a cleaned map. If there is any doubt about the base level - and this will practically always be the case - then the source will be measured relative to a local base level fitted to neighbouring portions of the map. Now imagine that a short spacing measurement is changed by an arbitrary amount; this will make no difference to the point source analysis since the local base level would simply be drawn differently. The large spacing measurements are in effect used to deduce what the contribution of the source to the short spacings ought to be, ignoring the measurements themselves.

To how large a region about the origin of the spacing domain does this argument apply? This clearly has something to do with how base levels are drawn in practice. It seems that spacings as large as 1/10 of the maximum spacing will often be ignored in such operations. Local non-zero baselevels are ascribed either to instrumental effects or to the presence of some extended region of emission in the area. The conclusion must be that the short spacing data should be excluded from the usual small-diameter clean operation: the information is ignored anyway and its inclusion only adds problems in the form of extended features which, according to (5), should not be cleaned with a high resolution weighting of the measured visibilities. The procedure, in effect, is similar to the earlier mentioned method of subtracting a rough model (the transform of the short spacing measurements) and let CLEAN work on the (fine structure) difference map.

Feature cleaning

The real sky contains resolved sources, 'features' of many kinds and shapes. CLEAN must represent all such features as compact clusters of point sources, an extremely unlikely configuration in any random distribution. Indeed, CLEAN has problems reproducing such distributions correctly. Furthermore, weak extended halo components for which there is in fact good evidence in the measurements may be overlooked as low level structures may become masked by the confusion noise during the iterations.

Equation (5) shows that the optimum set of weights will depend on what kind of feature we are looking for. Ideally we should search independently for all kinds of structure that the eye would single out as a 'feature' in the sense that it stands out from the rest as something united. One problem of course is that we do not know in detail what makes us describe some part of a complex structure as a feature separate from the rest. However, one strong - though by no means sufficient - indication is continuity, as exemplified by expressions such as 'blob', 'jet', 'arc' etc.

The standard weights corresponding to the usual full resolution cleaning are not well suited for such features because their transforms $f(u)$ will be close to zero over large portions of the spacing domain. However, various classes of feature cannot normally be separated because the algoritm works only with amplitudes and not with shapes. A possible modification would be to let the program make a local estimate of the shape of the feature responsible for the maximum on the dirty map and then to subtract the effects of a model feature derived from this estimate.

CONCLUSION

CLEAN has survived for a decade and is still one of the most commonly used algoritms for deconvolving synthesis measurements. This is to a great extent due to its inherent simplicity which has made it relatively easy to introduce modifications and improvements to the basic algoritm to meet the changing demands as synthesis telescopes develop.

REFERENCES

Bates,J.H.T.,Fright,W.R.,Millane,R.P.,Seagar,A.D.,Bates,G.T.H., Norton,W.A.,McKinnon,A.E. and Bates,R.H.T.(1982). Optik,62,333.

Clark,B.G.(1980).Astron.Astrophys.,89,377.

Clark,B.G.(1982), in Synthesis mapping: Proceedings of the NRAO-VLA Workshop, Chapter 10, National Radio Astronomy Observatory, A.R.Thompson and L.R.D'Addario (eds.).

Cornwell,T.J.(1982), ibid, Chapter 9.

Cornwell,T.J.(1983).Astron.Astrophys.,121,281.

Högbom,J.A.(1974).Astron.Astrophys.,Supplement,15,417.

Readhead,A.C.S.,Pearson,T.J.,Cohen,M.H.,Ewing,M.S.and Moffet,A.T. (1979).Astrophys.J.,231,299.

Readhead,A.C.S.and Wilkinson,P.N.(1978).Astrophys.J.,223,25.

Schwarz,U.J.(1978).Astron.Astrophys.,65,345.

Schwarz,U.J.(1979). Image formation from coherence functions in astronomy (C. van Schooneveldt ed). Astrophysics and Space Science Library, Vol 76,261.

DISCUSSION

J.R. FORSTER

It is sometimes claimed that CLEAN can give an estimate of the 'zero spacing' visibility. Can you explain how it does this?

J.A. HÖGBOM

A hole about the origin of the spacing domain is unique in that it makes you lose all information about certain very large smooth features that may be present. The contributions of such features, of course, cannot be estimated by CLEAN. However, as far as the analysed CLEAN components are concerned, their contribution to the zero spacing visibility will be retrieved.

I. KOCH

How do you choose your weights if you are looking for extended sources, i.e. do you use a priori information corresponding to the smoothing in pattern recognition?

J.A. HÖGBOM

The 'dirty map' can be produced in the normal way. However locally, in the surroundings, of a maximum, you can find out what the effects would be of different weightings: smoothing such a local region by e.g. a Gaussian is equivalent to changing the set of weights in the spacing domain in a controlled way. Comparing maps of the same region which have been smoothed by different Gaussians, you can decide on a reasonable shape to use as the component in the CLEAN iteration.

R.N. BRACEWELL

By smoothing the neighbourhood of a peak in the dirty map and fitting a suitably oriented Gaussian hogback you have in effect arrived at the three second moments (σ^2_x, σ^2_{xy} and σ^2_y) of that Gaussian function. Six adjacent values of the dirty map suffice to determine these moments; direct calculation thus may offer an alternative to determination of the parameters by fitting.

J.A. HÖGBOM

Six adjacent values are indeed enough to define a two dimensional Gaussian. However, the values on the dirty map will be corrupted by noise and by sidelobes from other sources. One way to find a suitable Gaussian shaped component would be to produce a set of maps of this particular region, each smoothed by a different Gaussian. This, of course, is equivalent to trying out different sets of weights w_k in the calculation of the dirty map (1). After deciding which of these show the most significant maximum, you can determine the parameters from six adjacent values on this smoothed map. The exact parameters are not very important, since the major purpose is to arrive at a component whose shape gives a better match to an extended emission region than does the usual 'point' component.

C.F. CHEN

Two comments: 1. An algorithm called CLEAN-TIDY has been developed in 1978. The speed gain over the conventional clean is between one and two orders of magnitude (Chen & Frater, this volume).

2. With hundreds or thousands of iterations, the accumulated digital quantization with shrinking dynamic range of the dirty map can have significant effect on the capability and reliability of clean (Chen, C.F., 1981, Proc. Astron. Soc. Aust., 4, 256).

7.2 THE RELIABILITY OF CLEAN MAPS AND THE CORRUGATION EFFECT

U.J. Schwarz
Kapteyn Laboratory, Groningen, The Netherlands

Abstract

In this paper we discuss the origin of corrugations in CLEAN maps. The period of the corrugation corresponds to non-observed spatial frequencies. Without *à priori* knowledge of the source structure, the corrugation cannot be avoided completely. However, some practical approaches are discussed which serve to diminish the effect.

The problem

Since the introduction of the CLEAN procedure (Högbom, 1974) it is known that occasionally 'clean' maps are not clean at all, but show stripes, or corrugations. Recently, this effect has been more frequently encountered; this is probably because faster computers make it possible to clean larger maps with thousands of components. The question is, is this a failure of CLEAN, or could the effect be avoided by conventional means (such as varying the number of iterations, loop gain, clean beam etc.)? The answer is that it is strictly speaking not a failure of CLEAN, but essentially an overinterpretation of the results.

The effect and its explanation

One may easily see what causes the effect: if one subtracts a component from an extended, smooth 'dirty' map, then the strongest negative sidelobe will - since it is subtracted - give rise to a secondary maximum (actually two maxima because the beam is symmetric). The next component will be put at this maximum (see Fig. 1). In this way a regular pattern will propagate. In the presence of noise this mechanism will work only if the amplitude of the sidelobe is significantly larger than the noise. The effect can also be suppressed by tapering the data, thereby reducing the negative sidelobe.

In the Fourier transform domain, the regular pattern will show up as a peak at high spatial frequencies. The components are convolved with a 'CLEAN' beam in order to reduce the effect of the highest spatial frequencies. A numerical 1-dimensional example, illustrated in Fig. 2. Fig. 2a shows one half of the beam pattern and Fig. 2b the clean map (full line) with the components as δ-functions; 500 iterations were used with a loop-gain of 0.1; the broken line gives the true distribution (a Gaussian convolved with the clean beam). The clean beam is a Gaussian of the same halfwidth as the dirty beam. A ripple is still clearly seen;

it has a period equal to the distance from the centre to the first negative sidelobe of the dirty beam. Fig. 2c displays the FT domain. The arrow indicates the spatial frequency corresponding to the period of the ripple; indeed, the FT of the components (full line) shows a strong maximum at this frequency. The true visibility is given by the broken line and the sampling function is illustrated as a series of delta functions. One can see that the extra peak falls completely outside the range of the sampling. CLEAN has done a good job, the FT of the 15 components is a good fit to the true ('observed') visibility, within the range of the sampling; CLEAN has done this in a least square sense (Schwarz, 1978). The ripple is caused by non-observed spatial frequencies.

A 2-dimensional example is shown in Fig. 3.

How large is the amplitude of the corrugation?

Since the FT of the ripple falls outside the observed range of sampling, only statistical predictions can be made. The amplitude can be estimated based on the predicted statistical errors of the visibility of the components.

In Fig. 2d the predicted errors for the example of Fig. 2b are displayed. One sees that the errors of the predicted visibility at places of 'gaps' in the sampling are small (comparable to the observational errors at the sampling points), but they rise dramatically (factor 1000) outside the range of observed spatial frequencies. The predicted errors depend on the distribution of positions of the components within the map; in the present case it is the extended source which causes this sharp increase of the errors. The errors have to be scaled with the rms deviations of the FT of the components from the observed visibility. The smaller the noise in the data and the better the fit (sufficient number of components and sufficient iterations) the smaller is the corrugation. In the limiting case of noise free data and a true distribution restricted to a part of the map, a perfect fit is obtained (under certain conditions, Schwarz 1978) and *no* corrugation will appear.

Is it possible to avoid the corrugation?

The question is, is it possible to let CLEAN run in such a way that no regular pattern of the positions of the components occurs? Or, expressed in the FT domain: is it possible to prevent the extra peak at high spatial frequencies? The question posed in this way makes it clear that one wants to solve a more delicate problem involving *à priori* restrictions on the solution; such restrictions could be positiveness, maximum smoothness or maximum entropy.

In this paper I wish only to mention one method provided by the conventional implementation of the CLEAN algorithm.

As can be seen easily from Fig. 1 one may avoid the propagation of the regular pattern by choosing a small loop gain; the smaller, the flatter the dirty map. However, the small loop gain results quickly in an even distribution of residuals, therefore sooner or later the corrugation pattern will re-emerge.

How to eliminate the corrugation.

A standard procedure in CLEAN is to convolve the derived components with a clean beam in order to eliminate the influence of the high spatial frequencies. The usual clean beam, a Gaussian, as used in the example of Fig. 2, is obviously not sufficient in this case.

What are the alternatives?

A quite radical approach was used by Rogstad and Shostak, 1971, in one of the very early applications of CLEAN. A clean beam was used whose transform was zero outside the largest observed spacings: they did not believe in a fictive 'Gaussian' telescope of infinite diameter. Such a beam has also strong negative sidelobes. Another solution would be to use a clean beam, which attenuates the large spatial frequencies more strongly than a Gaussian, but one which does not have negative sidelobes. Such a beam is, however, wider than the original dirty beam; in other words one smoothes the ripples away.

A statistically sound solution would be the use of an earlier proposed approach (Schwarz, 1978), namely to define a clean beam based on the predicted statistical errors of the visibility of the components. Thus, where the errors are large, one uses a small weight ($\propto 1/\text{error}^2$). The resulting beam in the present example (the FT of the inverse square errors) would be almost identical to the clean beam used by Rogstad and Shostak, discussed above. A disadvantage of this method is that it is practically impossible to meet the computational requirements to compute the errors of the predicted visibility for a large field, but from a theoretical point of view it shows nicely the limitations of CLEAN. A possible practical approach would be to use Monte Carlo calculations.

Fig. 1

Schematic representation of the mechanism, which produce the corrugations.
a) shows the dirty beam, b) the dirty map after subtraction of a component. The arrows point to the new maxima, which are at a distance of the negative side-lobe to the central point of the dirty beam. The new components will appear at these maxima.

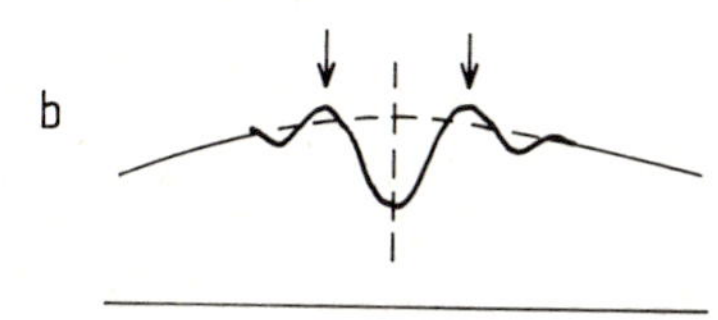

Fig. 2

Numerical 1-dimensional example of a model distribution.

a) Dirty beam (only one half) is displayed.

b) The result of CLEAN (full line) with components as δ-functions; the broken line shows the model smoothed with the clean beam.

c) The FT of b). The sampling (FT of the beam) is indicated by a series of delta functions. The period of the ripple seen in b) (two peaks are indicated by arrows) corresponds to the spatial frequency in the predicted visibility (c) at the position indicated by the arrow.

d) The error estimates of the visibility; the errors are normalised to standard deviation.

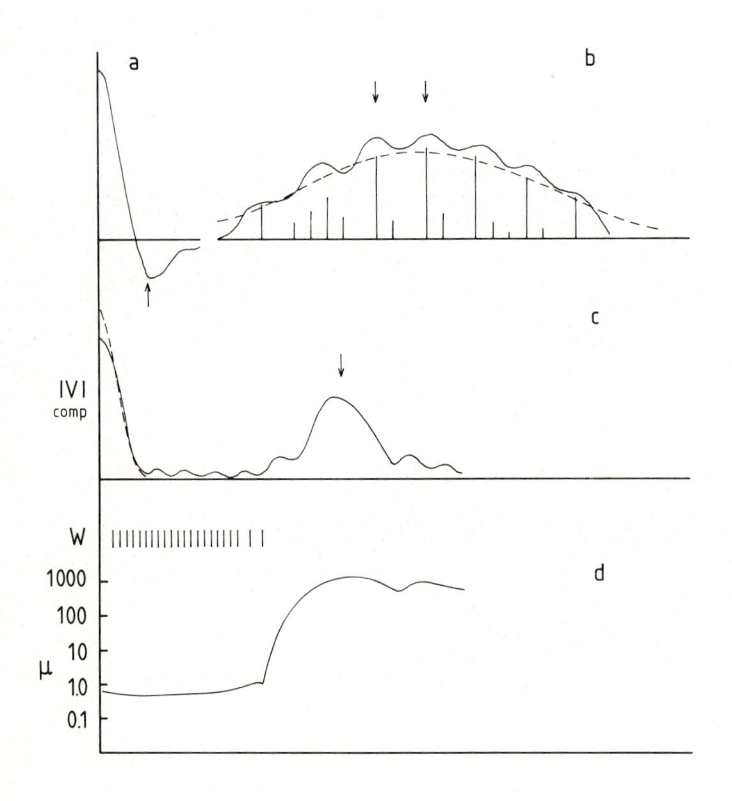

References

Högbom, J.A.: 1974, Astron. Astrophys. Suppl. 15, p. 417.
Rogstad, D.H., Shostak, G.S.: 1971, Astron. Astrophys. 13, p. 99.
Schwarz, U.J.: 1978, Astron. Astrophys. 65, p. 345.

Fig. 3

2-dimensional example. A completely flat distribution was cleaned, which resulted in the strong corrugation a); b) gives the FT of the clean map and c) the sampling. The contribution of the corrugation in the FT domain lies just outside the sampling function.

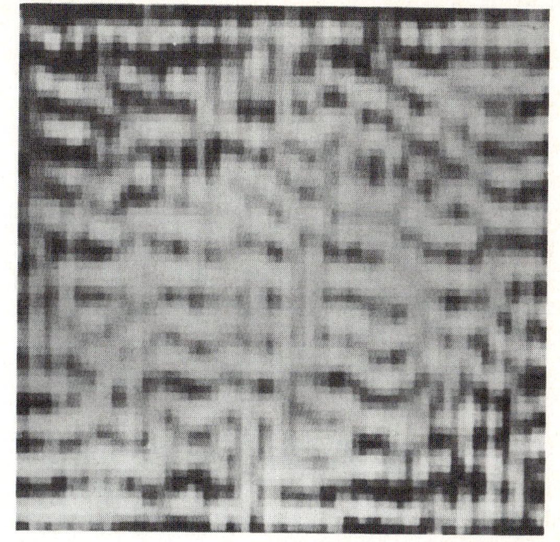

a

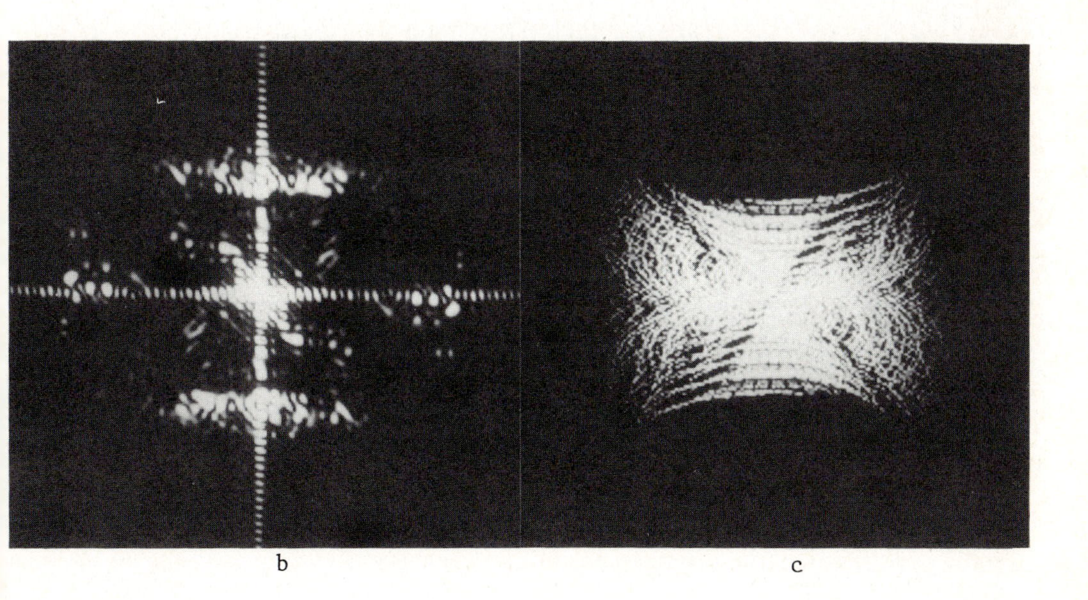

b c

DISCUSSION

J.A. HÖGBOM

Do I understand that the ripple is due to the first negative sidelobe and that the ripple always corresponds to spatial frequencies outside the surveyed area in visibility space? If so, would it help to use weights that suppress this lobe or put it further out?

U.J. SCHWARZ

Yes

R.H.T. BATES

The theory of cleaning extended objects is covered rigorously by the method called 'subtractive reblurring' (Bates, McKinnon & Bates, Optik, 61, 349 and 62, 1 and 333).

T.J. CORNWELL

The smoothness stabilized CLEAN (Cornwell, 1983, Astron. Astrophys., 121, 281) uses a very inexpensive and easily coded modification of CLEAN to prevent the occurrence of corrugations. Experience with the smoothness stabilized CLEAN is accumulating at various observatories and it seems that it is often very successful in preventing both corrugation and fragmentation of extended structure.

U.J. SCHWARZ

Indeed, your algorithm, which minimizes the sum of intensities squared, seems to prevent the occurrence of the corrugation. The algorithm is simple; one has only to add a δ-function of amplitude α to the centre of the dirty beam (normalized with maximum = 1). The working of the algorithm can be understood qualitatively easily in the FT-domain. There the δ-function has a constant amplitude throughout the whole FT-plane and therefore at each iteration it removes a certain contribution from every point in the FT-plane including the unwanted extra peak outside the sampled region.

There are however some precautions to be taken:

(i) A 'deep' CLEAN is required, because the mechanism responsible for the corrugation (Fig. 1) is only mildly affected, at least for small amplitudes α of the δ-function.

(ii) If no restriction on the searching area is applied, then the result will tend towards the principal solution (the dirty map). However if the search area is limited to a suitable window, this cannot happen and one obtains a good result.

7.3 PRE-CLEANING MOLONGLO SNAPSHOTS

M. Kesteven
Queen's University at Kingston

This paper addresses the question of whether it could be useful to operate the Molonglo Observatory Synthesis Telescope (MOST) in a sampled synthesis ("snapshot") mode. We conclude that it could be.

1. The motivation, and the problem.

The MOST (Mills, 1981) is the synthesis telescope obtained with the E-W arm of the Molonglo Mills Cross, reworked to 843 MHz. A contiguous array of 88 independent tracking units provides an effective collecting area (at the zenith) of about 16000 m^2, giving excellent sensitivity: 2 mJy RMS noise after 6 minutes. This makes the sampled synthesis mode of operation attractive for programs where detection is the principal objective, since a few short scans distributed over a 12 hour period will produce a beam of good resolution and sensitivity.

The difficulty - common to all E-W arrays - is the confusion produced by strong sources outside the field of view. By way of example, consider fig. 1, which contrasts 2 views of the same region: 1(a) is a 12 hour full synthesis, 1(b) is a sampled synthesis map. The problem is obvious. The cure is less so.

2. Diagnosis.

At least 2 remedies are available to a conventional E-W synthesis telescope:
- one could simply synthesize a larger field of view, and proceed to CLEAN the entire map;
- one could make a series of small maps, each approximately centred on a confusing source. The source's position would then be determined, and its influence removed from the U-V data stream. The final map would be made with the modified data stream.

The data processing scheme adopted for the MOST precludes both these approaches. A beam-forming matrix combines the outputs from the 88 receivers to produce 64 fan beams spaced approximately at the Nyquist interval. (This scheme is a hardware equivalent of Kenderdine's (1974) software solution for the Cambridge 5-km telescope). This scheme produces excellent 12 hour synthesis maps with high rejection of confusion by sources outside the field of view. However, for snapshots, there is a difficulty: a confusing source will appear in only a few scans. This means that only sources within the field of view will have the complete synthesized beam shape; confusing sources will appear as ridge lines crossing the entire map. It also means that the position of the confusing sources cannot be determined from the map.

In a conventional synthesis system we could locate the confusing sources with the full telescope resolution; but is this necessary? The manifestation of the confusing source - the extended ridge - is itself a beam, albeit somewhat elongated, with dimensions of order 1 degree by 22 arcsec. It should therefore be adequate to determine the position of the confusing sources to a comparable accuracy, in order to deal with them.

3. The proposed procedure.

The analysis consists of 3 steps:

(a). A map is produced from each 6 minute scan. This map has a resolution of 1 degree along the direction of the mean fan beam orientation, and 22 arcsec in the orthogonal direction. The synthesized beam in this frame is shown in fig. 2. (Note that in this figure the scale of the Y-axis has been compressed relative to the X-axis by a factor of 180, so that the beam appears roughly circular).

(b). This map is CLEANed. However, at the stage of restoring components to the residual map, we retain only those components which will lie within the final field of view. The point to note here is that with a reasonable signal to noise ratio we can expect to locate sources to within a fraction of a beamwidth. Thus there will be a relatively small area of ambiguity at the map border; for safety, one could include all sources within one standard error of the map boundary. (One might argue that a fully automatic program is too ambitious here, and that a small degree of supervision by the observer would help in resolving ambiguous cases). The depth of CLEANing is related to the number (N) of scans contributing to the final map. If S is the flux density of the weakest source we expect to detect in the final map, then we need to CLEAN to a level of NxS in each map.

(c). Only the central region of the map is retained for subsequent inclusion in the final map.

This scheme has been implemented at Sydney, and appears to work well. A reduction in the confusion "noise" of about one order of magnitude seems achievable. A further benefit of the scheme is worth noting: the computing load is reduced. Rather than attempting to synthesize the entire primary beam at full resolution, we restrict our efforts to a small number of maps, each quite modest in size. In essence, we match the CLEAN algorithm to the specific requirements of the problem.

4. References.

Kenderdine, S. (1974) Astr. and Ast. Suppl. Ser. 15, p 413
Mills, B.Y. (1981) Proc. ASA 4(2)

Figure 1a. A 12 hour synthesis of a 10 arcmin field in the SMC. The peak flux density is 40 mJy, and the contours are at 20, 40, 60 and 80% of the maximum.

Figure 1b. A 5 scan snapshot of the same field as in Figure 1a. The peak flux density is 54 mJy, and the contours are at -20, 20, 40, 60 and 80% of the maximum. (The maximum is now at the south of the map.)

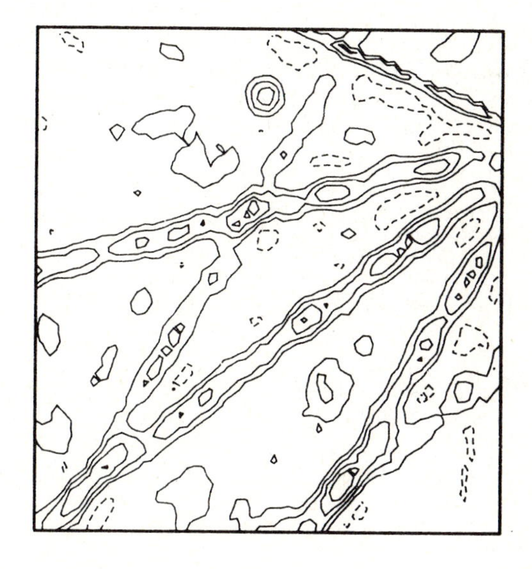

Figure 2. Synthesised beam for a 10 minute scan.

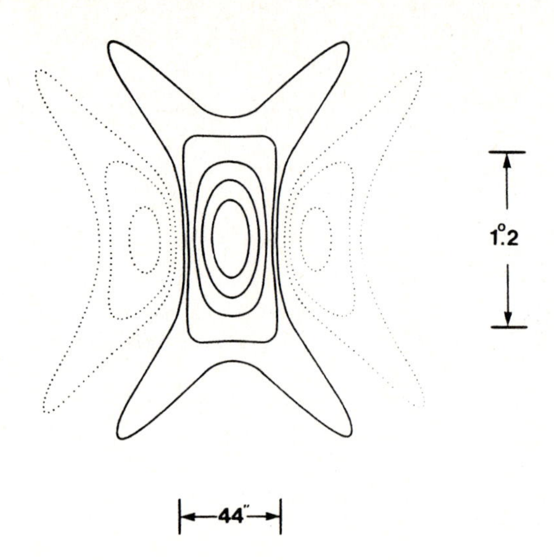

DISCUSSION

D.F. CRAWFORD

Procedures that implement the location of sources outside the field of view are in existence in Sydney. They work and can be used to 'clean' the raw data before synthesis.

R.N. MANCHESTER

Is it better to make two longer scans separated by time such that the two beams are approximately orthogonal or a larger number of shorter scans?

M.J. KESTEVEN

Three scans seems the best compromise. The factors to reconcile are these:

1. The synthesised dirty beam in the final map: improves as N increases.

2. The area of ambiguity (whether or not a source will lie within the final field of view): decreases as the scan duration increases.

3. The number of confusing sources that need to be removed: decreases as the scan duration increases.

8

Processing – maximum entropy

8.1 THE MAXIMUM ENTROPY METHOD

S.F. Gull
Mullard Radio Astronomy Observatory, Cavendish Laboratory,
Madingley Road, Cambridge CB3 OHE, England.

J. Skilling
Department of Applied Mathematics and Theoretical Physics,
Silver Street, Cambridge CB3 9EW, England.

Abstract We review the use of the maximum entropy principle in image reconstruction. We emphasise that entropy maximization is the only consistent regularisation technique for images: there is no room for any other variational method. This "fundamentalist" approach nevertheless gives complete freedom to use initial information to improve reconstructions. We also show how the MEP can now be properly extended to polarised images and to multi-channel spectral maps.

"So, are you faced with problems you can barely understand?
Do you have to make decisions though the facts are not to hand?
Perhaps you'd like to win a game you don't know how to play?
Just apply your lack of knowledge in a systematic way!"

Anonymous.

INTRODUCTION

Maximum entropy is being increasingly used as a technique of image reconstruction using data of many different types. Straightforward linear problems involving Fourier transform data, optical or X-ray convolutions, (Gull & Daniell 1978; Frieden 1972, 1980; Frieden & Burke 1972; Frieden & Wells 1978; Skilling *et al.* 1979; Daniell & Gull 1980; Bryan & Skilling 1980; Fabian *et al.* 1980; Willingale 1981; Burch *et al.* 1983) tomographic data (Minerbo 1979; Kemp 1980), etc., are now being supplemented by nonlinear applications such as crystallography (Collins 1982; Wilkins *et al.* 1983) and "blind" deconvolution (Skilling & Gull 1983). To illustrate dramatically the wide range of problems to which maximum entropy is now the everyday answer, we reproduce several practical examples from forensic imaging, radio astronomy, plasma diagnostics, medical tomography and blind deconvolution. The extensive demonstrable success of maximum entropy over and above conventional techniques (Burch *et al.* 1983, Burch 1980) demands a clear understanding of the fundamental theory underlying the technique.

The problem of image reconstruction is fundamentally ill-posed: only a selection of data from an external object is observed,

before after
Maximum entropy deconvolution
(UK Home Office)

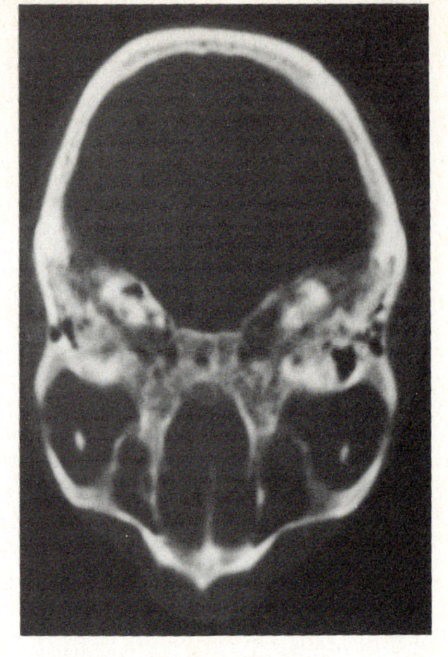

ME X-ray tomography
(skull in perspex, EMI Ltd)

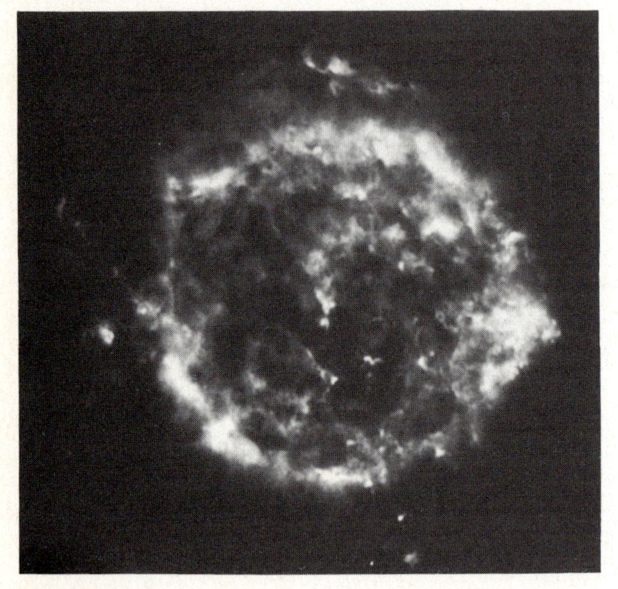

SNR Cas A at 5 GHz - 1024^2 ME image
(5-km telescope MRAO, Cambridge)

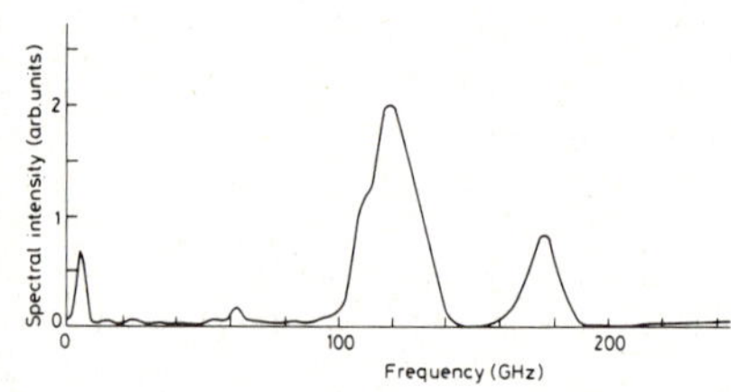

mm-wave Michelson interferometer spectrum of cyclotron emission from DITE tokamak
(Culham Laboratory)

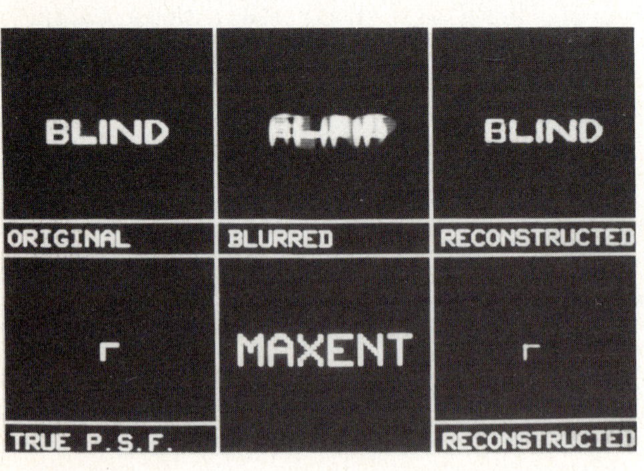

"Blind" deconvolution of unknown blurring.

(left) true image & blurring

(middle) data as given to ME program

(right) reconstructions
(T.J. Newton)

and even the data we do have are corrupted by noise. It follows that there are many different "feasible" images which are consistent with the data. This "feasible set" of images contains a degree of freedom for each independent unmeasured parameter and allows a range of values for each measured but noisy parameter.

For an image, this feasible set is far too large to comprehend; even if each image of the set were divided into finite pixels, there may be a million separate intensities to be determined. It is a matter of practical necessity to select a small number of images (ideally just one) from the feasible ones, with the open acknowledgement that we are making a deliberate selection.

The Maximum Entropy method (MEM) that we commend consists of choosing the single feasible image which has the greatest configuration entropy:

$$S = - \sum_i p_i \log (p_i/m_i),$$

where p_i = proportion of intensity originating in pixel "i" and m_i = corresponding measure or initial estimate.

THE PRINCIPLE OF MAXIMUM ENTROPY

Historically, the word "entropy" has been used for (at least) two different concepts. Clausius (1865) introduced the term as a state variable in thermodynamics, and Shannon (1948) used it as the (negative) information content of a probability distribution.

It is remarkable that there is any connection between these concepts (Jaynes 1983) and it is hardly surprising, therefore, that considerable confusion has resulted. For a full discussion of the origins of the word "entropy", the reader is referred to Jaynes (1980). The maximum entropy principle (MEP) relates to the informational entropy and is essentially due to Gibbs (1902), but its modern exposition and development is the work of Jaynes, whose pioneering papers are now available as a collection (Jaynes 1983). He suggested that, faced with the problem of determining probabilities given only partial constraints upon them, it is wise to employ that probability distribution for which this information entropy is maximal, whilst fitting the available constraints. In this way, we are maximally non-committal about unavailable data or constraints. Thus, given for example the average value of a quantity $\langle A\rangle = \sum_i p_i A_i$, we obtain by the method of Lagrange multipliers the solution $p_i = \frac{1}{Z(\lambda)} \exp(-\lambda A_i)$, where $Z(\lambda) = \sum_i \exp(-\lambda A_i)$ and the constraint is satisfied when λ obeys $\langle A\rangle = \frac{\partial}{\partial\lambda} \log Z(\lambda)$.

The principle described above is a powerful tool in thermodynamics and statistics, but its use in image reconstruction appears to require that we identify some quality of an image with a probability distribution. The obvious way to do this (Skilling & Gull 1983) is to remove the dimensionality of the image by concerning ourselves only with the dimensionless pattern of proportions of intensity $p_i = f_i / \sum_j f_j$, which then represents the configurational structure of the image, but not its overall intensity level. The connection with probability can then be made because $\{p_i\}$ are just the probabilities that any given photon originated from pixel "i". In view of the confusion that still exists, it is worth emphasising that this argument does not depend in any way on the real quantised nature of radiation as received, what is important is the location of sources in the external object. Interpreting our data as ensemble average constraints on this probability distribution, we then have the correct ingredients to use the MEP.

PROBABILITIES OR PROPORTIONS?

To some, this argument is intuitive and compelling, but it causes violent objections from a few (see the paper by Cornwell in this volume). The source of this objection is that, whilst all probability distributions are defined as proportions, the proportions of intensity in an image do not by themselves constitute a particularly convincing probability distribution. At first sight, one is apparently being asked to select "at random" a "small element of luminescence", a procedure which does indeed seem rather artificial - even if it could be done! However, the probability connection and the selection idea are quite unnecessary. They only appear because of our previous (non-Bayesian) indoctrination that probability distributions must necessarily be concerned with unpredictably changing physical variables. The idea of "entropy" applies equally to any set of proportions and so too does the principle of maximum entropy, without ever having to foist unnecessary interpretations upon them. Thus, the relative proportions of $\{p_i\}$ of letters of the alphabet used in a book have entropy: in fact that was almost the first example used to illustrate the information theoretic method! We do not have to imagine selecting a letter "at random" from the book. The proportions simply have entropy $S_I = -\sum_i p_i \log p_i$, and $\exp(S)$ has the interpretation of being the average number of letters used. Even (pro)portions of a cake have entropy in this sense. From this point of view the application of entropy to the pattern of proportions of intensity in an image seems very natural.

CONSISTENT REGULARISATION

How can we justify the claim made above that the MEP can be used directly to determine the proportions? We cannot appeal to the axioms of "information" - the word "information" seems inappropriate. Counting methods, based on teams of monkeys, convince some people, but we cannot justify the use of the MEP on the grounds that it gives more likely images unless the physical microstates are in fact made up in this way. For an image, they are certainly not.

Realising these difficulties, Shore and Johnson (1980) adopted a significantly different approach, and asked instead: "under what circumstances can a method of inference based on a variational principle be self-consistent?" They insisted merely that different ways of using the same information should yield consistent results. From axioms of transformation invariance and system and subset independence they show that the only consistent variational principles are those which maximise functionals (known as regularising functionals) of the form $S = -\sum p_i \log(p_i/m_i)$ or (trivially) monotonic functions of S. The following gives a physicists' perversion of the Johnson and Shore mathematical argument, but the reader is referred to the original article for a formal treatment.

THE KANGAROO PROBLEM

As Englishmen, we arrived in Australia in a state of ignorance concerning its marsupial wildlife. However, in the course of our travels we concluded the following;

Information: (1) One third of kangaroos have blue eyes.
(2) One third of kangaroos are left-handed.

Question: On the basis of this information alone, estimate the proportion of kangaroos that are both blue-eyed and left-handed.

The joint proportions of left-handedness and incidence of blue eyes can be represented as a little 2 by 2 contingency table, and we show three possible solutions to the problem; the independent case and those cases which display the maximum amount of positive and negative correlation.

Left-handed

Blue eyes	T	F
T	1/9	2/9
F	2/9	4/9

Uncorrelated

	T	F
T	1/3	0
F	0	2/3

Positive correlation

	T	F
T	0	1/3
F	1/3	1/3

Negative correlation

Suppose, though, that we have to choose only one answer - which is the best? Clearly, the answer we select cannot be thought of as being more likely than any other choice, because there may well be, and presumably is, some (small?) degree of genetic correlation between eye colour and handedness. But, to select either positive or negative correlations without having any relevant data is certainly nonsensical. The independent choice p(blue and left) = 1/9 is not any more likely, but is clearly to be preferred. It is this choice which is maximally noncommittal about unmeasured parameters. Of course, in this simple example, we do not have to make this choice; the problem is small enough to comprehend to full range of solutions. As pointed out earlier, for an image this is simply not possible, the megadimensional set of feasible images is far too large.

Suppose now, that we seek a solution by regularisation, maximising a functional $\sum_i F(p_i)$. The table shows the result:

Function	proportion (blue eyes and left-handed)			Correlation
$-\sum p_i \log p_i$	$1/9$	$\equiv$	.11111	Uncorrelated
$\sum p_i^2$	$1/12$	$\equiv$	.08333	Negative
$\sum \log p_i$	$\frac{\sqrt{17}-1}{24}$	$\equiv$	.13013	Positive
$\sum p_i^{1/2}$			.12176	Positive

With one notable exception, these results are very peculiar: the $\sum p_i^2$ form predicts, for example, that, although 1/3 of all kangaroos have blue eyes, only 1/4 of left-handed ones do. It yields similar predictions for any other quantity distributed in the same way, for instance, the handed-ness of kangaroos that drink Fosters ....

At this point, the reader may be forgiven for doubting the relevance of this example to astronomical imaging. Consider, though the following re-statement of the problem:

Information (Data) (1) One third of the intensity comes from the top half of (a region of) the sky.
(2) One third of the intensity comes from the left half of the sky

Question: What proportion of the flux comes from the top left quarter?

Suddenly, the contingency tables have become images, and the $\sum p_i^2$ form is now saying that, although 1/3 of the flux comes from the top half-sky, only 1/4 of the flux from the left half comes from the top. Without further data, it is difficult to imagine any sensible basis for this general decision. Once this is pointed out, we find it inconceivable that any rational user will continue to tolerate such behaviour in an image reconstruction technique!

When dealing purely with large images from a single type of complicated dataset, it is understandable that some image reconstructors have said "I can't see much difference between $f^{1/2}$ and flogf in practice" (Nityananda & Narayan 1983, Cornwell 1983) (though many practitioners would agree that $\sum \log f$ is worse). For this reason, it is easier to see what is going on in our simple example. All currently favoured functionals (excepting $-\sum p \log p$) fail to be consistent on even the simplest non-trivial image problem. One cannot suppose that they become more consistent for larger problems!

CONSISTENCY OF $-\sum p_i \log p_i/m_i$

Because it has been suggested that no "magical" functional exists for image regularisation (Högbom 1978, Nityananda & Narayan 1983) it is useful to point out explicitly where the "magic" property is hidden. Maximising $-\sum p_i \log p_i$ under constraint $\langle A\rangle$ yields, as before $p_i \propto \exp(-\lambda A_i)$. Under constraints $\langle A\rangle$ and $\langle B\rangle$ it yields

$$p_i \propto \exp(-\lambda A_i - \mu B_i) \equiv \exp(-\lambda A_i).\ \exp(-\mu B_i)$$

This is the magic property: when using Lagrange multipliers one *adds* the constraints linearly by maximising $S+\lambda\langle A\rangle+\mu\langle B\rangle$. When adding information one breaks the classes down into smaller sub-classes by *multiplication* of proportions. The exponential function does this for us, implying $\dfrac{\partial S}{\partial p_i} = \dfrac{\partial F}{\partial p_i} \propto \log p_i$ and $F \equiv -p_i(\log p_i - \text{constant})$. For a very much more careful discussion of the above points, the reader is again referred to Shore & Johnson (1980).

None of the above discussion should surprise devotees of the "information-theoretic" school. After all, by insisting that regularising techniques be self-consistent, we have merely retraced the steps (albeit in a different order) in the argument that led us to the proof of the uniqueness of $-\sum p_i \log p_i$ as an information measure! It does not matter then whether we justify the MEP on the basis of axioms of information, appeals to monkeys or as a consistent regularisation technique, all these arguments lead to the same conclusion. Equally, we must be clear as to what we achieve when we use the MEP in image reconstruction. The ME image is *no more likely* than any other feasible image; but with the available information we can make no other consistent choice.

PRIOR KNOWLEDGE *MUST* BE USED

Having concluded this rather lengthy sermon on the sins of using regularising techniques other than maximising the relative entropy $-\sum p_i \log(p_i/m_i)$, we now emphasize the enormous scope that this "fundamentalist" approach nonetheless gives us. The great power and flexibility of the formula arises from the appearance of the measure, or initial information $\{m_i\}$in the denominator. Without any data at all, the unconstrained MEM image defaults to $p_i \propto m_i$: for this reason we often refer to m_i as the "default level" for cell "i". By varying $\{m_i\}$ it is clearly possible to obtain any image that agrees with the data, in an infinite number of ways: so the formula is certainly flexible enough! But more importantly, the formula tells us how to encode initial information in a consistent manner. For example, if we observe a radio supernova remnant with background point sources we can use an initial flat $\{m_i\}$ to determine the positions and fluxes of the point sources and will obtain a picture of the SNR that is slightly corrupted by unavoidable sidelobes of the background sources. The MEP is surprised to see these sources, but the astronomer is not; armed with this knowledge, we can then use ($\{m_i\} \equiv$ flat + sources) to obtain a far superior image. We now make everyday use of $\{m_i\}$ in this way, to reduce grating responses for undersampled data, for VLBI to give initial phase estimates, etc. A particularly powerful example of "default level" manipulation has been given by Horne (1982) who reconstructed images of accretion discs in binary stars from eclipse data; with an $\{m(r,\theta)\} = \int_0^{2\pi} p(r,\theta)d\theta$ giving the most nearly circularly symmetric image consistent with his data. Another common form is to let the m_i depend on the neighbouring pixels: $m_i \equiv (p_{i+i}\ p_{i-i})^{\frac{1}{2}}$ in one-dimension.

These examples, showing that useful information can be encoded in a data-adaptive manner $\underline{m}(\underline{p})$ provide a glimpse of the present frontier of maximum entropy research. The message, though, is already clear: The MEP does not restrict your freedom: "Anything goes!"

We do however, advise caution: the MEP enables initial information to be used, but it is not fussy. The MEP does not by itself ask whether the information represents real knowledge or merely the guess of an idiot; in the latter case the results will probably be worse than if $\{m_i\}$ was not used at all.

MAXIMUM ENTROPY POLARISATION IMAGES

The application of the MEP to polarization data (Ponsonby 1977, Nityananda & Narayan 1983) requires considerable care in order to define properly the quantities that we wish to reconstruct. For total intensity maps this is the pattern of flux density proportions (emitters) on the sky: what is the similar quantity for polarised emitters? Partly polarised emission is characterized by the four Stokes parameter I, Q, U, U per pixel. In the quantum mechanical circular representation

$$\left|\begin{matrix}1\\0\end{matrix}\right\rangle = \text{L.H.} \qquad \left|\begin{matrix}0\\1\end{matrix}\right\rangle = \text{R.H.} ,$$

we have

$$I = \begin{pmatrix} 1 & 0 \\ 0 & 1 \end{pmatrix} = \text{total intensity}, \qquad V = \begin{pmatrix} 1 & 0 \\ 0 & -1 \end{pmatrix} = \text{net left-hand},$$

$$Q = \begin{pmatrix} 0 & 1 \\ 1 & 0 \end{pmatrix} = \text{net x-linear}, \qquad U = \begin{pmatrix} 0 & i \\ -i & 0 \end{pmatrix} = \text{net quadrature}.$$

Observations of I, Q, U, V fix the density matrix

$$\rho = \begin{pmatrix} I+V & Q-iU \\ Q+iU & I-V \end{pmatrix} /2I$$

This density matrix is the quantum mechanical generalisation of a probability distribution function, satisfying $\mathrm{Tr}(\rho) = 1$ and having entropy $S = \mathrm{Tr}(-\rho \log \rho)$.

Only in exceptional cases will the polarisation state be pure Q or U or V. More usually, it will be some partially polarised combination, best expressed in terms of the diagonal representation of the density matrix. This diagonal representation is

$$\begin{pmatrix} (1+\alpha)/2 & 0 \\ 0 & (1-\alpha)/2 \end{pmatrix}$$

where $\alpha = (Q^2 + U^2 + V^2)^{\frac{1}{2}}/I$

is the overall degree of polarisation, ranging from 0 to 1. The eigenvalues $(1 \pm \alpha)/2$ are the probabilities than an emitted photon would fall into either eigenstate. Correspondingly, the uncertainty in the polarisation state is given by the entropy

$$S^{pol} = \mathrm{Tr}(-\rho \log \rho) = -\frac{1+\alpha}{2} \log \frac{1+\alpha}{2} - \frac{1-\alpha}{2} \log \frac{1-\alpha}{2} .$$

Completely polarised emission ($\alpha = 1$) has $S^{pol} = 0$ because the polarisation state is fully predictable, whilst unpolarised radiation ($\alpha = 0$) has $S^{pol} = \log 2 = 1$ bit. With more than one pixel in the image, the density matrix diagonalises to

$$\rho = \mathrm{diag}\,(p_1(1+\alpha)_1),\ p_1(1-\alpha_1),\ p_2(1+\alpha_2),\ p_2(1-\alpha_2),\ \ldots)\ /2$$

where $p_j = I_j/\sum I$. The entropy

$$S = -\sum p_j(\log -_j + S_j^{pol})$$

measures the uncertainty in the answer to the question "What is the density matrix of the next photon to arrive?".

THE COLOUR PROBLEM - SPECTRAL INDEX MAPS

As our last illustration of the variety of questions which can be answered by the use of MEP, we turn to the problem of making images of the same object at two or more different frequencies. Suppose we have 2 data sets.

(1)	VLA 'D' array	:	Frequency 5.0 GHz	"Blue data"
(2)	5-km Telescope	:	Frequency 2.7 GHz	"Red data"

How do we combine these disparate observations into a useful map of spectral index? First let us mention what you must NOT do: make separate ME maps and take their ratio. The 2 different antenna arrangements give different sidelobe levels in different places and the result will of course be disastrous. However the "Red image" and "Blue image" made in this way do answer the question of "what does the source look like on the basis of the red/blue data alone". The answers are useful, telling us how much reliance can be placed on these data. But there is something else which we can loosely call a "colour" information channel. If we wish to make reliable colour images, we must display ones which have coloured features only if forced to have them by the data. We do this by "default level" manipulations, for example by setting:

$$m_{\substack{R\\B}}(\underline{x}) = p_{\substack{R\\B}} \cdot p(x)$$

where $$p_R = \sum_{\underline{x}} p_R(\underline{x}) \quad ; \quad p(\underline{x}) = \sum_{c=R,B} p_c(\underline{x}) \quad .$$

Thus, $\{p_R, p_B\}$ are the proportions of "red" or "blue" intensity independent of their position in the image, and $\{p(\underline{x})\}$ are the proportions in pixel $\underline{x}$, averaged over the colours.

In this way, by maximising $$-\sum_x \sum_{R,B} p_c(x) \log \frac{p_c(\underline{x})}{m_c(\underline{x})}$$

we obtain the "least coloured" image that is permitted by the data. This has the higher resolution of either dataset, yet shows spectral features only when there is definite evidence for them in the data.

The combination of "Red image", "Blue image", "Colour image" must all be displayed, and all are relevant.

This argument can of course be generalised to multiple-channel maps. Its practical implementation is relatively straightforward given data which allow ME "Blue" and "Red" images to be calculated.

CONCLUDING REMARKS

The principle of maximum entropy is alive and well, is based firmly on information theory and is thriving as a method for image reconstruction in astronomy and elsewhere. We have emphasised the uniqueness of the MEP as an inference method, but also stressed that the principle provides a natural mechanism for the inclusion of initial information. The thorny problems of polarised images and spectral index maps have yielded to the MEP during the last few years. We can modestly hope that the next five years will show similar progress on some other problems of interest to this audience: depolarisation and rotation measure maps and the separation of intensity and absorption effects from multi-channel data.

Whilst these new developments will require more robust algorithms and programming tricks, we confidently predict that one of the foremost guiding principles of inference will still be that of entropy maximisation. We leave the last word to Ed. Jaynes.

"For the principles of logic are the same in every field,
And regardless of your circumstances you always know they yield
What your information indicates, and (whether good or bad)
The best predictions one could make, from the data that you had"

ACKNOWLEDGEMENTS

We thank Rodney Johnson and John Shore for elucidating their consistency argument to us, Jon Ables for providing the initial "snake oil", Andy Coward of BBC Bristol for our marsupial friend, and Ed. Jaynes for everything else.

REFERENCES

Bryan, R.K. & Skilling, J.,(1980). Deconvolution by maximum entropy as illustrated by application to the jet of M87. Mon. Not. R. astr. Soc., 191, 69-79.

Burch, S.F., (1980). Comparison of image generation methods. UKAEA Harwell report AERE-R 9671.

Burch, S.F. & Gull, S.F., (1983). Image restoration by a powerful maximum entropy method. Comp. Vision Graphics Image Processing, 23, 113-128.

Clausius, L., (1865). Memoir read at Philos. Soc. Zürich, April 24. Pogg. Ann., 125, 353.

Collins, D.M., (1982). Electron density images from imperfect data by iterative entropy maximisation. Nature, 298, 49-51.

Cornwell, T. VLA Technical report (in preparation).

Daniell, G.J. & Gull, S.F., (1980). Maximum entropy algorithm applied to image enhancement. IEE Proc. 127E, 170-172.

Fabian, A.C., Willingale, R., Pye, J.P., Murray, S.S. & Fabbiano, G., (1980). The X-ray structure and mass of the Cassiopeia A supernova remnant. Mon. Not. R. astr. Soc., 193, 175-188.

Frieden, B.R., (1972). Restoring with maximum likelihood and maximum entropy. J. Opt. Soc. Am. 62, 511-518.

Frieden, B.R., (1980). Statistical models for the image restoration problem. Com. Graphics Image Processing 12, 40-59.

Frieden, B.R. & Burke, J.J., (1972). Restoring with maximum entropy. II: Superresolution of photographs of diffraction-blurred impulses. J. Opt. Soc. Am. 62, 1202-1210.

Frieden, B.R. & Wells, D.C., (1978). Restoring with maximum entropy. III: Poisson sources and background. J. Opt. Soc. Am. 68, 93-103.

Gibbs, J.W., (1902). Elementary principles of Statistical Mechanics. Yale University Press. Reprinted in "The Collected Works of J. Willard Gibbs" Dover 1960.

Gull, S.F. & Daniell, G.J., (1978). Image reconstruction from incomplete and noisy data. Nature, 272, 686-690.

Högbom, J.A., (1978). The introduction of a priori knowledge in certain processing algorithms; in "Image formation from coherence functions in astronomy", Groningen (D. Reidel).

Horne, K.D., (1982). Eclipse mapping of accretion disks in cataclysmic binaries. Ph.D. Thesis, California Institute of Technology.

Kemp, M.C., (1980). Maximum entropy reconstructions in emission tomography. Medical Radionuclide Imaging, 1, 313-323.

Jaynes, E.T., (1980). The Minimum entropy production principle. Ann. Rev. Physical Chemistry. (also reprinted in Jaynes 1983).

Jaynes, E.T., (1983). Papers on probability, statistics and statistical physics., ed. R.D. Rosenkrantz. Synthese Library Vol. 158, D. Reidel.

Minerbo, G., (1979). MENT: A maximum entropy algorithm for reconstructing a source from projection data. Comp. Graphics Image Processing, 10, 48-68.

Nityananda, R. & Narayan, R., (1983). Maximum entropy image reconstruction - a practical noninformation theoretic approach. J. Astrophys. Astron., (in press).

Ponsonby, J.E.B., (1977). An entropy measure for partially polarised radiation. Mon. Not. R. astr. Soc., 163, 359-380.

Shannon, C.E., (1948). A mathematical theory of communication. Bell System Tech. J., 27, 379-423 and 623-656.

Shore, J.E. & Johnson, R.W., (1980). Axiomatic derivation of maximum entropy and the principle of minimum cross-entropy. IEEE Trans. IT-26, 26-37.

Skilling, J., Strong, A.W. & Bennett, K., (1979). Maximum entropy image processing in gamma-ray astronomy. Mon. Not. R. astr. Soc., 187, 145-152.

Skilling, J. & Gull, S.F., (1983). The entropy of an image. American Mathematical Society (in press).

Willingale, R., (1981). Use of the maximum entropy method in X-ray astronomy. Mon. Not. R. astr. Soc., 194, 359-364.

Wilkins, S.W., Varghese, J.N. & Lehmann, M.S., (1983). Statistical Geometry. I: A self-consistent approach to the crystallographic inversion problem based on information theory. Acta. Cryst. A39, 49-60.

8.2 MAXIMUM ENTROPY-FLEXIBILITY VERSUS FUNDAMENTALISM

Ramesh Narayan
Raman Research Institute, Bangalore 560080, India.

Rajaram Nityananda
Raman Research Institute, Bangalore 560080, India.

Abstract: Maximum Entropy (ME) image reconstruction has traditionally been viewed in information theoretic terms. We present an alternative interpretation which emphasises that the method essentially fits the data with a non-linear transform of a band-limited function. We show that the widely discussed functions, f(B) = lnB and -BlnB, are just two members of a continuous family of "entropy" functions having the key properties $d^2f/dB^2 < 0$ and $d^3f/dB^3 > 0$. The best choice of f in a particular application depends on the user's requirements and not on fundamental issues. We recognise the sensitivity of the reconstruction to the DC level of the map and use this in a flexible numerical scheme permitting control over the resolution of the restoration and sensitivity to noise in the data. Situations where standard ME fails are easily understood and corrected with our interpretation. For instance , we find that ME should be "inverted" for absorption maps and that the band-limited function should be chosen with care when the data are awkwardly sampled. We generalise ME with an arbitrary f to polarisation maps in radio astronomy and present numerical simulations.

INTRODUCTION

Maximum Entropy (ME) image reconstruction (Burg 1967; Ables 1972; Gull & Daniell 1978) has not come into widespread use in astronomy, perhaps because of the controversy over the form of "entropy" to be used. Two schools of thought, backing the lnB and -BlnB entropies where B(x,y) is the "brightness" at the map point (x,y) have strongly defended their particular choice through information theory, statistical thermodynamics and combinatorial reasoning. Practical discussions of the properties of ME reconstructions have been rare and this, in our view, is a serious gap.

In this paper we focus on the maximisation conditions that the ME reconstruction satisfies. In what follows, the term "entropy" function does not have any thermodynamic or information theoretic connotation but just means the function f(B) whose integral is being maximised. By a simple geometrical argument we show that both the lnB and -BlnB entropy functions can be regarded as members of a family of functions which implicitly make a similar a priori assumption regarding the map viz., that it consists of isolated sharp features separated by flat extended regions. Where this assumption is valid all "entropies" (including novel forms like $B^{\frac{1}{2}}$, $-1/B$ and $-\exp(-\alpha B)$) produce "good" reconstructions, while where it is wrong (we give examples) none of the forms work. Further, we can control the resolution of the reconstruction and prescribe modifications of the basic ME scheme in some cases where it fails. Finally, we present a natural generalisation of ME with arbitrary f to polarised brightness maps and show, with examples, that all our results continue to be valid for this case.

NON-LINEAR TRANSFORM OF A BAND-LIMITED FUNCTION

Consider the problem of estimating the sky brightness distribution $B(x,y)$, $0 \leq x,y \leq 1$, given a partial set of its Fourier coefficients $\rho_{mn}, m,n \in K$, where K denotes the "known" set. The principal solution $B_p(x,y)$ is obtained by setting the unknown coefficients to zero.

$$B_p(x,y) = \sum_{m,n \in K} \rho_{mn} \exp[2\pi i(mx+ny)] \quad (1)$$

It is known that the "band-limited" function $B_p(x,y)$ is unsuitable for astronomical purposes because of the excessive termination ripple which swamps weak features in the map. The ME method estimates the unknown Fourier coefficients by maximising the integral (over the map) of a suitable function f(B) subject to the measurements as constraints, i.e.,

$$\text{maximise } E\left[B(x,y)\right] = \iint f(B)\, dxdy \quad (2)$$

It is easy to show (e.g., Nityananda & Narayan 1982, NN) that this variational scheme leads to

$$f'\left[B(x,y)\right] = \sum_{m,n \in K} \sigma_{mn} \exp[2\pi i(mx+ny)] \quad (3)$$

where f' is the derivative of the function f. In other words, f'(B) is a band-limited function whose Fourier coefficients σ_{mn}, $mn \in K$, which can be viewed as Lagrange multipliers, have to be determined to fit the data. Denoting the function inverse to f' by g, we thus have

$$B(x,y) = g\left[f'(B)\right] = g\left[\text{band-limited function}\right] \quad (4)$$

If the transformation $g(\equiv f'^{-1})$ is linear in its argument, then we clearly get back the principal solution $B_p(x,y)$. Thus the virtues of ME arise from the non-linearity of f' or g.

Fig. 1 is a schematic representation of the non-linear transformation f'(B) as a function of B. Since E(B) has a unique maximum only when $f'' < 0$

(NN 1982) we plot $f'(B)$ with a negative slope. Further, the non-linearity of the transformation implies a non-zero f''' and we choose a positive sign for f'''; this choice is crucial. In the ME method the function $f'[B(x)]$ is band-limited (we consider a one-dimensional "map" $B(x)$ though the argument is quite general) and thus typically has rounded peaks and excessive ripple as in Fig. 1. For the particular transformation with $f''' > 0$, the reconstructed $B(x) \equiv g\,[f'(B)]$ will clearly have reduced ripple and sharpened peaks. Thus the choice of an "entropy" $f(B)$ with $f''' > 0$ is equivalent to the *a priori* assumption of a map with sharp features and a flat baseline. This particular assumption is very natural and valid in most applications in radio astronomy aperture synthesis and spectral line analysis (however see the next section) and is also the basis of the CLEAN algorithm. Both the "entropy" functions $\ln B$ and $-B\ln B$ have $f''' > 0$ for positive B. However, they are but two special cases of a family of functions $f(B)$ having

$$f''(B) \propto -B^{-n} \quad , \quad n \geq 1 \tag{5}$$

all of which (i) enforce positivity of B since $|f'(B)| \to \infty$ as $B \to 0$ (Högbom 1978), (ii) have $f'' < 0$ (for uniqueness) and (iii) have $f''' > 0$ (to introduce the right *a priori* input). Non-standard members of this family such as $f(B) = B^{\frac{1}{2}}$ and $-1/B$ (having n=3/2 and 3) should also produce good reconstructions. We have verified this elsewhere (NN 1982; see also the section on polarisation). The "entropy" $B^{\frac{1}{2}}$ has n midway between those of $\ln B$ and $-B\ln B$ and could be a bridge between the warring camps! The argument of Fig. 1 shows that $f''' < 0$ leads to reconstructions with a flat top and pointed troughs (next section).

Since a linear transformation f' produces the principal solution, the degree of improvement that ME produces depends on the degree of non-linearity. For the family of entropy functions defined by (5) a simple measure of non-linearity is

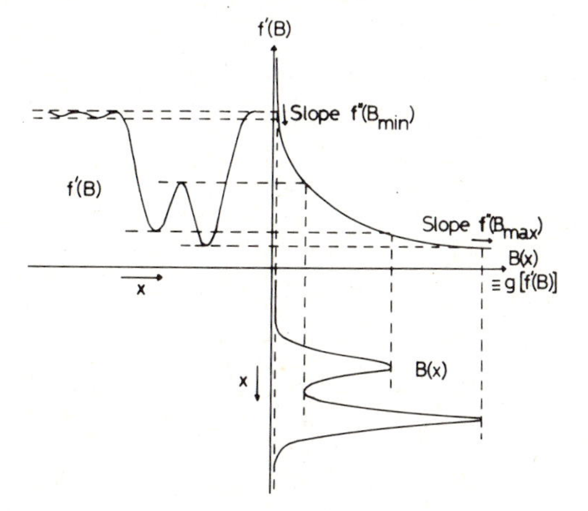

Fig. 1. The ME map $B(x)$ is a non-linear transform g of a band-limited function $f'(B)$ (eqn 4). The figure illustrates how a suitable choice of g can flatten the baseline and sharpen the peaks of B relative to those of $f'(B)$. The ratio R of the slopes of f' at the highest and lowest levels in the map measures the non-linearity of the transformation.

$$R = f''(B_{min})/f''(B_{max}) = (B_{max}/B_{min})^n \qquad (6)$$

where B_{min} and B_{max} are the minimum and maximum brightness values in the map. This explains a peculiarity of ME that was noted by Bhandari (1978). When the map has a high DC background, $(B_{max}/B_{min}) \sim 1$ and $R \sim 1$. Hence we essentially have a linear transformation and obtain only the principal solution. As the DC is reduced R increases and ME progressively modifies the principal solution to more acceptable maps. However, when the DC is very low R tends to very high values and the extreme non-linearity may sometimes produce unnatural features like peak-splitting (Bhandari 1978; Komesaroff, Narayan & Nityananda 1981, KNN). Some of these effects also depend on the "entropy" function and the dimensionality of the map (NN 1982).

Thus there is an intermediate range of R (say from 10 to 10^3) where ME produces good reconstructions without spurious features. It is important to work within this range. In our numerical implementation we add a suitable constant C to all brightness values in the map to attain the required value of R. By this technique which we call "float" we essentially operate with the entropy function $f(B + C)$. We feel that "floating" is an important and necessary aspect of ME. If instead one chose to regard the zero Fourier coefficient as a piece of data that should not be modified, aperture synthesis reconstructions would depend on such incidental parameters as the intensity of the cosmic microwave background and spectral reconstructions would depend on the intensity of the continuum background or the system temperature.

In our experience R=100 is an optimal choice. At a given value of R, to first order, reconstructions by different "entropy" functions are essentially similar. As R increases the baseline becomes flatter and the peaks sharper (as can be deduced from Fig. 1). At a given R, low values of n give flatter baselines and blunter peaks while higher n reconstructions have sharper peaks and more ripple. This second order effect can also be understood from Fig. 1 by considering the detailed variation of $f'(B)$ for different values of n (NN 1982).

MAPS IN ABSORPTION

Suppose we are interested in a map having a uniform constant background with superposed absorption features. Cases are known in aperture synthesis (e.g. low frequency observations of HII regions) and spectral analysis (e.g. 21 cm absorption). From the earlier discussion it is clear that the standard ME would be particularly bad since it is built to reconstruct a flat baseline and sharpened peaks. Fig. 2b shows an example where the $-B\ln B$ reconstruction at R=25 of the model absorption map of Fig. 2a is worse than the principal solution. As R increases the reconstruction becomes even worse.

Fig. 2. (a) A model map with absorption features seen against a background of height 1000. (b) A ME reconstruction using data with $|u|$, $|v| \leq 3$. (3) A modified ME reconstruction in which the sign of B is changed and its origin shifted to give R = 100. Note the improvement over (b). The broken lines represent the contours above 1000 in the sequence 1010, 1030, 1050, 1100, 1200. The full lines represent contours below 1000 in the sequence 990, 970, 950, 900, 800, 700, 500, 300, 0.

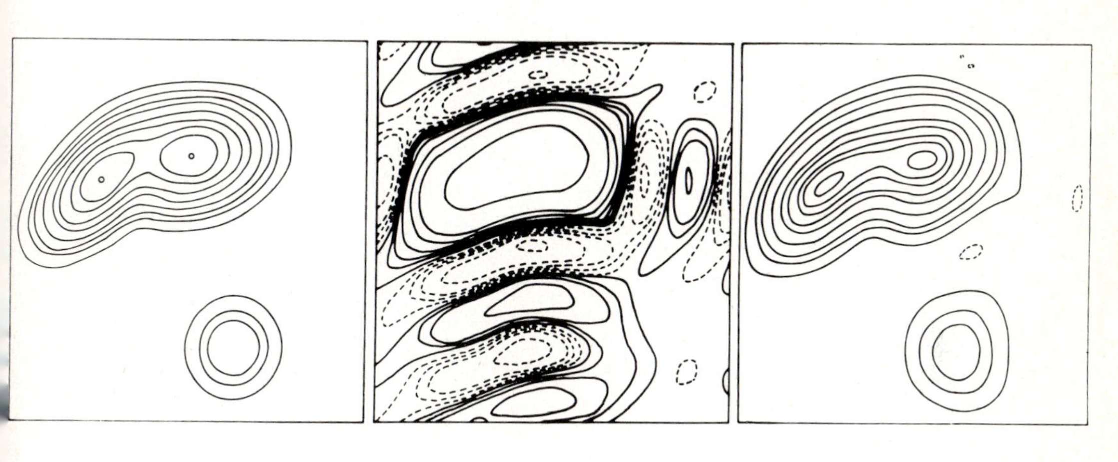

a b c

When one wants a flat top and sharpened troughs the sign of f''' should be negative. This is easily arranged by using the entropy function f(C-B) where C is a suitable "floating" constant whose value is adjusted to produce the required non-linearity:

$$R = (C - B_{min})^n / (C - B_{max})^n \qquad (7)$$

Fig. 2c shows the reconstruction when $\int \ln (C-B)$ is maximised at R = 100. Similar excellent results are obtained by maximising $-\int (C-B) \ln (C-B)$ or $\int (C-B)^{\frac{1}{2}}$ or any of the other "entropy" functions of eqn(5). Our experience with this and other cases, where the entropy function has to be modified to get good results, suggests that the ME principle may not be fundamental or unique in some deep philosophical sense. Statements that ME reconstructions are maximally noncommittal, are as featureless as the data allows, etc., are frequently made. The absorption map discussed here adds to the list of counterexamples (Bhandari 1978; KNN 1981; NN 1982).

NON COMPACT DATA COVERAGE

Another difficult case is when the data coverage in the uv plane is not compact but has holes or is in sectors. Now in addition to extrapolating the Fourier information outwards from a central measured region the ME method also has to interpolate in regions close to the origin. The principal solution has very high amplitude sidelobes. From the argument of Fig. 1 we see that the degree of non-linearity needed to flatten the baseline is much higher. For instance, even at R = 100 the -BlnB entropy reconstruction from data in sectors (Fig. 3a) is rather bad (other entropy functions produce similar results). One remedy would be to increase R to very large values (NN 1982). But the computation time required for convergence to the ME solution goes up rapidly at large R. Also, in one-dimensional problems it is known that as $R \rightarrow \infty$, the map tends towards a set of δ-functions (KNN 1981), the reason for the peak-splitting known in spectral analysis (Bhandari 1978). In two and higher dimensions, the $R \rightarrow \infty$ map depends on the entropy function that one chooses. Preliminary results suggest that -BlnB is stable in this limit (private communication from Dr. J.Skilling) but more studies are needed. On the other hand lnB certainly seems to be bad at large R (see also Appendix B of NN 1982).

An alternative, admittedly ad hoc, approach could be the following. Eqn (4) shows that B(x,y) is the non-linear transform g of a band-limited function whose band extends over the same region of uv space as the measured data. If one ignores for the moment that this result

Fig. 3. (a) ME reconstruction with f=-BlnB, R=100. The original model is essentially the negative of Fig. 2a, having emission features on a zero background. The data occupy a sector in the uv plane as shown at top centre. The full lines represent the contour levels 10, 30, 50, 100, 200, 300, 500, 700, 1000. The dashed contours correspond to the negative values -10, -30, -50, etc. Note the heavy ripple remaining in the ME map because of the non-compact uv coverage. (b) Reconstruction from the same data as in (a), fitting them to a non-linear transform (exponential) of a band-limited function with the compact uv coverage shown at bottom centre. Note the significant improvement over (a) because of the more physical model being fitted to the data.

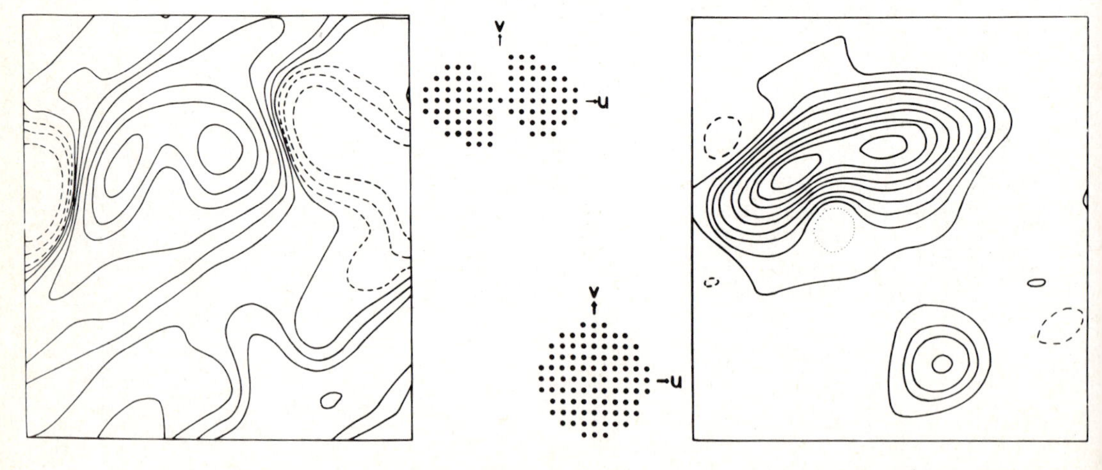

follows from maximising eqn (2) one could view it as a form of model-fitting with adjustable parameters σ_{mn}. It then appears unnecessary to take an inherently bad uv coverage for the σ_{mn} just because the data are awkwardly sampled. One could instead take a more compact coverage so that f'(B) would have weaker sidelobes and a moderate R would be sufficient to produce an acceptable map. Fig. 3b shows a sample calculation with a more reasonable band shape at R=100 with B = exp (band-limited function), which corresponds to -BlnB "entropy". There is significant improvement over Fig.3a. Note that we still fit all the data. The parameters σ_{mn} in this case were refined to fit the measured ρ_{mn} best in a least squares sense. Being non-linear least squares the solution is not unique and it is necessary to start with a map not too far from the true solution.

The point here and in the previous section is that once we look at the equations which define the reconstruction we often get new insight in problem areas, not provided by discussions of the foundations of ME. The problems can often be solved with a little flexibility in approach.

NOISY DATA

With noisy data a least squares form of ME (Ables 1972) has gained wide acceptance. The data are not fitted exactly but the ρ_{mn} in the model are permitted to deviate by an amount determined by the known noise level. One is essentially trying to solve for the noise. Bryan and Skilling (1980) noted that the residuals obtained do not have a gaussian distribution. We have shown (NN 1982) that the residuals are to first order uncorrelated with the actual noise present in the data but depend in a completely predictable way on features in the map. Thus, the refined ρ_{mn} at the measured locations have rms deviations from the <u>true</u> values $\sqrt{2}$ times greater than those of the original data. In view of this increased deviation from the truth, the usefulness of least squares ME is debatable. We feel that any modification of the data must be attempted only if one expects to come closer to the true values. We prefer to fit the data exactly and "float" suitably to get the resolution appropriate to the noise level.

MAXIMUM ENTROPY FOR POLARISED MAPS

Ponsonby (1973) extended ME with the lnB form of entropy to polarised brightness distributions and suggested maximising

$$E = \iint \ln \left[I^2(x,y) - Q^2(x,y) - U^2(x,y) - V^2(x,y) \right] dxdy \qquad (8)$$

where I,Q, U and V are the usual Stokes parameters. We have generalised this expression for an arbitrary form f(B) of the entropy function (Nityananda & Narayan 1983, NN) and find that one should maximise

$$E = \iint \mathrm{Trace} \left[f \{ \underline{\underline{B}}(x,y) \} \right] dxdy \qquad (9)$$

where $\underline{\underline{B}}(x,y)$ is the correlation matrix given by

$$2\underline{\underline{B}} = \begin{bmatrix} I + Q & U + iV \\ U - iV & I - Q \end{bmatrix} \qquad (10)$$

In the diagonal representation of $\underline{\underline{B}}$, the diagonal elements of the matrix $f(\underline{\underline{B}})$ are given by $f(B_{ii})$. Physically one interprets eqn. (9) by uniquely decomposing the radiation at each point into two orthogonally polarised, mutually incoherent components whose intensities are given by the eigenvalues λ_1 and λ_2 of $\underline{\underline{B}}$. The "entropy" at each point (x,y) is taken to be $f(\lambda_1) + f(\lambda_2)$ and the integral over the map is maximised. The numerical algorithms used for scalar brightness maps can be easily generalised to the polarisation case by replacing operations on B(x,y), ρ_{mn}, σ_{mn}, etc. by equivalent matrix operations on $\underline{\underline{B}}(x,y)$, $\underline{\underline{\rho}}_{mn}$, $\underline{\underline{\sigma}}_{mn}$, etc. Again the solution is unique if $f'' < 0$ (NN 1983).

Fig. 4 shows a ME reconstruction of a model polarised map from limited Fourier information on its Stokes parameters. For convenience in displaying the polarisation vectors we have taken V to be zero though we have confirmed that the method works well even with non-zero V. We display the $B^{\frac{1}{2}}$ "entropy" reconstruction (R=100) to emphasise that our earlier results on the similarity of different forms of entropy continue to be valid. We have verified that the -BlnB and lnB reconstructions are equally good. The reconstructions from different entropies at a given value of R are to first order qualitatively similar. However, as in the scalar brightness case, the "entropy" -BlnB with n=1 produces the smallest ripple on the baseline while lnB with n=2 has more super-resolution (peak sharpening). There is remarkable improvement in the reconstructions of Fig. 4c over the principal solutions of Fig. 4b. It is notable that ME has considerably improved the Q and U maps though there is no direct positivity constraint. To our knowledge Ponsonby's (1973) entropy measure (8) has not been tested so far. Our studies show that eqn (8) as well as our generelisation eqn (9) work very well.

An interesting feature of eqn (9) is that if the correlation measurements were performed with rotated dipoles, corresponding to a rotation of the data in the QU plane, the reconstruction is similarly rotated. The usual procedure of cleaning Q and U maps separately would not satisfy such a rotational invariance requirement for a general rotation angle.

CONCLUDING REMARKS

Without needing to establish the foundations of ME with great rigour, we understand the simple 'mechanics' of the way it works - non-linear transform of a band-limited function - and why and where it fails. Our work is an extension of the ideas of Högbom (1978) and Subrahmanya (1978) who atribute the successes of ME to the *a priori* information built in, particularly the penalty against negative values. Even if the search for the perfect entropy fails, we feel the method will still have a role to play in image reconstruction problems whereever the *a priori* information is appropriate.

Fig. 4. (a) Model of a source with partial linear polarisation. The three maps show the Stokes parameters I,Q and U. The I map has straight line segments showing the direction and degree of linear polarisation. The contour levels are the same as in Fig. 3. (b) The dirty maps (I,Q & U) obtained by restricting the data to $|u|$, $|v| \leq 3$. (c) Result of a ME restoration with $f = \underline{\underline{B}}^{\frac{1}{2}}$, R = 100. Note the suppression of spurious features in all the Stokes parameters.

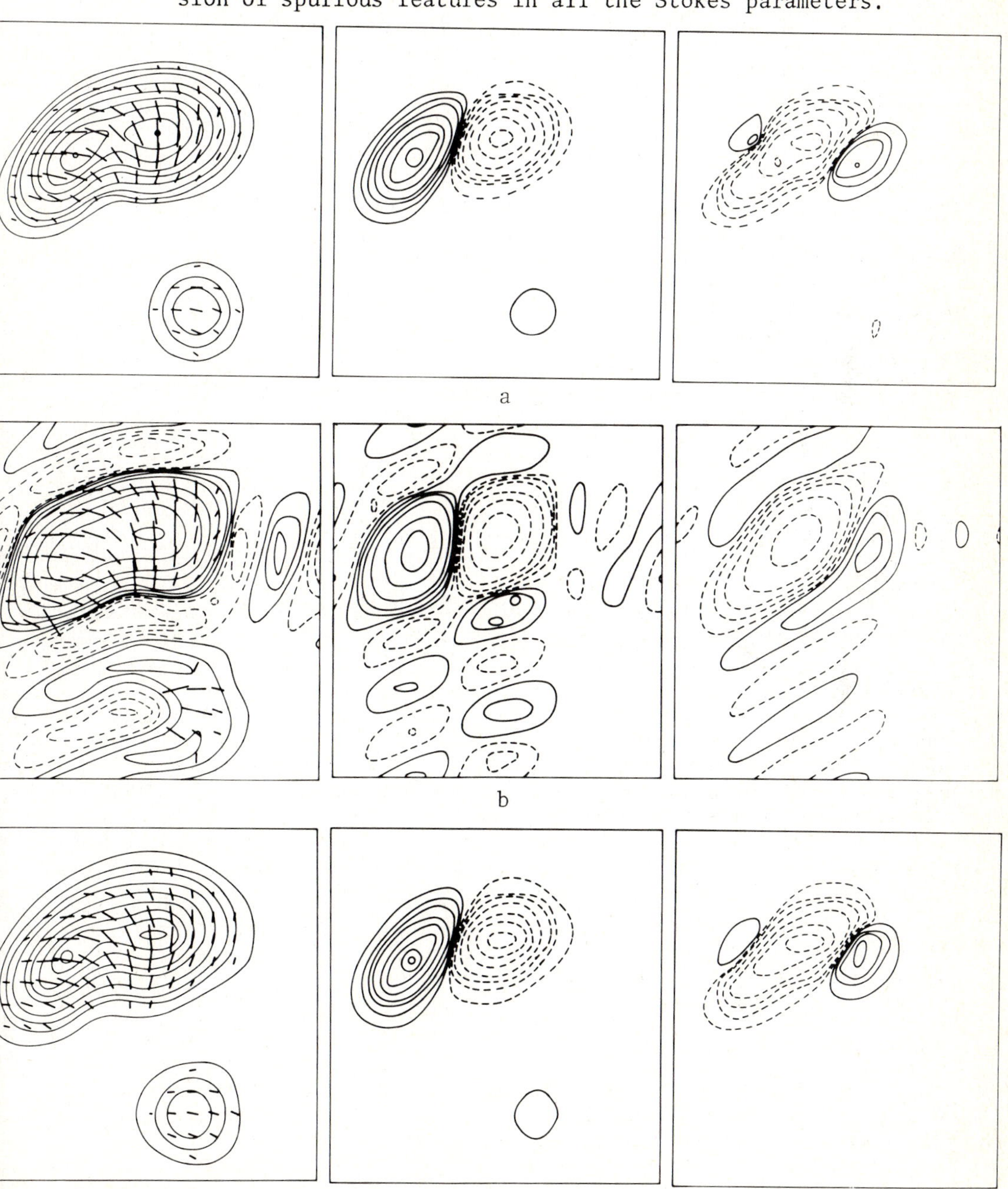

REFERENCES

Ables, J.G. (1972). Symposium on the Collection and Analysis of Astrophysical Data, published as Astron. Astrophys. Suppl., 15, 383 (1974).

Bryan, R.K. & Skilling, J. (1980). Monthly Notices Roy. astron. Soc., 191, 69.

Burg, J.P. (1967). Paper presented at the Annual International Society of Exploration Geophysicists Meeting, Oklahoma City. Reprinted in "Modern Spectrum Analysis", ed. D.G. Childers, IEEE Press (1978).

Gull, S.F, & Daniell, G.J. (1978). Nature, 272, 686.

Högbom, J.A (1978). In Image Formation from Coherence Functions in Astronomy, ed. C. van Schooneveld, p. 237, Reidel, Dordrecht, Holland.

Komesaroff, M.M., Narayan, R. & Nityananda, R. (1981). Astron. Astrophys., 93, 269 (KNN 1981).

Nityananda, R. & Narayan, R. (1982). J. Astrophys. Astron., 3, 419 (NN 1982)

Nityananda, R. & Narayan, R. (1983). Astron. Astrophys., 118, 194 (NN 1983).

Ponsonby, J.E.B. (1973). Monthly Notices Roy. astron. Soc., 163,369.

Subrahmanya, C.R. (1978). In Image Formation from Coherence Functions in Astronomy, ed. C. van Schooneveld, p. 287, Reidel, Dordrecht, Holland.

DISCUSSION

S.F. GULL

How do you test that you have achieved the unique maximum of entropy in your simulations?

R. NARAYAN

We have carried out maximum entropy refinement starting from different initial maps, say the principal solution and the true map, and confirmed that the refined maps are essentially the same. We believe all our maps have converged, and correspond to the value of the non-linearity parameter R that we specify.

T.W. COLE

The infinity of possible solutions amongst which we select one by an ad hoc process horrifies me. Each would redistribute the uncertainty due to noise and (e.g.) missing u-v samples differently. Should we not avoid problems by choosing one (or two?) and learning to live with their (eventually) well known properties?

R. NARAYAN

I disagree. I think there is a great deal to be learned by looking at several different reconstructions, all of which are consistent with the data and incorporate the same a priori information.

8.3 IS JAYNES' MAXIMUM ENTROPY PRINCIPLE APPLICABLE TO IMAGE CONSTRUCTION ?

T.J. Cornwell,
National Radio Astronomy Observatory, Socorro, N.M.

1 INTRODUCTION

The Maximum Entropy Principle due to Jaynes (1957a,b) has proved to be a very powerful and elegant solution to the problem of assigning a form to a probability distribution on the basis of incomplete information usually expressed as ensemble averages of functions. A rather complete description and justification of the MEP is given by Jaynes in a lengthy review article published in 1978 (ed. Tribus 1978). The attraction of the MEP is that it provides the least informative or maximally non-committal answer; (the reader is referred to Jaynes' article for replies to the number of objections that can be raised to such a statement - we will assume hereafter that the reader is familiar with the contents of that article.) Since ill-posed problems are abundant in physics the MEP has been applied to a wide range of different problems (see e.g. Tribus 1978, Smith 1983, Jaynes 1978). For this one must re-pose the ill-posed question so that it refers to a probability distribution and so that the constraints are in the form of averages of functions; having done this the MEP formalism is then easily applied (see the next section). In most cases the first part of this procedure is nearly always the most difficult conceptually since the object of interest may have no natural interpretation as a probability distribution.

Since image reconstruction nearly always constitutes an ill-posed problem considerable efforts have been made to utilise the MEP. As always, the major obstacle is the identification of one aspect of the problem with a probability distribution. In this paper we will consider a number of different approaches that various authors have made to the problem of image reconstruction. Firstly, we give a brief discussion of the practical use of the MEP.

2 THE MAXIMUM ENTROPY PRINCIPLE

Suppose that we wish to assign a form to a probability distribution $\{p_i | i=1,2,..N\}$ of a discrete variable x given constraints of the form :

$$\Sigma_i f_k(x_i).p_i = F_k \qquad k=1,2,...K$$

Given certain axioms about lack of predictive power it may be shown that the entropy $H(\{p_i\})$ of a given probability assignment is a unique measure of its lack of predictive power (Jaynes 1957a,b).

$$H(\{p_i\}) = - \Sigma_i \, p_i.\log(p_i)$$

The Maximum Entropy Principle due to Jaynes (1957a,b) asserts that given incomplete information we should choose that probability assignment which obeys the above constraints and has maximum entropy or minimum predictive power. By the use of Lagrange multipliers the MEP probability distribution may easily be calculated to be :

$$p_i = Z^{-1}.\exp(-\Sigma_k \, \lambda_k.f_k(x_i))$$

where the partition function is :

$$Z = \Sigma_i \, \exp(-\Sigma_k \, \lambda_k.f_k(x_i))$$

and the Lagrange multipliers can be found from the relation :

$$F_k = - (\partial/\partial\lambda_k) \log Z$$

Constraints which add nothing to our knowledge of the probability distribution will automatically be assigned a potential λ_k equal to zero.

As an example, we note that the normal distribution is the ME probability distribution for constraints of the first and second moments. Jaynes (1978) has provided a beautiful and simple application of the MEP to a more complicated problem : that of assigning a probability distribution to represent our knowledge of the properties of a die which was tossed 20,000 times by R.Wolf in the period 1850-1890. Jaynes showed that the frequency distribution for this set of random trials is inconsistent with an unbiased die. By making simple guesses about the physical causes of the bias he was able to reconcile the frequency data with that expected from a die having two imperfections : missing mass due to the placement of the dots and an elongation of one axis. (Technically, these imperfections can be introduced by two special functions f_2 and f_3 which measure the level of the relevant effect). Of course, we might expect other imperfections to be present, such as non-orthogonal axes, but the point is that the data do not force the introduction of such a possibility. This die example shows in a very clear way how the MEP can be used to introduce physics to solve problems parsimoniously.

3 THE MEP AND IMAGE RECONSTRUCTION

Image reconstruction is a classic ill-posed problem : we wish to reconstruct the "true" image f from a noisy and incomplete data set g where

$$g = h*f + n$$

h represents the point spread function, * represents convolution and n represents noise. Since h is, in general, singular or at least ill-conditioned there exist homogeneous solutions or ghosts z such that

$$h*z = 0 \qquad \text{or} \qquad h*z \sim n$$

Consequently there are usually many solutions to the measurement equation; further constraints are required to define a unique solution. In this rather general formulation the h may represent a point spread function due to either a filled aperture or an array of interferometers.

The main conceptual difficulty encountered when we try to apply the MEP to image reconstruction lies in identifying a probability distribution connected with a single image f. Of course, if we had access to a collection of images for which ensemble averages were known then we could apply the MEP to estimate the probability that we should assign to a given image. This p.d. could then be used with Bayes' theorem to find the a posteriori probability of a single image given some extra data (D'Addario 1976). Normally, however, this approach is not appropiate because either the ensemble data are not available or they cannot be reduced to a suitably simple form. Thus we must confine our considerations to a single image and we must identify a probability distribution connected with that image.

The first attempt is based upon the observation that the brightness distribution in an image f(x,y) predicts either the rate of emission of photons or the strengths of incoherent oscillators, depending upon the physical model preferred for the emission process. D'Addario (1976), among others, showed that the entropies for these two models are, neglecting some constants :

$$H_{waves} = \int_{sky} \log(f(x,y))\, dx.dy$$

$$H_{photons} = - \int_{sky} f'(x,y).\log(f'(x,y))\, dx.dy$$

where, in the latter, f'(x,y) is the brightness normalised by the total power. The Maximum Entropy images are found by maximising one of these quantities constrained by the observed data. In the discrete approximation where there are N pixels in the images

$$g_i = \Sigma_j\, h_{i,j}.f_j + n_i \;,\; i=1,2,\ldots N$$

then

$$\text{Waves} : f_i = (\Sigma_m\, \lambda_m.h_{i,m})^{-1}$$

$$\text{Photons} : f_i = \Gamma.\exp(-\Sigma_m\, \lambda_m.h_{i,m})$$

where, in the latter,

$$\Gamma = \exp(\, \Sigma_i\, (f_i/(\Sigma_j\, f_j)).\log(f_i))$$

and the Lagrange multipliers are chosen so that the constraints are obeyed.

The physical meaning of these forms is that they are the brightness distributions having the least predictive power concerning the electric fields emitted at the true object and the position of emission of the next photon respectively (see e.g. D'Addario 1976). However, rather than the details of the emission processes we are concerned with the image itself so, although these forms are ME solutions to a well-posed problem, they are of little use in solving our problem. Before moving on we should note one remaining ambiguity of these forms : that is, when is a given emission model appropiate ? Clearly, the division must depend upon experimental details such as bandwidth,aperture size and source brightness and so, consequently, neither of the two forms is to be preferred universally.

Skilling and Gull (1983a,b), in an attempt to justify general use of the photon measure, have asserted that the fundemental question defining the problem of image reconstruction is "Where would the next photon come from ?". They then go on to claim that the question of photon degeneracy is irrelevant since we are only interested in a single photon. However, in our view, it is meaningless to ask where the next photon will come from if the radiation is degenerate since we cannot then distinguish one photon from another. In other terms, since the minimum unit of time for distinguishing photon arrivals is the inverse bandwidth and, in degenerate radiation, many photons arrive in an inverse bandwidth, we have no way of measuring the arrival direction of the "next photon". Skilling and Gull (1983a,b) object to this argument on the grounds that it forces the ME image to depend upon "accidents of observation" such as bandwidth,aperture size,etc. On the contrary, it seems to us to reflect a fundemental physical distinction between classical and quantum descriptions. Finally, we assert that the MEP cannot be used to escape the necessity of asking physically meaningful questions.

In another justification of their use of the photon measure Skilling and Gull (1983a,b) note that since the set of proportions or normalised pixel brightnesses, $\{p_i = f_i/(\Sigma_j f_j)\}$, obeys the Kolmogorov axioms of probability it can be viewed as being a probability distribution, and so the Shannon/Jaynes form of entropy applies directly

$$H(\{f_i\}) = - \Sigma_i\, p_i . \log(p_i)$$

In other words, since the set of proportions is isomorphic to a probability distribution it must itself <u>be</u> the probability distribution we would attach to a set of events. This may be true but there seems no reason why this should be the case or, if so, that the set of events thus described should be meaningful for the discussion of image reconstruction. Gull (private communication, 1983) has suggested one appropiate set of events : suppose that the source is composed of a large number of <u>distinguishable</u> incoherent emitters and consider the problem of locating one such incoherent emitter; given no extra prior information we clearly must equate the probability of the incoherent emitter being in pixel k with the normalised brightness in that pixel p_k. The lack of predictive power in a brightness distribution concerning the location of an incoherent emitter is then uniquely measured by the above form of entropy. Hence, if one <u>defines</u> the central question of image reconstruction as "Where is a given incoherent emitter in the field ?" then the Skilling and Gull form is appropiate and useful. We should note that this whole argument rests upon the distinguishability of the incoherent emitters; if this does not apply then it is clearly nonsense to ask the position on any emitter. We can see no meaningful way of distinguishing emitters and therefore reject this argument.

In contrast to Skilling and Gull we view the most basic question in image reconstruction as "What is the true image ?". In the general spirit of subjective probabilities we can assign a degree of belief or probability to any possibility f as P(f). (We emphasise that this only reflects a degree of belief not an actual frequency.) The entropy implicit in our probability assignment is

$$H(P(f)) = - \int_{\text{all } f} P(f).\log(P(f)/Q(f))\, df$$

where Q(f) is an appropiate measure function representing any prior knowledge we have about the true image (see e.g. Jaynes 1968). To use the MEP to assign a form to P(f) we need constraints of the form

$$< F_k > = \int_{\text{all } f} F_k(f).P(f)\, df$$

The ME form for P(f) is then

$$P(f) = Q(f).Z^{-1}.\exp(-\Sigma_k\ \lambda_k.F_k(f))$$

where Z is the partition function

$$Z = \int_{\text{all } f} \exp(-\Sigma_k\ \lambda_k.F_k(f))\, df.$$

To introduce the measurements, while taking into consideration noise, we define, in the discrete approximation,

$$< F > = N.\sigma^2 \text{ where } F = \Sigma_{k,l}\ e_k.W_{k,l}.e_l.$$

where W^{-1} is the covariance matrix of the noise of which $N.\sigma^2$ is the expected value. The errors have the form

$$e_i = g_i - \Sigma_j\ h_{i,j}.f_j$$

We find

$$P(\{f_i\}) = Q(\{f\}).Z^{-1}.\exp(-\lambda.(\Sigma_{k,l}\ e_k.W_{k,l}.e_l))$$

where λ is a Lagrange multiplier chosen so that the constraint on the data is obeyed. Thus for a uniform prior (expressing complete ignorance), we will obtain a normal form for P({f}). This approach only yields a probability distribution expressing our degree of belief in an image; however, we can use all the tools of statistics to extract simpler information such as definite predictions in the form of the mean, median or mode. The last is nearly always easiest to calculate : choosing {f} to maximise P({f}), we find that the solution must obey

$$(\partial/\partial f_i)\ [\log(Q(\{f\})) - \lambda.\ \Sigma_{k,l}\ e_k.W_{k,l}.e_l] = 0$$

We believe that this is a more satisfactory solution to the image construction problem and should be used in place of the maximisation of the entropy of an image. Thus the entire problem can be reduced to finding a prior Q({f}) to represent our state of a priori knowledge about the true image. Given no extra data, for example from the class of images to which f is thought to belong, the MEP can provide no further help : we must use other methods to assign forms to the prior Q (see e.g. Jaynes 1968, 1978 for a discussion of the techniques available).

4 CONCLUSIONS

We have demonstrated that, by itself, the MEP is powerless to provide an answer to what we believe to be the central question in image reconstruction "What is the true image ?"; further information in the form of a prior is required.

5 ACKNOWLEDGEMENTS

I thank Steve Gull for some interesting discussions and Larry D'Addario for his criticism of a draft of this paper.

6 REFERENCES

D'Addario,L.R., (1976). "Maximum Entropy Imaging : theory and philosophy", Conf. Proc. on Image Analysis and evaluation, Toronto, 1976.

Jaynes,E.T., (1957). "Information theory and statistical mechanics", Phys. Rev., 106, 620, and 108, 171, 1957.

Jaynes,E.T., (1968). "Prior probabilities", IEEE SSC-4, 227-241.

Jaynes.E.T., (1978). "Where do we stand on Maximum Entropy ?", published in Tribus, (1978).

Skilling,J. and Gull,S.F., (1983). "Algorithms and applications", preprint.

Skilling,J. and Gull,S.F., (1983). "The entropy of an image", preprint

Smith,C.J., (1983). Proceedings of the first and second workshops on Maximum Entropy and Bayesian Methods in Applied Statistics, Laramie, 1981,1982, to be published 1983.

Tribus,M, (1978). editor "The maximum entropy formalism",MIT press.

8.4 MAXIMUM ENTROPY - SOME RESERVATIONS

J.D. O'Sullivan* and M.M. Komesaroff
Division of Radiophysics, CSIRO, Sydney, Australia

Abstract. The maximum entropy method (MEM) has been used in a number of problems in applied physics, including the problem of correcting spurious effects due to sidelobes in radio astronomy maps made using the aperture synthesis technique. It appears to have a firm mathematical foundation and to be related to well-understood methods used in classical statistical mechanics. Jaynes (1981, 1982) has given a systematic and logical account of MEM, indicating the various assumptions going into its derivation. One of these assumptions seems questionable, and the purpose of the present communication is briefly to review Jaynes's presentation, to indicate this questionable assumption, and to show that this not only poses philosophical problems, but that it is also related to shortcomings of MEM. A possible solution to some of the problems raised is proposed.

1 INTRODUCTION

In any problem in applied physics - for example, image reconstruction - the measured data, designated by

$$D \equiv \{d_1, d_2 \text{ --- } d_m\} \quad ,$$

are frequently insufficient by themselves to yield a unique solution. Jaynes (1981, 1982) points out that in addition to D we often have prior information about the problem and that this should be taken into account in determining a solution. He designates the prior information by I. (For example, in image reconstruction, one simple, but very important, piece of prior information is that the image intensity should nowhere be negative.)

2 FORMULATION OF THE MEM PROBLEM

2.1 General considerations

The Bayesian approach is central to Jaynes's treatment of the problem. His introduction to the Bayesian method is based on some rudimentary Boolean algebra. He denotes various propositions by A, B

---

*Formerly at Netherlands Foundation for Radioastronomy.

and C and the proposition 'A and B' by AB. He denotes the probability that AB is true given that C is true by $p(AB|C)$, and writes as an axiom

$$p(AB|C) = p(A|C) \cdot p(B|AC) \ .$$

On the basis of straightforward algebraic manipulation he applies this result to the problem of image reconstruction. He considers the problem of constructing an image, comprising n pixels, designating the image by

$$F \equiv \{f_1, f_2 \text{ --- } f_n\} \ , \tag{1}$$

where f_i is proportional to the intensity of pixel i, using data D measured with r.m.s. uncertainty σ, and also using the prior information I. He demonstrates that the posterior probability $p(F|D\sigma I)$, is related to the prior probability $p(F|I)$ by

$$p(F|D\sigma I) \propto p(F|I) \ p(D|F\sigma I) \tag{2}$$

and selects that F which maximizes the posterior probability as the solution to the problem.

2.2 The prior probability

It is the choice of the prior probability function for MEM which has led to the greatest degree of discussion.

Jaynes (1981, 1982) envisages N 'elements of luminance' distributed among the n pixels so that the number of elements of luminance in the first pixel is $N_1 \equiv Nf_1$, in the second $N_2 \equiv Nf_2$, and so on. The number of ways of forming a particular image F is W(F), given by

$$W(F) = \frac{N!}{(Nf_1)!(Nf_2)!\text{----}(Nf_n)!} \ . \tag{3}$$

If all the N_i are very large we can apply the approximation to Stirling's formula, and derive

$$\ln W(F) = -N \sum_{k=1}^{n} f_k \ln f_k \ . \tag{4}$$

Jaynes takes the prior probability to be proportional to W(F). Thus we may write

$$p(F|I) \propto W(F) = \exp - \left\{ N \sum_{k=1}^{n} f_k \ln f_k \right\} \ . \tag{5}$$

2.3 The elements of luminance, the number N and the equi-probability assumption

We interrupt this resumé of Jaynes's argument here to stress the following points.

(1) Jaynes makes it clear that in general the elements of luminance are not photons as that term is usually understood.

(2) In general N is a finite number. We cannot allow it to become infinite and expect to obtain useful results.

(3) Taking the prior probability to be proportional to the multiplicity W(F) implies that the probability of an element of luminance entering a pixel is the same for all pixels.

The last of these points is a questionable basic assumption which leads to the effects noted in the second point.

2.4 The zero-error case

If our measurements could be made with complete accuracy the problem would become a straightforward one of constrained maximization and the number N would then present no difficulty. For example, consider the aperture synthesis problem of radio astronomy. There the data d_j $(1 \leqslant j \leqslant m)$ are a subset of the Fourier components. The solution necessarily has a Fourier transform in agreement with the given data d_j and thus satisfies the convolution conditions precisely (but also contains additional Fourier components). There are no unphysical regions of negative intensity and in practice it is found that sidelobes are either absent or greatly reduced relative to the principal solution.

2.5 The effect of measuring error - the Bayesian approach

When measuring error is taken into account, the full Bayesian approach becomes necessary. This involves evaluating $p(D|F\sigma I)$ (see equation 2).

The d_j corresponding to a particular F are given by

$$d_j = \mathrm{IFT}_j\{f_1, f_2, \text{---}\, f_n\} + \varepsilon_j \quad , \tag{6}$$

where ε_j is the jth measuring error and IFT_j represents the jth inverse Fourier component. Thus

$$\varepsilon_j = d_j - \mathrm{IFT}_j(f_1, f_2, \text{---}\, f_n) \ . \tag{7}$$

If each of the errors ε_j has independent Gaussian distributions with the same standard deviation σ, the probability distribution of the errors $p(\varepsilon_1, \varepsilon_2, \text{---}\, \varepsilon_j)$ is proportional to $\exp\left\{\frac{1}{2\sigma^2} \sum_{j=1}^{m} \varepsilon_j{}^2\right\}$. Thus we may write

$$P(D|F\sigma I) = \exp\left\{- \frac{1}{2\sigma^2} \sum_{j=1}^{m} |d_j - \hat{d}_j|^2\right\} , \tag{8}$$

where the $\hat{d}_j$ are estimates of the d_j derived from the estimated image $\hat{F}$,

$$\hat{d}_j = \mathrm{IFT}_j(\hat{f}_1, \hat{f}_2, \text{---}\, \hat{f}_n) \ .$$

Thus from equations (2), (5) and (8), apart from a proportionality constant, $p(F|D\sigma I)$ is given by

$$P(F|D\sigma I) = \exp\left\{-N \sum_{k=1}^{n} f_k \ln f_k\right\} \exp\left\{-\frac{1}{2\sigma^2} \sum_{j=1}^{m} |d_j-\hat{d}_j|^2\right\}. \quad (9)$$

In Jaynes's formulation the problem then becomes one of maximizing this function subject to the constraint $\sum_{k=1}^{n} \hat{f}_k = 1$, which is equivalent to maximizing the function

$$Q = -N \sum_{k=1}^{m} \hat{f}_i \ln \hat{f}_i - \frac{1}{2\sigma^2} \sum_{j=1}^{m} |d_j-\hat{d}_j|^2 + \lambda \sum_{k=1}^{m} f_k ,$$

λ being a Lagrange multiplier. The form of the solution is

$$\frac{\partial Q}{\partial f_k} = 0; \quad 1 \leqslant k \leqslant n ,$$

which becomes

$$-N(1+\ln \hat{f}_k) - \frac{1}{2\sigma^2} \sum_{j=1}^{m} \frac{\partial |d_j-\hat{d}_j|^2}{\partial f_k} + \lambda = 0 . \quad (10)$$

3 THE DIFFICULTIES

The first term, $-N(1 + \ln f_k)$, on the left-hand side of equation (10) comes from the assumed prior probability distribution. If the errors σ are sufficiently large, the second term may become insignificant compared with the first, and equation (10) becomes

$$-N(1 + \ln f_k) + \lambda \approx 0 ,$$

corresponding to a uniformly grey image. As the errors become smaller the second term, representing the observations, comes to play a larger part in determining the solution. This seems to be in accordance with reasonable expectations. The difficulty arises in connection with N, the number of elements of luminance. We should expect this number to increase as we increase the duration of observations, but if N becomes large the first term on the left-hand side of equation (10) will dominate, yielding a uniformly grey image.

There is another difficulty associated with the number N. It has been shown by Schwarz (1978), O'Sullivan (1981) and Cornwell (1983) that

$$\sum_{j=1}^{m} \frac{\partial |d_j-\hat{d}_j|^2}{\partial f_k} = 2(\mathcal{f}_k-\hat{\mathcal{f}}_k) .$$

Here $\mathcal{f}_k$ is intensity of the kth pixel of the principal solution - that is, the Fourier transform of the data values which is equivalent to our estimate of the true distribution convolved with the beam - and $\hat{f}_k$ is the kth pixel of the MEM solution, after convolution with the same beam.

From equation (10) the MEM solution then has the form

$$\ln f_k = \frac{1}{N\sigma^2}\{f_k - \hat{f}_k\} + \lambda/N - 1 \quad . \tag{11}$$

Now if the result satisfied the convolution equation precisely, as in the case of zero errors, we would have $f_k - \hat{f}_k = 0$ for $1 \leqslant k \leqslant n$. In fact the residuals $(f_k - \hat{f}_k)$ have a systematic dependence on the image pixel intensities $\hat{f}_k$. They do not have a Gaussian distribution, which is at variance with a basic assumption. That is, although $(d-\hat{d})$ in equation (8) was assumed to be Gaussian-distributed, the residuals $f_k - \hat{f}_k$, given by the inverse Fourier transform of $(d-\hat{d})$, are not Gaussian-distributed.

To examine the consequences of this consider a very simple distribution consisting of a flat background with a single compact source close to the resolution limit superimposed on it. The source could, for example, be a circular Gaussian. Let p be the height of the true peak above the baseline, and $\hat{p}$ the height of the peak in the MEM solution above its baseline. We would expect that for such a simple distribution the same peaks after convolution with the beam, would satisfy

$$P \approx K\, p \quad \text{(principal solution)},$$

$$\hat{P} \approx K\, \hat{p} \quad \text{(convolved MEM)}.$$

For simplicity, we make this assumption. If b and $\hat{b}$ are the corresponding base levels, we can write (from equation 10)

$$\frac{P}{\hat{p}} = 1 + \frac{N\sigma^2}{K}\,\frac{\ln(1+\hat{p}/\hat{b})}{\hat{p}} \quad . \tag{12}$$

The larger the term on the left-hand side, the greater the 'squashing' of peaks in the MEM solution. All terms on the right-hand side are necessarily positive and $\ln\{1+\hat{p}/\hat{b}\}/\hat{p}$ is a monotonically decreasing function of $\hat{p}$. It therefore follows that the height of small peaks in the MEM solution are reduced relative to stronger peaks, and the lower the base level the more marked the effect. Furthermore, the magnitude of the distortion as predicted by equation (12) is proportional to N, the number of elements of luminance.

There is another, apparently paradoxical, consequence of equation (12). If we made identical sets of measurements on two successive days we should expect the value of σ^2 to be halved. However, we should also expect N, the number of elements of luminance, to be doubled, leaving $N\sigma^2$ unchanged. Thus the degree of systematic distortion would be independent of measuring error. Not only is this a surprising result, but it also flatly contradicts the result asserted in Section 2.4 - namely, that as the measurement error tends to zero the distortion should also tend to zero.

4 A POSSIBLE SOLUTION TO THE DIFFICULTIES

The problems with MEM as formulated in the previous section are both philosophical and practical. Firstly, we are obliged to cling to the rather nebulous concept of luminance elements in order that the measured data be able to influence the prior assumptions. This is in spite of the fact that a clear candidate for a luminance element is present - the photon. Secondly, the fitting of data to within some noise criterion introduces bias in the result. The estimate is on the average incorrect!

As an alternative we suggest that the probability that a luminance element enters a pixel should be taken as proportional to the intensity of that pixel. Making this suggestion highlights a further paradox in the MEM philosophy as previously described. If the restored image $\{I_i\}$ is non-uniform then a better estimate of the probability p_i of a luminance element entering pixel i is given by $I_i/\Sigma I_j$. This is at variance with the implicit prior assumption $p_i = 1/n$.

We might then seek as a goal of the restoration, not the most *a posteriori* probable set of luminance element counts, but an estimate of the set of probabilities $\{p_i\}$, and hence $\{I_i\}$.

5 ACKNOWLEDGMENTS

The argument outlined in this paper emerged from discussions between the authors when JO'S was at the Netherlands Foundation for Radioastronomy and MMK was receiving financial assistance from the Foundation.

References

Cornwell, T.J. (1983). A method of stabilizing the clean algorithm. Astron. Astrophys., 121, 281.

Jaynes, E.T. (1981). Where do we go from here? First Workshop on Maximum Entropy Methods, University of Wyoming.

Jaynes, E.T. (1982). The rationale of maximum entropy methods. Proc. IEEE, 70, 939.

O'Sullivan, J.D. (1981). Modified CLEAN - generalised constrained optimisation of maps. Netherlands Foundation for Radio Astronomy Note 352.

Schwarz, U.J. (1978). Mathematical-statistical description of the iterative beam removing technique (method CLEAN). Astron. Astrophys., 65, 345.

DISCUSSION

J.A. ROBERTS

In a practical case what is the value of N?

J.D. O'SULLIVAN

This appears to be the crux of the matter. We are suggesting that no physical significance can be attached to the number of luminence elements and their relevance is unclear in the Jaynes formulation.

8.5 EXPLICIT AND IMPLICIT SOLUTIONS IN THE MAXIMUM ENTROPY METHOD

Nai-Long Wu
School of Electrical Engineering,
University of Sydney, N.S.W. AUSTRALIA

1 *INTRODUCTION*

The Maximum Entropy Method (MEM) is a nonlinear technique for estimating power density spectra or for image reconstruction. Its merits are determined on the basis of the positiveness of solution, enhanced resolution, suppression of noise and so on.

It is known that there are two schools of thought in the MEM divided by the definition of entropy:

$$H1 = \int \log S(f)df \quad \text{(entropy of p.d.f. of the time series)}$$

and

$$H2 = -\int S(f)\log S(f)df \quad \text{(configurational entropy).}$$

The former was proposed by Burg (1967) for spectral estimation, and the latter was employed by Frieden (1972), Gull & Daniell (1978), Bryan & Skilling (1980) and others for image reconstruction. In extrapolating the autocorrelation function (ACF) beyond the measure limit H1 assumes the maximal flexibility of the predicted ACF while H2 makes a maximally non-committal prediction of the shape of the image.

It has been argued that in image reconstruction, entropy of the form H2 is superior to H1 since one should consider the image as a whole rather than be interested in the individual samples. On the basis of the above understanding and since MEM with entropy of the form H1 (called MEM1) has been extensively studied, this presentation will concentrate on solutions of MEM with entropy of the form H2, called MEM2.

The problem can be stated as follows: Given measured data (the original signal $x(n)$, or its ACF $R(n)$, $n=0,\ldots m$), estimate the spectrum (or image) which maximizes H2 subject to the constraints

$$\int_{-\frac{1}{2}}^{\frac{1}{2}} S(f)\exp(2\pi ifn)df = R(n), \quad n=0,\ldots,m. \tag{1}$$

By using the Lagrange undetermined multiplier (LUM) method it can be shown that the solution is

$$S(f) = \exp\{-1 + \sum_{k=-m}^{m} \lambda_k \exp(2\pi ifk)\}. \tag{2}$$

Putting (2) into the form

$$\lambda_k = \delta_k + \int_{-\frac{1}{2}}^{\frac{1}{2}} \exp(2\pi ifk)df \log \sum_{n=\infty}^{\infty} R(n)\exp(-2\pi ifn) \quad (3)$$

one can recognize the undetermined multipliers λ_k as the cepstrum of R(n) (except for the additional term δ_k). How to determine them will be shown later.

2 *SOLUTIONS USING THE CEPSTRUM ANALYSIS METHOD*

Because of the intrinsic nonlinearity in MEM2 it is believed that an explicit solution will be subject to very restrictive conditions. Indeed, if we assume that the signal x(n) of one-dimension is real, causal and minimum-phase, an explicit solution and data extension can be determined by virtue of the cepstrum analysis method (CAM) as shown by Wu (1983). If the conditions of causality and minimum-phase are not satisfied, one needs to solve iteratively the following simultaneous equations in N/2 unknowns $\lambda_1,\ldots,\lambda_m$, R(m+1),..., R(N/2):

$$R(n) = \sum_{k=-m}^{m} (k/n)\lambda_k R(n-k), \quad n=1,\ldots,N/2, \quad (4)$$

where N, the length of the Fourier transform to be used, can be arbitrarily large. The MEM2 spectrum then can be determined by the correlogram method or (2).

A numerical example will be illustrated in Section 5. At this point, however, it is interesting to notice the n in the denominators in (4). This implies that the extrapolated R(n), $|n| > m$ will vanish rapidly as n increases. That is, the data extension in MEM2 is 'conservative'. Therefore, we cannot expect too much as far as improvement of resolution is concerned.

3 *SOLUTION BY R-λ PROCEDURE*

By inspection of (1) and (3) one can see that MEM2 spectral estimation is equivalent to the following.

Given the ACF, R(n), for $n \in D$, estimate the spectrum S(j) such that:
(i) The constraints IFT[S(j)] = R(n) for $n \in D$ are satisfied, and

(ii) $$\lambda_k = \delta_k + \text{IFT}[\log S(j)] \begin{cases} \text{to be determined for } k \in D \\ =0 \text{ (desired by MEM2) otherwise,} \end{cases}$$

where the discrete form of S is used and IFT denotes the inverse Fourier transform. Note that the above problem statement applies to the support D of any dimension and shape, and to complex as well as real R(n).

An iterative procedure to determine S(j) can be established in light of (i) and (ii). In particular, start with a uniform estimate of S(j), repeat the correction of IFT[S(j)] with the known R(n) for $n \in D$ and truncation of λ_k for $k \notin D$ until (i) is satisfied within an error limit. This R-λ procedure for MEM2 is similar to that developed by Lim & Malik (1981) for MEM1. Some considerations in the implementation of this kind of algorithm can be found in their paper. A numerical example will be shown in Section 5.

4 *REVISED GULL AND DANIELL ALGORITHM*

In all the algorithms shown so far no errors in the data were considered. Taking errors into account one should tolerate misfit between the calculated ACF (the LHS of (1)) and the measured data (the RHS of (1)) in maximizing the entropy. By using the χ^2-statistic and introducing an LUM Gull & Daniell (1978) found a solution of the form

$$S(j) = A \exp\{-1 - (\lambda/2)\partial C(S(j))/\partial S(j)\}, \tag{5}$$

where

$C = \sum_{n \in D} |R(n) - IFT[S(j)]|^2/\sigma_n^2$, λ is the LUM and σ_n^2 is the noise variance of $R(n)$. A is a scale factor which will be discussed later.

For a particular λ (5) can be solved, starting with a uniform image, according to the following iterative formula

$$S_{(j)}^{(m+1)} = A \exp\{-(\lambda/2)\partial C(S_{(j)}^{(m)})/\partial S(j)\} \tag{6a}$$

or

$$S_{(j)}^{(m+1)} = A \exp\{\lambda\ FT[(R(n) - IFT[S_{(j)}^{(m)}])\ W(n)/\sigma_n^{\ 2}]\}, \tag{6b}$$

where m denotes the m-th iterate and the window $W(n)=1$ for $n \in D$, $W(n)=0$ otherwise.

Increase λ to decrease C until it falls just below the expected value (the number of datum points). One then obtains the desired MEM2 image.

The attractive feature of the Gull-Daniell (GD) algorithm is the positiveness of iterates and rapid development of peak values due to the exponential function. However, it is also due to the exponential function that the behaviour of the GD algorithm is erratic and unstable, especially at high values of λ. Sometimes arithmetic floating point overflow occurs on a computer.

To avoid this one must average successive iterates to ensure convergence by setting

$$S_{(j)}^{(m+1)} = (1-\beta)S_{(j)}^{(m)} + \beta\ A \exp\{-(\lambda/2)\partial C(S_{(j)}^{(m)})/\partial S(j)\}.$$

For a large λ, the memory coefficient β could be as small as 0.01; thus, the iteration effectively stops.

In order to overcome these drawbacks of instability and speed of convergence, the two following techniques have been introduced to revise the GD algorithm.

(I) Use of an Optimal Scale Factor

If we choose the scale factor A to minimize C the convergence will be speeded up. In particular, λ starts with zero so that A starts with the initial value

$$A^{(o)} = \text{maximum modulus of ACF/size of image.}$$

After the m-th iterate let

$$\partial C/\partial \alpha = \frac{\partial}{\partial \alpha} \sum_{n \in D} |R(n) - \alpha \mathrm{IFT}[S^{(m)}_{(j)}]|^2/\sigma_n^{\ 2} = 0$$

and determine:

$$\alpha = \frac{\sum_{n \in D} \mathrm{Re}(\mathrm{IFT}[S^{(m)}_{(j)}]R^*(n))/\sigma_n^{\ 2}}{\sum_{n \in D} |\mathrm{IFT}[S^{(m)}_{(j)}]|^2/\sigma_n^{\ 2}}$$

where * denotes the complex conjugate. The scale factor to be used in the (m+1)-th iterate is set to be

$$A^{(m+1)} = \alpha\, A^{(m)}.$$

Experience has shown that α approaches unity after several iterations and the computational time can be reduced by one-third or so.

(II) Use of the Second-Order Solution

(5) is an exact solution but (6) is not. Apparently, the following solution is exact:

$$S^{(m+1)}_{(j)} = A \exp\{-(\lambda/2)\, \partial C(S^{(m+1)}_{(j)})/\partial S(j)\}.$$

Note $S^{(m+1)}_{(j)}$ on the RHS.

$\partial C(S^{(m+1)}_{(j)})/\partial S(j)$ can be expanded in a Taylor series:

$$\frac{\partial C(S^{(m+1)}_{(j)})}{\partial S(j)} = \frac{\partial C(S^{(m)}_{(j)})}{\partial S(j)} + \sum_{\ell} \frac{\partial^2 C(S^{(m)}_{(j)})}{\partial S(j)\, \partial S(\ell)} (S^{(m+1)}_{(\ell)} - S^{(m)}_{(\ell)}) + 0 \qquad (7)$$

The third and higher order terms vanish since C is a quadratic form in S(j). The first-order approximation gives (6), which is denoted by $S^{(m+1)}_{(j)}$ (GD).

Having calculated the second-order derivatives in (7) one can find

$$S^{(m+1)}_{(j)} = S^{(m+1)}_{(j)}(\mathrm{GD})\, \exp\{-\lambda(S^{(m+1)}_{(j)} - S^{(m)}_{(j)}) * w(j)\} \qquad (8)$$

where * denotes convolution and $w(j) = \mathrm{FT}[|W(n)|^2/\sigma_n^{\ 2}]$.

In some cases, e.g., in radioastronomy, w(j) is close to the δ-function, then the following would be a good approximation for (8):

$$S^{(m+1)}_{(j)} = S^{(m+1)}_{(j)}(\mathrm{GD})\, \exp\{-\lambda(S^{(m+1)}_{(j)} - S^{(m)}_{(j)})w(0)\}. \qquad (9)$$

Furthermore, take approximation $e^x \approx 1+x$, then $S^{(m+1)}_{(j)}$ in (9) can be solved:

$$S^{(m+1)}_{(j)} = \frac{1 + \lambda w(0) S^{(m)}_{(j)}}{1+\lambda w(0) S^{(m+1)}_{(j)}(\mathrm{GD})}\, S^{(m+1)}_{(j)}(\mathrm{GD}). \qquad (10)$$

This is the formula used in our iterative algorithm. The factor before $S^{(m+1)}_{(j)}$ (GD) on the RHS of (10) will be greater or less than unity depending on whether $S^{(m+1)}_{(j)}(GD)$ is less or greater than $S^{(m)}_{(j)}$. The big jump from $S^{(m)}_{(j)}$ to $S^{(m+1)}_{(j)}(GD)$ will be compensated by this factor, $S^{(m+1)}_{(j)}$ being upper-bounded by $S^{(m)}_{(j)}+1/(\lambda w(0))$. As a consequence, the stability is remarkably improved. In all the numerical examples tested so far the memory coefficient β is at least 0.25 so that the convergence rate is also higher.

5 *EXAMPLES*

5.1 *The Cepstrum Analysis Method and R-λ Procedure*

Fig. 1(a) shows the true ACF and ACF extrapolated from the first 16 true ACF points by the CAM. The true ACF is of the form $R(n) = a^2\text{Cos}(\omega n) + \sigma^2\delta_n$, $|n| = 0,1,2,\ldots$. The signal-to-noise ratio is defined by $\text{SNR(db)} = 10 \log(a^2/\sigma^2)$.

In Fig. 1(b) the line labelled "true" is an isolated peak plus flat background. The first 16 true ACF values were used in computing the other spectra. "FT" was computed by the correlogram method with the Bartlett window. "CAM" was computed in two ways: by the correlogram method with the extrapolated ACF in (a), and according to (2) having determined λ_k in (4); "R-λ" was computed by the R-λ procedure. From the figure we can see the self-consistency in the CAM and the agreement between the CAM and R-λ procedure: the three curves are indistinguishable. The sidelobes of about -10 db are plausible due to the isolated peak in the true spectrum and the 'conservative' extension in MEM2.

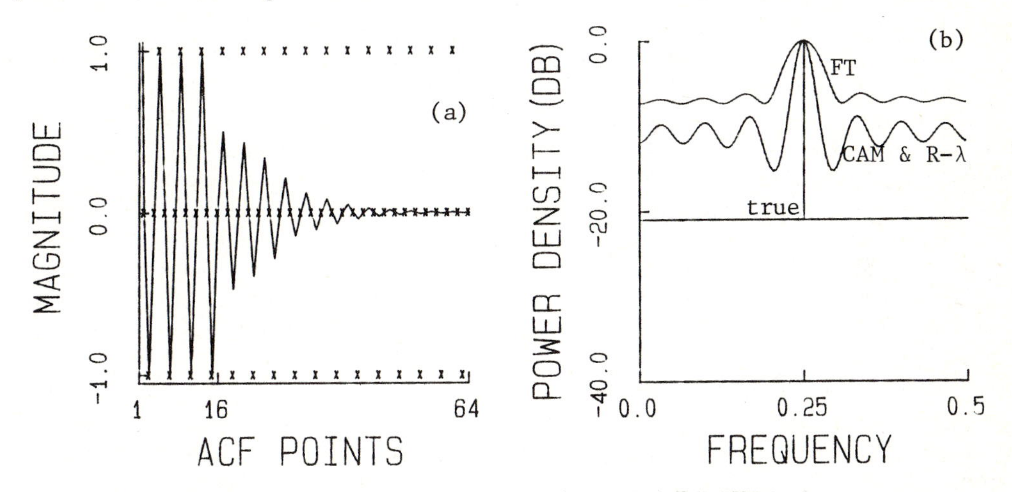

Fig. 1(a) x shows the first 64 true ACF points, $\omega/2\pi=.25$, SNR = -3db. The actual DC term is 3. The oscillating curve shows the extrapolated ACF by CAM. The negative halves are not shown. (b) The spectra computed from the true ACF; by CAM, R-λ procedure and FT.

5.2 *The revised Gull and Daniell algorithm*

Some visibility functions from the Fleurs Synthesis Telescope (Sydney University) were used for testing. Fig. 2(a) is a map deduced by the regridding-FFT method. Fig. 2(b) shows its MEM2 map generated by the revised GD algorithm. It can be seen from Table 1 that the revised GD algorithm shows a large improvement over the original version in its stability and convergence speed.

Table 1: The original and revised GD algorithm

Map size: 256 x 256 points			Computer: VAX 11/780	
	No. iterations	No. FFT	β	CPU time (min)
Original	48	96	0.03	60
Revised	24	48	1.0	30

(a) (b)

Fig. 2(a) The map of radio source PKS 1733-56 deduced by the regridding-FFT method. (b) The MEM2 map.

6 *CONCLUSION*

We have introduced four solutions and illustrated some numerical examples in MEM2. The revised GD algorithm has been shown to be very useful in image reconstruction in radioastronomy. The others can be used to study the properties of MEM2. In this respect more theoretical work and numerical examples are needed. For instance, the conditions for the existence and uniqueness of the solutions are still unknown. Comparison between the properties of MEM2 and MEM1 is also to be done.

ACKNOWLEDGEMENT

I would like to thank Professor T.W. Cole for instructive discussion, and Mr. R. Sault for his help in software. This research work was supported by a University of Sydney Postgraduate Research Award.

REFERENCES

Bryan, R.K. & Skilling, J. (1980), Mon.Not.R.Astr.Soc., 191, pp. 69-79.
Burg, J.P. (1967), Maximum Entropy Spectral Analysis, presented at the 37th Ann.Int.Meet.Soc.Explor.Geophys., Oklahoma City, OK.
Frieden, B.R. (1972), J.Opt.Soc.Amer., Vol. 62, pp. 511-518.
Gull, S.F. & Daniell, G.J. (1978), Nature, Vol. 272, pp. 686-690.
Lim, J.S. & Malik, A. (1981), IEEE Vol. ASSP-29, pp. 401-413.
Wu, N.L. (1983), IEEE Vol. ASSP-31, pp. 486-491.

8.6 X-RAY STRUCTURE DETERMINATION USING INFORMATION THEORY - THE STATISTICAL GEOMETRIC APPROACH.

Stephen W. Wilkins
CSIRO, Division of Chemical Physics, P.O. Box 160, Clayton, Victoria, Australia 3168

Joseph N. Varghese
CSIRO, Division of Protein Chemistry, Royal Parade, Parkville, Victoria, Australia 3052

Stig Steenstrup
CSIRO, Division of Chemical Physics, P.O. Box 160, Clayton, Victoria, Australia 3168
(on leave from: Physics Laboratory II, University of Copenhagen, Denmark.)

ABSTRACT

The foundations of a general information-theoretic based approach to X-ray structure determination are outlined and some simulated structure refinements for a simple one-dimensional model presented.

1. INTRODUCTION

The outstanding problem in X-ray crystallography is the determination of molecular structure inside the unit cell from the measured integrated reflectivities of the Bragg diffraction peaks and any other available information. More specifically, one wishes to determine the electron density in the unit cell:

$$\rho(\underset{\sim}{r}) = V^{-1} \sum_{\underset{\sim}{k}} F(\underset{\sim}{k}) exp\{-2\pi i \underset{\sim}{k}.\underset{\sim}{r}\}, \quad (1)$$

where i) the summation is over all possible reflexions, $\underset{\sim}{k}$, ii) V is the volume of the unit cell and iii) $F(\underset{\sim}{k})$ is the structure factor - in general complex - for reflexion $\underset{\sim}{k}$. In practice, one has only measurements (with errors) of structure amplitudes (without phases) and even then only for a subset of possible reflexions, thus direct inversion of $F(\underset{\sim}{k})$ via (1) to obtain ρ is ruled out. A number of numerical methods aimed at solving crystal structures with such diffraction data have evolved over the past 30 years and are known as "direct methods" (e.g., Sayre, 1982). Most of these methods assume structures consisting of identical distinct atoms, and work very well for fewer than, say, 100 atoms, but are not applicable to the important area of biological macromolecules (especially below atomic resolution). On the experimental side, methods exist for obtaining some information on phases (e.g., Blundell & Johnson, 1976), however, most of these methods depend on the presence (or isomorphous substitution) of heavy atoms in the structure and, even then, give only partial information. Thus, the problem of X-ray structure determination is inevitably a problem in statistical inference.

A very powerful technique for statistical inversion involves the use of the information-theoretic procedure of maximum entropy (ME) developed by Jaynes (1957, 1968, and 1982). In a recent paper, Wilkins, Varghese and Lehmann (1983), hereafter WVL, have examined the application of information theory to the crystallographic inversion problem and outlined a general method of approach which we termed statistical geometry. Our work is an extension, in particular, of the ideas of Frieden (1972), Wernecke (1977) and Gull & Daniell (1978) who applied these principles to the corresponding problems in 2 dimensions in optics and astronomy. These methods hold the exciting prospect of providing a method for crystal structure determination which is conceptually simple and numerically tractable and, moreover, one which is also optimal in a clearly defined statistical sense.

In the present article we briefly outline some of the salient ideas in our approach to crystallographic inversion using information theory, and present some illustrative results for a simple 1-dimensional structure. The basic principles and equations of the method are independent of the dimensionality of the system, while the practical task of numerical solution depends approximately linearly on the number of pixels in the structure and the number of reflexions considered.

2. INFORMATION THEORY FUNDAMENTALS AND THE ROLE OF PRIOR KNOWLEDGE

The key concept and starting point for an information theoretic approach to a problem in statistical inference is the concept that there exists a *unique and consistent measure of the amount of information*, I (= const. $-S$, the entropy), in a (discrete) probability distribution, $P(\underset{\sim}{x})$, and that it is given by (Shannon, 1948)

$$I[P] = const. -S[P] = \sum_{\underset{\sim}{x}} P(\underset{\sim}{x}) \ln[P(\underset{\sim}{x})/P^{o}(\underset{\sim}{x})], \tag{2}$$

where $\underset{\sim}{x}$ is a random vector and $P^{o}(\underset{\sim}{x})$ denotes the prior probability distribution on the $\underset{\sim}{x}$-space as introduced in the present context by Kullback (1959) and Jaynes (1968). The information-theoretic criterion for statistical inference is to choose the probability distribution, $P^{ME}(x)$, which minimizes I (i.e., maximizes S), subject to *all* the available information (Jaynes, 1957). Such information need not only be in the form of constraint relations but may also be introduced via $P^{o}(x)$ and implicitly via the specific choice of the $\underset{\sim}{x}$-space (see Steenstrup & Wilkins, 1983, for a discussion of prior information in the present context).

For the X-ray structure determination problem, the quantity to estimate is the electron density in the unit cell, $\rho(\underset{\sim}{r})$. For simplicity, only the one-dimensional case is considered (although all results hold for arbitrary dimension) and we work with the discrete and normalized density, $\underset{\sim}{\rho} = \{\rho_1,\ldots,\rho_N\}$, ρ_j being the normalized density in the jth pixel leading to the unitary structure factor

$$P_k = |P_k|\exp\{i\phi_k\} = \langle \sum_{j=1}^{N} \rho_j \exp\{2\pi ijk/N\}\rangle = \sum_j p_j \exp\{2\pi ijk/N\}, \qquad (3)$$

where $p_j = \langle\rho_j\rangle$ is the expectation value taken w.r.t. P.

Assembling our prior knowledge about $\underset{\sim}{\rho}$, we proceed as follows: n identical discrete units of structure, corresponding to elements of resolution in ρ-space rather than to actual physical entities such as atoms or electrons (although they are not perforce excluded), are assumed distributed at random over the N pixels with a prior probability distribution $\underset{\sim}{q}$, so that the probability of a given state, $\{n_1,\ldots,n_N\}$, is the multinomial distribution

$$P^o(n) = \frac{n!}{n_1!\ldots n_N!} \prod_{j=1}^{N} q_j^{n_j}. \qquad (4)$$

Extending some results given by Levine (1980), minimization of the information, I, (2) subject to the N constraints (3) and the prior distribution (4) yields the following derived expression for the (structural) information in this special case:

$$I[\underset{\sim}{p}] = const. - S[\underset{\sim}{p}] = n \sum_{j=1}^{N} p_j \ln[p_j/q_j], \qquad (5)$$

where $p_j = \langle n_j\rangle/n = \langle\rho_j\rangle$, and leads to a ME distribution of multinomial form so that, e.g., the variance of n_j is given by $\langle n_j^2\rangle - \langle n_j\rangle^2 = n\, p_j(1 - p_j)$ and allows an estimate for the error in a ME map to be made when n can be given some physical meaning.

To summarize, the information function (5) is *derived* from (2) and embodies the very specific prior information:
I1: positive integers $n_j = 0,1,\ldots,n$
I2: prior probabilities for states given by (4)
I3: fixed total number of scattering units, i.e. $\Sigma\, n_j = n$
I4: constraints (3).

In the context of the crystallographic inversion problem, these mathematical assumptions have the immediate consequence that:

A1: only those structures are considered where the total number of electrons exactly equals the total number of electrons (Z) in the unit cell and
A2: only structures with positive discrete bounded density $\underset{\sim}{\rho}$ are considered.
A3: prior information about the structure may be injected into $\underset{\sim}{p}$ via $\underset{\sim}{q}$.

It is perhaps worth pointing out that other recent approaches to the crystallographic inversion problem (e.g., Britten & Collins, 1982; Piro, 1983) tend to violate one or both assumptions A1 and A2 above (for further discussion of this, see Steenstrup, 1983; Steenstrup & Wilkins, 1983).

Subject to the inbuilt prior information discussed above, (5) may be used as a basis for further inference by minimization of $I[\underset{\sim}{p}]$ w.r.t. $\underset{\sim}{p}$ subject to any further information available such as the diffraction data (which will be discussed in the next section), and constitutes an example of sequential inference using ME, as discussed by Levine (1980).

3. INFORMATION FROM DIFFRACTION DATA

Conventional crystallographic data is usually in the form of structure amplitudes with phases which are unknown or estimated by techniques such as isomorphous replacement, anomalous dispersion, dynamical scattering effects, direct methods, etc. In keeping with the approach outlined in Gull & Daniell (1978) and WVL, we intend to introduce the information from measurement into $\underset{\sim}{p}$ using information theory via weak constraints, since this will help to guarantee a solution to the ME determination of structure and provide a structure consistent with the accuracy of the data. The constraints which we consider are:

Phases approximately known (constraint 1)

If there are N_1 distinct measured unitary structure factors U_k (i.e., $U_0 = 1$), whose phases are assumed known, with estimated standard deviations $\sigma_{k,1}$, then constraint 1 is

$$f_1(\underset{\sim}{p}) \equiv \frac{1}{2N_1} \sum_{k \varepsilon D_1} |P_k - U_k|^2/\sigma^2_{k,1} = C_1, \tag{6}$$

where the summation is over the set D_1 of all measured reflexions including Friedel pairs (i.e., $\pm k$) and P_k is the model unitary structure factor corresponding to $\underset{\sim}{p}$ and is given by (3). Expression (6) represents the reduced χ^2 distribution and should be *s.t.* $C_1 \simeq 1$ when the structure fits the data within experimental errors.

Phases unknown (constraint 2 or 3)

In this case only the $|U_k|$ are assumed known, and the constraint used is

$$f_2(\underset{\sim}{p}) \equiv \frac{1}{2N_2} \sum_{k \varepsilon D_2} |P_k - |U_k| \exp\{i\phi_k\}|^2/\sigma^2_{k,2} = C_2, \tag{7}$$

where there are $2N_2$ reflexions with unknown phases belonging to the set D_2 (including Friedel pairs) and ϕ_k is the phase of P_k. This term does not have a χ^2 distribution, but if N_2 is sufficiently large, we can assume a normal distribution with $C_2 \simeq 1$.

An alternative constraint to (7) for including phaseless data which may have some advantages, especially if data contains superimposed reflexions, is to work instead with the intensities, $I_k = |U_k|^2$, viz.

$$f_3(\underset{\sim}{p}) \equiv \frac{1}{2N_3} \sum_{k \in D_3} (P_k^* P_k - I_k)^2 / \sigma_{k,3}^2 = C_3. \tag{8}$$

This form of constraint for introducing phaseless data also has better mathematical properties than $f_2(\underset{\sim}{p})$ (see Table 1 in WVL).

4 THE FUNDAMENTAL EQUATIONS AND THE PROBLEM OF NUMERICAL SOLUTION

Minimization of the structural information (5) subject to normalization of $\underset{\sim}{p}$ and the set of constraints

$$\underset{\sim}{f}(\underset{\sim}{p}) = \underset{\sim}{C} \tag{9}$$

yields the fundamental equations of the SG method (see Gull & Daniell, 1978, and WVL):

$$p_j = q_j \exp\{-\lambda_o - \underset{\sim}{\lambda} \cdot (\partial \underset{\sim}{f}(\underset{\sim}{p})/\partial p_j)\}, \tag{10}$$

where λ_o and $\underset{\sim}{\lambda}$ are the Lagrange multipliers associated with the normalization and constraints (9), respectively. The required $\underset{\sim}{p}$, i.e. electron density or maximum entropy structure (ME structure) is that which solves (10) with λ_o and $\underset{\sim}{\lambda}$ chosen so that the relevant constraints are satisfied.

The task of actually solving these equations for constraints of the type given in §3 is, in general, a non-trivial task and represents the major obstacle to practical application of the statistical geometric method for large-scale systems (i.e., N large). The problem has been approached in different ways (see, e.g., Wernecke, 1977; Wernecke & D'Addario, 1977; Gull & Daniell, 1978; Skilling, 1981; Skilling & Bryan, 1983; Skilling, 1983; and Wilkins, 1983a,b). However, even the most powerful of these methods falter in the case of the phaseless problem.

Our methods for tackling the numerical solution of (10) are described in WVL and Wilkins (1983a,b), and have been used in obtaining the results described in §5.

5 SIMULATED STRUCTURE REFINEMENT

For an illustration of a stucture refinement by the statistical geometric method, we have taken a simple one-dimensional model structure consisting of three Gaussian peaks with different heights and half-widths (see Figs. 1 and 2) and assumed a uniform prior for $\underset{\sim}{q}$ (q_j=1/N). Fig. 1 shows (in direct space) the result of a simulated structure refinement involving one observed structure factor (with error) in constraint 1 leading to a ME structure shown as the dashed curve, followed by the additional inclusion of 11 amplitudes (with errors) in constraint 2 leading to the ME structure shown as a thin solid curve. This refinement involved no assumed phase (since the origin is arbitrary) and leads to a ME structure with strong centrosymmetric character due, undoubtedly, to the limited information and the strong smoothing influence of the information function. This case helps to highlight the interpretation one should place on ME structures, namely that they portray values for mean occupation levels of pixels in terms of the units of resolution or structure. Thus, one should bear in mind the fact that ME structures are essentially probabilistic in nature and cannot immediately be interpreted in terms of atoms, etc.

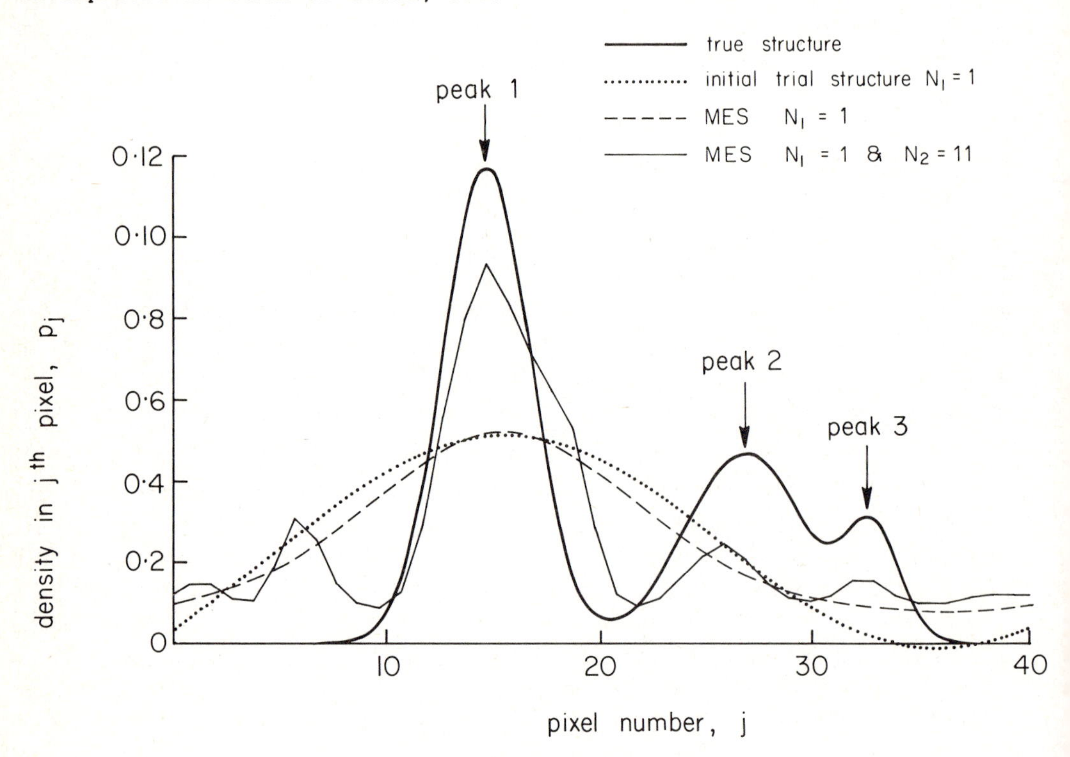

Fig. 1.
Simulated structure refinement with no assumed phase information.

Addition of one extra structure factor into constraint 1 ($N_1 = 2$) and inclusion of errors in the amplitudes and phases leads to the corresponding results shown in Fig. 2, where the full ME structure now has peaks closely matching those of the model structure and demonstrates good resolution and low noise in the background.

Fig. 3 shows the progress of the ME refinement towards the final ME structure for the same data as Fig. 2, but with different numbers of pixels, N. Starting from a flat map ($\underset{\sim}{\lambda}$ = 0), the constraint functions are seen to decrease monotonically as λ_1 is increased until $f_1(\underset{\sim}{p}) = 1$ at which point the second constraint is switched on and both λ_1 and λ_2 are simultaneously adjusted until $f_1 = f_2 = 1$. Also shown in Fig. 3 is the information content of the structure, I, as a function of λ_1 and λ_2 in the refinement, which can be seen to increase monotonically. The slightly larger values of I for the N = 10 case reflect the additional information implied by linear interpolation used (only) for this case.

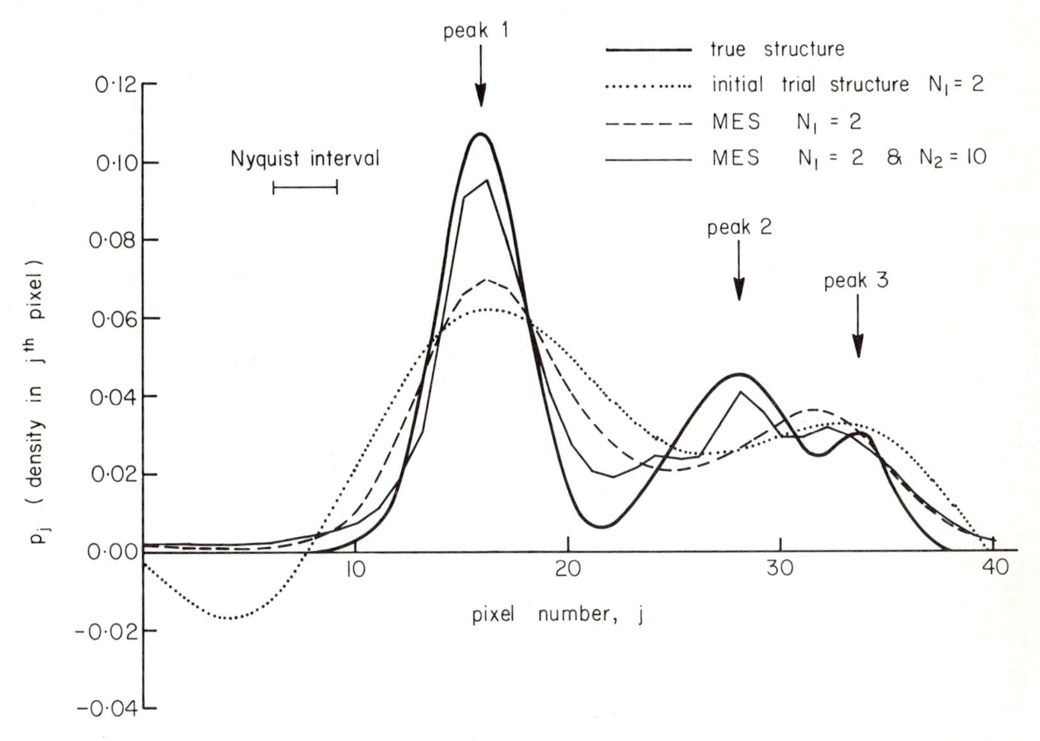

Fig. 2.
Simulated structure refinement with 1 assumed phase angle. (Based on Figure 1, in Wilkins, Varghese & Lehmann, 1983).

CONCLUSIONS

We have outlined an approach to crystallographic inversion based on information theory which offers a powerful method of 3-dimensional macromolecular structure determination, provided efficient and reliable numerical methods for solving the fundamental equations can be developed and coded when N (the number of pixels) and $N_1;N_2$ (the number of reflexions) are large. The method has the advantage over conventional direct methods that the concept of atomicity is *not* a prerequisite and so can apply to structure refinement below atomic resolution. Moreover, it is a self-consistent method in the sense that data and prior information of the types discussed can be exactly built into the structure and the structural estimate, $\underline{p}$, made least biased with respect to missing information.

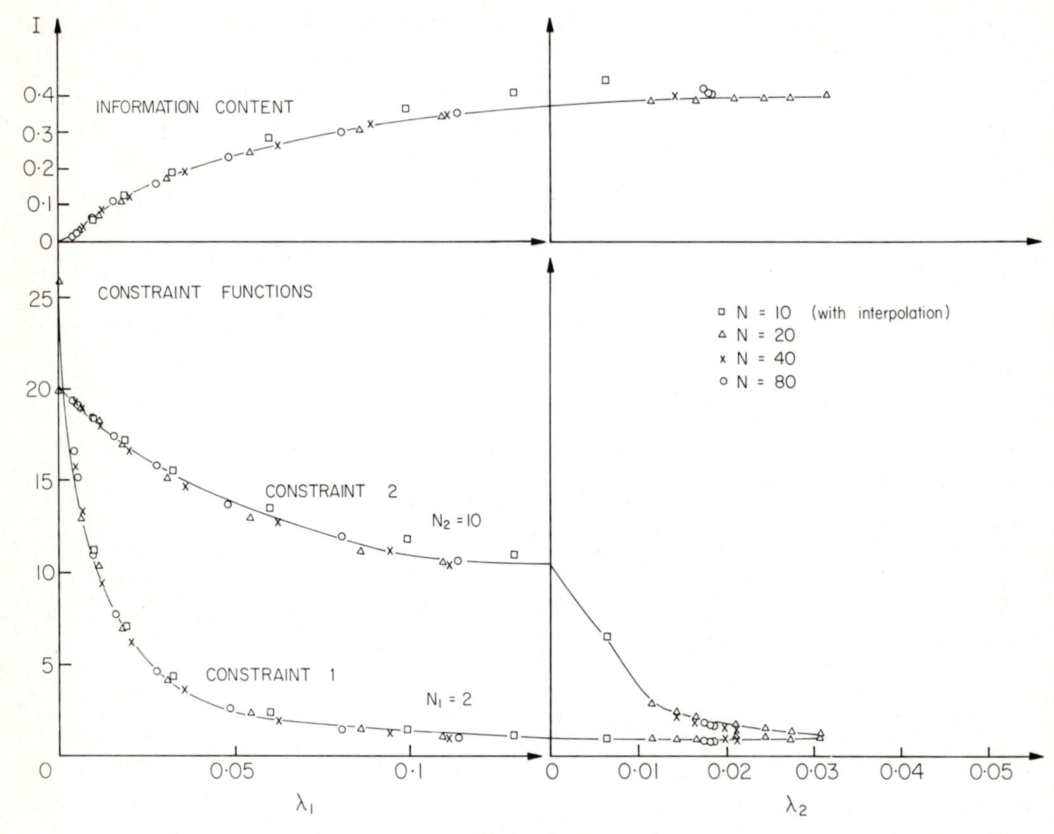

Fig. 3.
Information and constraint values as a function of $\underline{\lambda}$ *showing progress of refinement corresponding to fig. 2 for different values of the number of pixels* (N).

ACKNOWLEDGEMENTS

We are grateful to Dr. J. Skilling for helpful discussion and preprints of papers prior to publication. One of us (SS) wishes to thank the CSIRO for hospitality and the Danish Science Foundation for financial assistance.

REFERENCES

Blundell, T.L. & Johnson, L.N. (1976). Protein Crystallography, London: Academic.

Britten, P.L. & Collins, D.M. (1982). Acta Cryst. A38, 129-132.

Frieden, R. (1972). J. Opt. Soc. Am. 62, 511-518.
Gull, S.F. & Daniell, G.J. (1978). Nature (Lond.) 272, 686-690.
Jaynes, E.T. (1957). Phys. Rev. 106, 620-630.
Jaynes, E.T. (1968). IEEE Trans. SSC-4, 227-241.
Jaynes, E.T. (1982). Papers on Probability, Statistics and Statistical Physics. Dordrecht: D. Reidel.
Kullback, S. (1959). Information Theory and Statistics. N.Y.: Wiley.
Levine, R.D. (1980). J. Phys. A: Math. Nucl. Gen. 13, 91-108.
Piro, O.E. (1983). Acta Cryst. A39, 61-68.
Sayre, D. (1982). Computational Crystallography. Oxford: Clarendon Press.
Shannon, C.E. (1948). Bell Syst. Tech. J. 27, 379 and 623-656.
Skilling, J. (1981). Paper presented at Workshop on Maximum Entropy Estimation and Data Analysis, Univ. Wyoming, Laramie, Wyoming.
Skilling, J. & Bryan, R. K. (1983). Preprint.
Skilling, J. (1983). Preprint.
Steenstrup, S. (1983). Acta Cryst. A39, in the press.
Steenstrup, S. & Wilkins, S.W. (1983). Submitted to Acta Cryst. A.
Wernecke, S.J. (1977). Radio Sci. 12, 831-845.
Wilkins, S.W., Varghese, J.N. & Lehmann, M.S. (1983). Acta Cryst. A39, 47-60.
Wilkins (1983a). Acta Cryst. A39. In the press.
Wilkins (1983b). Acta Cryst. A39. In the press.

DISCUSSION

R. NARAYAN

You mention that the approach of Britten and Collins (1982) is wrong. How do you account for the fact that Collins (1982) has obtained interesting results with his maximum entropy on real protein data?

Your phaseless maximum entropy reconstruction of a one-dimensional model was nearly centrosymmetric although the model is non-centric. We have found that this effect is a function of the value of the zero Fourier coefficient, F_0. At large values of F_0, ME reconstructs a centric structure while at low F_0 it produces a non-centric one.

S.W. WILKINS

The remark that Britten and Collins are wrong refers to their paper: Acta Cryst. A38, 129-32 and concerns the argument relating the maximum entropy principle to the maximum determinant method. For a fuller discussion see Steenstrup and Wilkins: "On Information and Complementarity in Crystal Structure Determination" to appear in Acta Cryst. A.

8.7 WORKSHOP ON MAXIMUM ENTROPY METHODS

Chairman: R.D. Ekers
Reporter: J.A. Roberts

GRIDDING AND CONVOLUTION FOR MEM

Practical applications of MEM in radio astronomy use the Fast Fourier Transform. Hence the measured visibilities, which do not lie on a rectangular grid, must be approximated by values on a rectangular grid. As, for example, described by Schwab in this Symposium the measured values are convolved with a suitable function to produce a continuous function of u and v which is then sampled on a regular rectangular grid. The point-source-response ('beam') is found by replacing all measured visibilities by unity.

J.A. Roberts described the experiences of a group at the CSIRO Division of Radiophysics (see Dulk et al., this Symposium) in using the Cambridge Maximum Entropy Data Consultants package. Satisfactory results were obtained when a single cell 'boxcar' convolving function was used, taking as the MEM 'data file' the gridded visibilities divided (where they were non-zero) by the gridded point-source-response visibilities (i.e. the data file was the transform of a 'uniformly weighted' map), and as the 'accuracies file' a multiple of the gridded point-source-response visibilities (i.e. the transform of a 'naturally weighted' beam). However no satisfactory results had been obtained when, to reduce problems of aliasing, a Kaiser-Bessel convolving function was used. With this convolving function bright regions in the MEM restoration were surrounded by high level rings similar to the sidelobe responses in the 'dirty' map.

S.F. Gull and T.J. Cornwell each reported they had used MEM successfully with data convolved with functions designed to reduce aliasing (T.J. Cornwell using his own MEM package). S.F. Gull gave as his prescription for using the Cambridge M.E. Data Consultants package:

(i) Produce 'naturally weighted' maps and beams using a convolving function designed to reduce aliasing.

(ii) Correct these for convolution by dividing the map or beam by the transform of the convolving function.

(iii) As the 'data file' use the transform of the convolution-corrected naturally-weighted map divided by the transform of the convolution-corrected naturally-weighted beam.

(iv) As the 'accuracies file' use the transform of the convolution-corrected naturally-weighted beam multiplied by the accuracy of an individual visibility value.

DIFFERENT MEM ALGORITHMS

T.J. Cornwell described an algorithm (called VM - for 'variational method') which is available in the current version of the NRAO AIPS package (NRAO VLA Technical Memorandum to be published). Like the Cambridge (Skilling/Gull) algorithm this takes ~100 FFTs to converge. Discussion suggested that for general reconstruction work it was probably not as robust as the Cambridge algorithm, but was probably quite adequate for radio astronomy applications and is simple to code.

WINDOWING

S.F. Gull suggested 'windowing' should be used with MEM as it is with CLEAN, i.e. parts of the map believed to be at the background level should be set to be identically zero. Discussion emphasized that this differed from using windowing in CLEAN where, because the residuals are added to the map made by convolving the CLEAN components with the CLEAN beam, the map is not forced to be zero outside the window.

RESIDUALS

It was suggested that the residuals from the maximum entropy process should be displayed to provide a means of assessing the reliability of the map. S.F. Gull favoured studying the residuals in the u-v plane while others suggested that a map of residuals defined by (dirty map - dirty beam * MEM map) should be displayed, or added to the MEM map. It was noted that with the present MEM algorithms the residuals are not randomly distributed but are closely anti-correlated with the map intensities - a fact which made some participants even more emphatic that the residuals should be displayed.

SOURCE FLUX DENSITIES

S.F. Gull and T.J. Cornwell agreed that, even for 'unresolved' sources, the <u>peak</u> intensities in a MEM map do not bear a linear relation to the source flux densities. The MEM map is normalized so that intensities are in units of flux density/pixel, but the width of the effective point-source-response depends on the signal/noise. Gull and Cornwell suggested that flux densities were better estimated by integrating the MEM map over a region larger than the 'dirty beam' but, because of the non-random residuals, even these integrated values are not linearly related to the flux densities.

S.F. Gull favoured determining a flux density or other specific parameter by using some optimum procedure to estimate that parameter from the original data, e.g. in a least squares sense determining the flux density of a source at the location given by the MEM map (model fitting).

U.J. Schwarz argued that for flux density this was equivalent to summing the CLEAN components over the area of the source. This has been shown to yield an unbiased least squares estimate of the deconvolved source parameters (Schwarz 1979).

S.F. Gull stated that MEM maps should not be used for quantitative analysis, 'they are for looking at the data'. However he suggested that if an MEM map was made using a uniform default value (DEF) and then the intensities in this MEM solution were used as the default values for a few further MEM iterations the map so obtained would be a largely unbiased estimate (cf. O'Sullivan & Komesaroff, these Proceedings).

A CONSTANT POINT-SPREAD FUNCTION

R.D. Ekers and J.A. Roberts suggested there was a need for an algorithm which yielded a map which

(i) fitted the data within the errors,
(ii) was the true intensity distribution convolved with an invariant point-spread function, and
(iii) was 'best' in some defined sense.

T.J. Cornwell argued that this result could be approximated by convolving the MEM map with a 'clean' beam. Not everyone agreed.

THE D.C. VALUE

S.F. Gull agreed that Narayan and Nityananda's method (these Proceedings) of adjusting the maximum to minimum ratio in the map was certainly appropriate in some cases.

OTHER FORMULAE FOR 'ENTROPY'

S.F. Gull maintained that $\Sigma\, p \log(p/w)$ could be made to take any form by suitable choice of w!

J.G. Ables suggested that a proper extension of information theory would provide a measure of the information content of each of a set of possible images and thus enable the image with minimum information to be selected. The functional form for the information content of an image should, he believed, have the following properties:

(i) It should depend only on the image (not on the way it was produced).

(ii) It should be invariant to

(a) the addition of a d.c. level
(b) rescaling of either x, y or the intensity
(c) rotation of the image.

(iii) The information content of two images side by side should be the sum of the information content of each separately.

Reference

Schwarz, U.J. (1979). The method CLEAN - use, misuse and variations. In Image Formation from Coherence Functions in Astronomy, ed. C. van Schooneveld, pp. 261-266. Dordrecht, Holland: Reidel.

9

Processing – other topics

9.1 MAXIMUM RELIABLE INFORMATION IN LIMITED ANGLE TOMOGRAPHY

R.A. Niland
Wills Plasma Physics Department, School of Physics,
University of Sydney, N.S.W. 2006, Australia

Abstract. Suppose that we have m views equispaced in angle, of an object confined to the unit disk but otherwise unknown. Clearly the m projections contain some information about the object, but not all of it. Elementary considerations lead to the fact that the reconstruction will have a resolution cell size of about $^1/m$. This result was made precise in a previous paper (Niland, 1982) where it was shown that m^2 real quantities, generalized moments of the object with respect to a set of special functions, can be obtained without aliasing ambiguity from the m projections.

We extend this work to the case of m views constrained to lie in a limited angular range. For special choices of the directions of the views this reduces to the equispaced case and hence m^2 quantities can again be obtained. The physically intuitive fact that reducing the range of views reduces the collected information is reflected in the poorer condition of the reconstruction process. Numerical results are given to illustrate this effect. We can completely bypass the ill-condition of limited angle tomographic reconstruction by restricting ourselves to retrieval of those moments of the object which are known to be (a) uncontaminated by aliasing, (b) of reasonable condition.

INTRODUCTION

It is well known that prior knowledge such as positivity, or known support, is important in tomographic reconstruction. A special form of this is the knowledge (based on the geometry of the projections) of what information can, in principle, be retrieved from the data. Again, the condition in ill-conditioned problems is improved by attempting to retrieve less information. This is exploited by Davison and Grunbaum (1979) in their technique of "mollification". We exhibit the set of retrievable moments of an unknown 2D function given certain special projections of it. This allows us to have both the above benefits and, in addition, to easily incorporate positivity and support constraints.

EQUISPACED VIEWS

In (Niland, 1982) called (I) hereafter, the following fact is developed. We represent the object $\psi(r,\theta)$ contained in the disk of unit radius by its expansion into the complete set of nonorthogonal functions

$$r^k e^{i\ell\theta} \qquad k = 0,1,2 \ldots \quad \ell = 0,\pm1,\pm2, \ldots$$

so that ψ is fully described by a table of complex moments

$$Q_{\ell k} = \int_o^1 rdr \int_o^{2\pi} d\theta \; r^k e^{i\ell\theta} \psi$$

We take ψ real so that only $\ell \geqslant 0$ needs to be considered. The Q table can essentially be retrieved from the complete set of projections $G(t,\Phi)$, $0 \leqslant t \leqslant 1$, $0 \leqslant \Phi < 2\pi$ where $G(t,\Phi)$ is the integral along the line $y \cos \Phi - x \sin \Phi = t$, and hence ψ may be retrieved from $G(t,\Phi)$. This is already well known and the Radon transform is an explicit formula for ψ in terms of (all) its projections. We have introduced the Q-table because it has a simple relation to tomography with a finite number m views. From (I)

$$Q_{\ell k} = \frac{1}{\Gamma_{\ell k}} \int_o^1 dt \; t^k \int_o^{2\pi} d\Phi \; e^{i\ell(\Phi+\pi/2)} \; G(t,\Phi)$$

relating the Q table to the projections. $\Gamma_{\ell k}$ is a coefficient.

Suppose now that $G(t,\Phi)$ is known for m views only, i.e. $\Phi_\nu = \nu\pi/m$ $\nu = 0,1, \ldots m-1$, $-1 \leqslant t \leqslant 1$. The Fourier integral above must be replaced by a discrete Fourier transform with its accompanying aliasing errors:-

$$Q^*_{\ell k} = \frac{\pi}{2m\Gamma_{\ell k}} \int_{-1}^1 dt \; t^k \sum_{\nu=0}^{m-1} e^{i\ell(\Phi_\nu+\pi/2)} \; G(t,\Phi_\nu)$$

The main result in (I) is that $Q^*_{\ell k} = Q_{\ell k}$ for $\ell = 0,1, \ldots m-1$ and $k = \ell, \ell+2, \ldots, 2m-\ell-2$ since for these special values the aliasing error is absent because $G(t,\Phi_\nu)$ lacks the relevant t-moments (a property of the projection operator: $\psi \to G$). Note that there are m^2 real quantities retrievable from m equispaced views. They lie in a triangle of the Q-table and these triangles are nested as m increases.

For computational stability one can orthogonalize the m^2 weight functions (i.e. functions of the form $r^k e^{i\ell\theta}$ for the above ℓ and k) yielding an orthonormal set $\{\phi_i\}$, $i = 1,2, \ldots m^2$ whose span we call S. A particular example of such an orthonormal set are the functions $R_\ell^\nu(r)\, e^{i\ell\theta}$ on the disk, where R_ℓ^ν are the Zernike polynomials. The corresponding elements in the projection space are $U_{\ell+2\nu}(t)\, e^{i\ell(\Phi+\pi/2)}$ where the U_k are Chebyshev polynomials of the second kind.

LIMITED ANGLE TOMOGRAPHY

Assume that we know the projections of ψ at a set of angles α_i (i = 1,2, ... m) which are the images of the equiview angles Φ_i under some linear map (which will preserve parallelism of lines). For definiteness take $\alpha_1 = \alpha$, $\alpha_m = -\alpha$ and the mapping to be $x = kx$, $y = y$, where $k = \tan\alpha \tan \pi/2m$ (Fig. 1). A view at angle θ maps to one at θ' where $\tan\theta' = \frac{1}{k}\tan\theta$. This determines the value of α_i, i = 2 ... m-1. The unit disk is compressed to an ellipse boxed by $(0,\pm 1)$ $(\pm k,0)$ and each view is reduced in width by the factor

$$w(\theta') = \sqrt{\cos^2\theta' + (k \sin\theta')^2}$$

and its amplitude multiplied by k/w. Notice that as m increases the mapping packs the inner views towards $\theta = 0$ rather than space them uniformly in $[-\alpha,\alpha]$. This effect will tend to offset the advantages in increasing m. Note also that we have chosen the symmetric member of the one-parameter family of linear mappings that open out $[-\alpha,\alpha]$ to the desired angle.

We have now regained the equiview situation with a new object $\psi(x/k,y)$ and hence can ascertain m^2 moments with respect to functions $\phi_i(x,y)$ orthonormal on the disk. One source of loss of condition is immediately evident: the functions ϕ_i will not in general be orthonormal on the ellipse (for $k < 1$) so the m^2 moments are referred to a potentially very skew set of basis functions. This means the information about the object will be encoded in the extreme right hand decimal digits of the derived moments and so be vulnerable to noise. We show this phenomenon effectively reduces the number m^2.

CONDITION OF LIMITED ANGLE TOMOGRAPHY

If $\phi_i(x,y)$ are the orthonormal basis functions for m-view tomography we now know that only moments of our original object ψ with respect to the set $\{\phi_i(kx,y),\ i = 1,2,\ \dots\ m^2\}$ can be retrieved and investigate the condition of this retrieval. Adopting the notation of Davison and Grunbaum (1979) define a map

$$P : S \rightarrow \bigoplus_{i=1}^{m} L^2\left([-1,1],\ 2\sqrt{1-t^2}\right)$$

from S to the space of projection data (m-tuples of functions of $t : -1 \leqslant t \leqslant 1$). P is the Radon transform sampled in angle and restricted to S. For $\psi \in S$, $P\psi = (G(t,\alpha_1), \dots, G(t,\alpha_m))$. We have chosen the weight on the disk = 1 and the weight on PS its image $2\sqrt{1-t^2}$ as seems natural. On PS define the inner product as

$$(\underset{\sim}{\beta},\underset{\sim}{\gamma}) = \frac{1}{m} \sum_{i=1}^{m} \int_{-1}^{1} dt \left[2\sqrt{1-t^2}\right]^{-1} \beta_i(t)\, \gamma_i(t)$$

where $\underset{\sim}{\beta} = (\beta_1(t), \beta_2(t), \ldots , \beta_m(t))$ etc.

Now the adjoint P^t satisfies

$$P^t \underset{\sim}{\beta} = \frac{1}{m} \sum_{i=1}^{m} \frac{\beta_i(-x \sin \alpha_i + y \cos \alpha_i)}{2\sqrt{1 - t^2}}$$

i.e. a weighted average of the back projections.

The adjoint enters naturally because we want to determine the allowed moments (ψ,ϕ_i) of the object ψ from the data $P\psi$ using a relation like $(\psi,\phi_i) = (P\psi,\phi_i')$ where in general $P^t\phi_i' = \phi_i$.

Let us choose a new orthonormal basis $\{\hat{\phi}_i\}$ of S whose corresponding set $\{\hat{\phi}_i'\}$ is orthonormal. This ensures optimum condition and will occur when $P^tP\hat{\phi}_i = \lambda_i^2\hat{\phi}_i$, i.e. the $\{\hat{\phi}_i\}$ are the singular value functions of the operator P with the above defined domain and weights. Now $(\psi,\hat{\phi}_i) = (P\psi,\hat{\phi}_i')/\lambda_i$ so only if λ_i is not too small is the moment accurately determined by the data.

NUMERICAL WORK AND DISCUSSION

We chose m = 4 views. The functions spanning S are the Zernike functions

$$R_\ell^\nu(r)\, e^{i\ell\theta} \qquad \ell = 0,1,2,3$$

$$\nu = 0,1, \ldots 3-\ell$$

with $x \to kx$. The unit disk was represented by a polar-gaussian grid of 20 × 20 points so inner products could be accurately calculated. The above functions were orthogonalized and four projections calculated at angles which are the k-mappings of $(-3,-1,1,3) \times \pi/8$. The projections were represented by their Chebyshev expansions to facilitate back projection in the formation of the adjoint. We computed then ϕ_i (orthonormal basis function), $P\phi_i$, $P^tP\phi_i$ and $(P^tP\phi_i,\phi_j) = M_{ij}$ for $i,j = 1,2, \ldots 16$ and finally found the eigenvalues of M. This procedure was repeated for $\alpha = 67\frac{1}{2}$, 50, 40, 30, 20, 10, 5 degrees and the results shown in Fig. 2. As expected, as α decreases the condition of retrieval of the moments drops away so that with even small noise we cannot retrieve the full m^2. An unexpected finding was the 'saturation' of the seven largest eigenvalues as $\alpha \to 0$. This reflects the fact that as $\alpha \to 0$, S contains functions which are essentially polynomials in y only and so can be retrieved very well from a single view. The above analysis which deals only with S and its image PS does not fully take into account the special properties of S which is not merely a subspace of functions which can be retrieved from their projections (e.g. any set of ridge functions along one of the views would qualify for that) but has the further property that the projection onto S of any unknown function ψ can be retrieved from $P\psi$. Formally S has a very special relationship to the operator P in that there exists another operator, $\bar{P}$ say, such that $\bar{P}P = P_S$ where P_S is the projection operator on to S.

We can modify the numerical work to treat S as part of a larger space as follows. From the equiview theory, and using the k map, it can be shown that we can retrieve $P_S\psi$ from $P\psi$ by forming $(P\psi, v_{\ell k})$ where

$$P\psi = \begin{pmatrix} G(t,\theta_1) \\ G(t,\theta_2) \\ G(t,\theta_3) \\ G(t,\theta_4) \end{pmatrix}, \text{ and } v_{\ell k} = 2\sqrt{1-t^2}\, \frac{4\pi i^\ell t^k}{2m\Gamma_{\ell k}} \begin{pmatrix} w(\theta_1')^k\, e^{i\ell\theta_1'} \\ w(\theta_2')^k\, e^{i\ell\theta_2'} \\ w(\theta_3')^k\, e^{i\ell\theta_3'} \\ w(\theta_4')^k\, e^{i\ell\theta_4'} \end{pmatrix}$$

$\theta_i' = (2i-5)\pi/8$, $\ell = 0,1,2,3$ and $k = \ell, \ell+2, \ldots 2m-\ell-2$, $m = 4$.

Write the projection of $P\psi$ onto V the span of $\{v_{\ell k}\}$ as $P_V P\psi$. As before we seek two orthonormal systems but now $\hat{\phi}_i \in S$ and $\hat{\phi}_i' \in V$. This requires

$$P_S P^t P_V P\hat{\phi}_i = \lambda_i^2 \hat{\phi}_i$$

so we must include an extra projection on to V. Notice if V is spanned by (any) orthonormal set $\{v_i\}$ and $(\phi_i, P_S P^t P_V P\phi_j) = M_{ij}$ then

$$M = BB^T \quad \text{where} \quad B_{ij} = (\phi_i, P^t v_j) = (P\phi_i, v_j)$$

Results are shown in Fig. 3.

Notice that the result is similar to the previous result - however these eigenfunctions define a space onto which the projection of any other function can be retrieved (in principle) from the data. We point out that the condition is not catastrophic and also that the eigenfunctions are quite non radial as α becomes small. This represents a "mollification" in a way entirely native to the problem. There is a great loss of two dimensional information which can presumably be quantified by isolating the rotation invariant part of the space S. One expects as $\alpha \to 0$ to regain the single quantity predicted by the equiview theory for m=1.

ACKNOWLEDGEMENTS

Financial support for this work was provided by the Australian Research Grants Scheme, NERDDC, AINSE, and the Science Foundation for Physics within the University of Sydney.

Reference list

Davison, M.E. & Grunbaum, F. Alberto (1979). Convolution algorithms for arbitrary projection angles. IEEE Trans. Nucl. Sci., NS-26, 2670.

Niland, R.A. (1982). The maximum reliable information obtainable by tomography. JOSA, 72, 1677.

Fig. 1. Geometry of Limited Angle tomography.

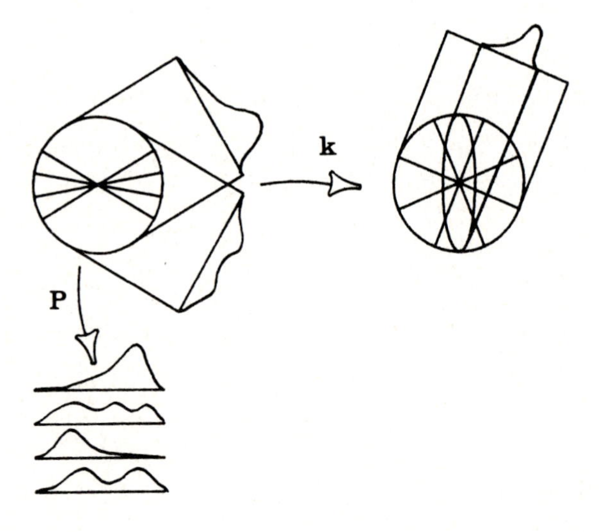

Fig. 2. Spectrum of $P_S P^t P$

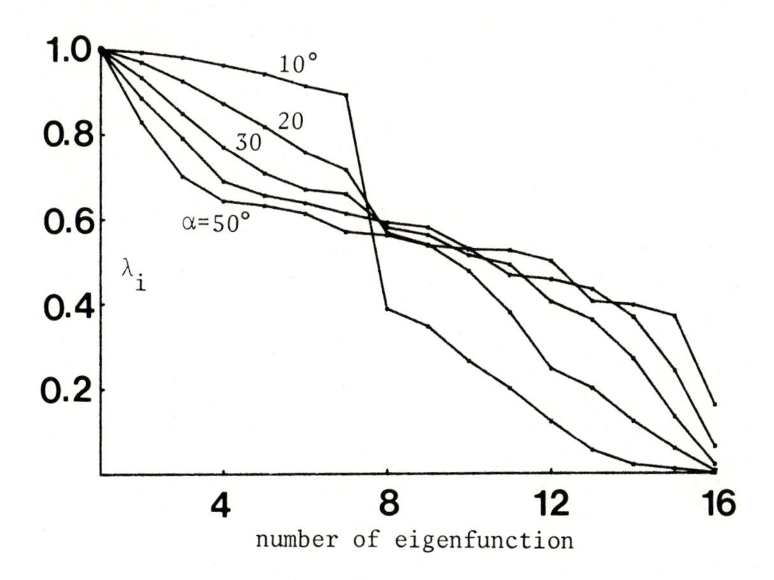

Fig. 3. Spectrum of $P_S P^t P_V P$

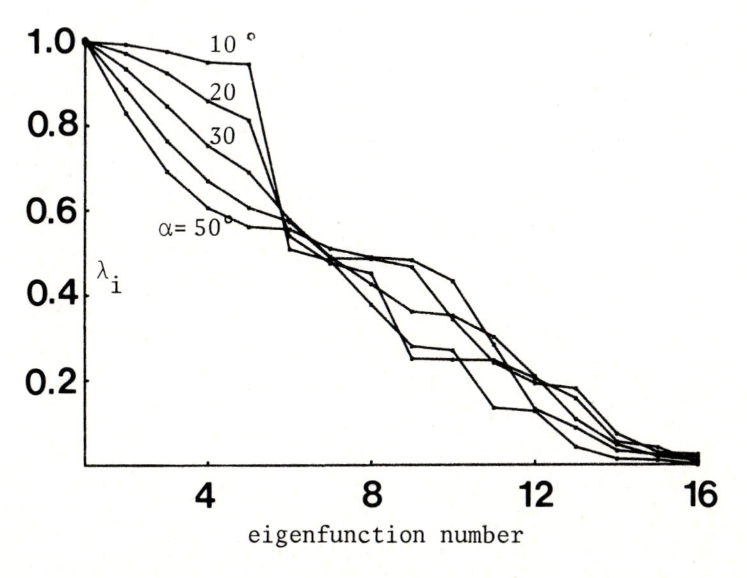

DISCUSSION

F.R. SCHWAB

Frank Natterer and M. Zuhair Nashed have done related work on the ill-conditioned nature of limited-angle reconstruction. Natterer's results, I think I recall, also worked out in terms of the distribution of the singular values of a discretized problem.

R.A. NILAND

I think the difference lies in the discretization. The main point of my paper is that for m views either equispaced in angle or specially chosen within the prescribed limited angle then there is a privileged m^2 dimensional subspace and the projections of objects on to this subspace can be retrieved from the tomographic views. This is modified by limited angle tomography so the effective dimension m^2 drops down.

9.2 OPTIMAL GRIDDING OF VISIBILITY DATA IN RADIO INTERFEROMETRY

F. R. Schwab
National Radio Astronomy Observatory, Charlottesville, Va.

Abstract. The obvious criterion for choosing the gridding convolution function (g.c.f.) in radio mapping leads to 0-order prolate spheroidal wave functions (p.s.w.f.'s). One can do "better", though. Borrowing from work of D. R. Rhodes, I introduce weighted optimality criteria. By de-emphasizing aliasing suppression around map edges, greater effect can be achieved in the center, near the object of interest. Spheroidal functions (including the p.s.w.f.'s) then are the optimal g.c.f.'s. Approximations to these are presented, and applications to wide-field mapping and confusion-limited self-calibration are outlined.

1 INTRODUCTION

In radio mapping, ordinarily the irregularly sampled visibility data are convolved with a suitably chosen function C of finite extent (i.e., of bounded support) in order that the data may be re-sampled over a rectangular grid and, hence, that the fast Fourier transform algorithm (FFT) then may be used to approximate the radio brightness. In addition to its use for smoothing and interpolating the data prior to re-sampling, C can be chosen in order to suppress the aliasing induced by the uniform spacing of the grid. The topic considered here is optimal selection of this *gridding convolution function* (g.c.f.), C, to yield the most effective aliasing suppression.

The importance of selecting an effective g.c.f. is worth emphasizing. For though it may happen that near the region of interest there are no sources that are likely to cause serious aliasing problems (owing to their responses having been suppressed by instrumental effects: finite integration time, finite bandwidth, antenna primary beam attenuation, etc.), aliasing of a source's own sidelobes, too, is a potentially serious problem. An illustration of this effect, discussed at length by Greisen [4], is shown in Figure 1. The map in Figure 1a is a CLEAN map which was derived using a "pillbox" g.c.f. (i.e., the gridding was by cell-averaging), whereas in Figure 1b the map was produced using one of the better g.c.f.'s which are described below. (The intensity scale has been exaggerated---the photos are over-exposed.) In this example, there are no aliased sources. Rather, all of the clutter in the central region of Figure 1a is due to aliases of sidelobes (sidelobes, lying outside of the field of view, of the three point sources within the map). Since none of the usual mapping/deconvolution algorithms can cope with aliased responses, whether due to "confusing" sources or to sidelobes, use of a good g.c.f. is a necessity for high dynamic range mapping.

It is well known, following the work of Slepian, Landau, & Pollak in the 1960's on functions which simultaneously are essentially band-limited

Figures 1a-b. Comparison of clean maps made with a poor g.c.f. (left) and a better one (right). The two g.c.f.'s are a 1 cell x 1 cell "pillbox" and a 6 cell x 6 cell spheroidal function, ψ_{10} (see §4). All of the "clutter" in the central region at left is due to aliased sidelobes, and not to "confusing" sources. A good g.c.f. is a necessity for high dynamic range mapping.

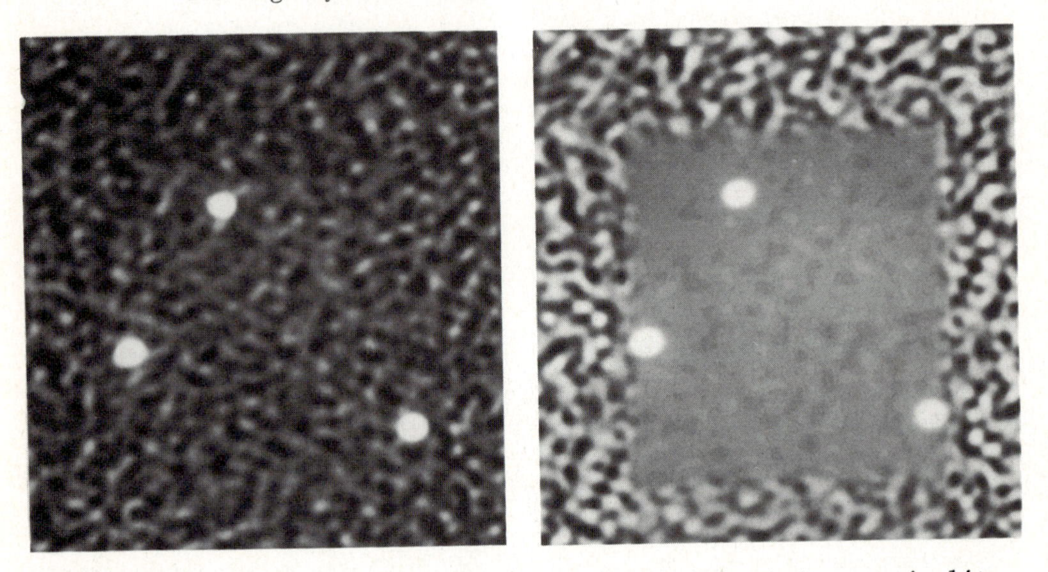

and essentially time-limited [13, 9], that the most obvious optimality criterion for choice of a g.c.f. leads to the selection of a 0-order prolate spheroidal wave function (p.s.w.f.) (cf. Brouw's 1975 review article [1]). Even so, this g.c.f., or one approaching its effectiveness, usually has not been used in the standard data reduction packages ---mainly because of lack of ready access to simple numerical approximations.

The main point of this talk is that one can make even a bit "better" choice than the 0-order p.s.w.f. But since that choice already is an optimal one, we need here a new definition of "optimal". Borrowing from work of Rhodes [10] and Fedotowsky & Boivin [3], I introduce a continuum of optimality criteria parametrized by a real number $\alpha > -1$. α determines a weight function. The weighting can be used to de-emphasize the importance of aliasing suppression around the map edges, so that significantly greater suppression can be achieved in the center, near the object of interest. In this case one chooses $\alpha > 0$; choosing $\alpha < 0$ has the opposite effect.

The g.c.f.'s that are optimal according to the criteria introduced here are known as *spheroidal functions* (see Stratton [16]). This class of functions includes the p.s.w.f.'s, which correspond to the case ($\alpha = 0$) of no weighting.

In most applications the p.s.w.f.'s, or more ordinary functions, are "good enough" g.c.f.'s. So, the main contribution here is only to "tidy-up" the subject a bit. I have computed a few simple rational approximations to selected spheroidal functions---including some of the ordinary 0-order p.s.w.f.'s---which are potentially useful g.c.f.'s. These approximations are tabulated in §4. In addition to the application to the standard mapping problem, two specialized applications---to the problems of wide-field mapping and confusion-limited self-calibration---are given in §5.

2 OPTIMALITY CRITERIA

Let us denote the u-v sampling distribution by S---i.e., if $\{(u_i,v_i)\}_{i=1}^{n}$ denotes the u-v coverage, then S is the generalized function given by $S(u,v) = \sum_{i=1}^{n} \delta(u-u_i,v-v_i)$---, and denote the visibility function by V. We may consider the sky to be flat. The "gridding" operation is simply one of pre-filtering and re-sampling---i.e., convolving the measurement distribution (SV + error) with some given function C, of compact support (i.e., 0 outside of some bounded region), and then re-sampling the convolution over a rectangular grid. Denoting by R that comb-like re-sampling distribution which is supported on (i.e., confined to) the lattice points of the grid, these operations can be represented as $R(C * (SV + \text{error}))$. The purpose of the gridding convolution is twofold: (i) to interpolate and smooth the measurements so that, by re-sampling, the FFT may be used to approximate the brightness distribution, and (ii) to reduce the intensity of spurious responses that are due to sources and to sidelobes lying outside of the map field. These spurious features, or aliases, appear in the map because the Fourier transform (FT) of R is again a comb-like distribution with peaks (or teeth) whose separation is equal to the linear dimensions of the map. The effect of the convolution, in addition to interpolating and smoothing, is to pre-multiply the "dirty map" $\widehat{SV}$ by the FT, $\hat{C}$, of C. Judicious choice of C greatly suppresses the aliased responses.

C always is taken to be real and symmetric about the origin, and $\hat{C}$ inherits these properties. Usually C is separable and confined to a rectangle. The computing time required for gridding is proportional to the area of the rectangle; so, for computational economy, the sides of the rectangle seldom measure more than six times the grid spacing.

Let $\eta = (\eta_1,\eta_2)$ denote the sky coordinates, and suppose that C is confined to a prescribed region B; i.e., that $C \equiv 0$ outside of B, or $\text{supp}(C) = B$. Let A denote the region of interest on the sky. When A coincides with the region being mapped, the part of $\widehat{SV}$ falling outside of A appears as the aliased response, but it is tapered by $\hat{C}$. So a reasonable optimization goal is to maximize the quantity

$$R = \frac{\iint_A |\hat{C}(\eta_1,\eta_2)|^2 \, d\eta_1 d\eta_2}{\int_{-\infty}^{\infty} \int_{-\infty}^{\infty} |\hat{C}(\eta_1,\eta_2)|^2 \, d\eta_1 d\eta_2},$$

which measures the extent to which $\hat{C}$ is concentrated in A. One also might wish to include a weight function in the definition of R:

$$R = \frac{\iint_A |\hat{C}(\eta_1,\eta_2)|^2 \, w(\eta_1,\eta_2) \, d\eta_1 \, d\eta_2}{\int_{-\infty}^{\infty} \int_{-\infty}^{\infty} |\hat{C}(\eta_1,\eta_2)|^2 \, w(\eta_1,\eta_2) \, d\eta_1 \, d\eta_2}$$

The problem, then, is to find a function C, with w and $\text{supp}(C) = B$ both given, that maximizes R.

We shall restrict our attention to the case of A and B rectangular, with sides aligned parallel to the coordinate axes. Without loss of generality, we may assume that A is the square $|\eta_1| \leq 1$, $|\eta_2| \leq 1$. We shall take B to be the rectangle of linear dimensions m_1 and m_2 times the respective grid spacings, Δu and Δv (m_1 and m_2 needn't be integral). And, in order to apply Rhodes' results, we shall choose a separable weight function given by $w(\eta_1,\eta_2) = w_1(\eta_1)w_2(\eta_2)$, where $w_i(\eta) = |1 - \eta^2|^{\alpha_i}$ with $\alpha_i > -1$ given. The solution is a separable function equal to $w_1 w_2$ times the product of the solutions to two 1-dimensional problems of the form:

"Find f which maximizes $R_\alpha(c) = \dfrac{\int_{-1}^{1} |F_\alpha(c,\eta)|^2 (1 - \eta^2)^\alpha \, d\eta}{\int_{-\infty}^{\infty} |F_\alpha(c,\eta)|^2 |1 - \eta^2|^\alpha \, d\eta}$,

where $F_\alpha(c,\eta) = \displaystyle\int_{-1}^{1} e^{ic\eta t} (1 - t^2)^\alpha f(t) \, dt$,

and where c is related to the m_i by $c_i = \pi m_i/2$."

(Here, $(1 - t^2)^\alpha f$ ought to be identified with C; and F_α, whose concentration is being maximized, with $\hat{C}$. By using normalized coordinates (up until §4), and omitting the factor of 2π from the argument of the exponential kernel of the FT, the notation is simplified. Thus, $\eta = \pm 1$ at the map edges and at the boundary of the support of the g.c.f.) For the original 2-dimensional problem, the maximal R is given by $R^{\text{opt}} = R^{\text{opt}}_{\alpha_1}(c_1) R^{\text{opt}}_{\alpha_2}(c_2)$.

A fair body of literature, much of it reviewed and extended in Fedotowsky & Boivin [3], deals with this type of maximal concentration problem.

Certain cases in which A and B both are rectangular, circular, or ellipsoidal, reduce to nice problems whose solutions satisfy well-studied second order ODE's. Results pertaining to rectangular domains A and B, in the absence of weighting, are due to Slepian, Pollak, & Landau [13, 9]; those pertaining to rectangular domains with the weight function used here are due to Rhodes [10]; and results pertaining to circular domains, in the absence of weighting, are due to Slepian [14]. Jarem in [8] gives a numerical procedure for constructing solutions to 1-dimensional problems of the form considered here, but allowing arbitrary weight functions; but I know of no analytic solutions more general than the spheroidal functions. Grünbaum [5] has derived discrete analogs of the p.s.w.f.'s and describes in [6] an application to tomographic reconstruction. [15] is an up-to-date review of much of this work.

3 THE OPTIMAL GRIDDING FUNCTIONS

The function f that maximizes $R_\alpha(c)$ is among those solutions of the differential equation $(1 - \eta^2)\, f'' - 2(\alpha + 1)\eta f' + (b - c^2\eta^2)f = 0$ which remain finite at $\eta = \pm 1$. These bounded solutions of the DE are termed *spheroidal functions* (see [16]). For each α and c there exist countably many solutions which are distinct up to an arbitrary normalization factor, and which, in the literature, commonly are denoted by $\psi_{\alpha n}(c,\eta)$, $n = 0,1,2,\ldots$. They correspond to eigenvalues $b_{\alpha n}(c)$ of the DE which may be arranged in ascending order, $0 < b_{\alpha 0}(c) < b_{\alpha 1}(c) < \ldots$. The solution $\psi_{\alpha 0}(c,\eta)$, corresponding to the smallest eigenvalue, maximizes $R_\alpha(c)$. More generally, for each α and c, and $N \geq 1$, the first $N + 1$ of the spheroidal functions are the $N + 1$ linearly independent functions of the form $F_\alpha(c,\eta)$ which are the most concentrated on $[-1,1]$, according to our criterion. The DE, above, is equivalent to the integral equation $\nu f(\eta) = \int_{-1}^{1} e^{ic\eta t}(1 - t^2)^\alpha f(t)\, dt$. So $\psi_{\alpha n}(c,\eta)$ is, apart from a constant factor (the eigenvalue $\nu_{\alpha n}(c)$ of the integral equation), its own finite, weighted-kernel Fourier transform. For these results, and the further properties summarized below, see Rhodes [10] from which all of the results of this section have been borrowed.

The spheroidal functions have a number of interesting properties. For fixed α and c, the $\psi_{\alpha n}$ are simultaneously orthogonal over two domains ---both over the interval $[-1,1]$, and over the whole real line (with respect to an inner product weighted by $|1 - \eta^2|^\alpha$). They are complete in the functions on $[-1,1]$ square-integrable with weight $(1 - \eta^2)^\alpha$, and complete in the space of band-limited, filtered functions of the form $F_\alpha(c,\eta)$. $\psi_{\alpha n}$ is real for real η, has exactly n zeros in the interval $[-1,1]$, is nonzero at $\eta = \pm 1$, is even or odd depending on the parity of n, and its analytic extension to complex η is entire. Certain of the spheroidal functions are named special functions. The relation to the p.s.w.f.'s (in "Flammer's notation") is $\psi_{mn}(c,\eta) = (1 - \eta^2)^{-m/2} S_{m,m+n}(c,\eta)$, $m = 0,1,2,\ldots$ (i.e., for each integral α, the optimal g.c.f. is simply related to an α-order p.s.w.f.). $\psi_{\pm\frac{1}{2},n}$ are related in a simple manner to the periodic Mathieu functions ce_n and se_{n+1}.

4 APPROXIMATIONS

So, having decided upon the use of a separable gridding convolution function $C(u,v) = C_1(u)C_2(v)$ supported in a rectangle whose sides in the directions of the coordinate axes measure m_1 and m_2 times the respective grid spacings, Δu and Δv, and having chosen exponents α_1 and α_2 in the separable weight function, the optimal g.c.f. is of the form $C(u,v) = |1 - \eta_1^2(u)|^{\alpha_1}\psi_{\alpha_1 0}(c_1,\eta_1(u))|1 - \eta_2^2(v)|^{\alpha_2}\psi_{\alpha_2 0}(c_2,\eta_2(v))$, where $\eta_1(u) = 2u/m_1\Delta u$, $\eta_2(v) = 2v/m_2\Delta v$, and $c_i = \pi m_i/2$. Thus, for gridding, we need to approximate $\psi_{\alpha 0}(c,\eta)$ for η in the range [0,1], α perhaps in the interval [0,2], and, assuming $m \leq 8$, for $c \leq 4\pi$. The map plane taper is proportional to $\psi_{\alpha_1 0}(c_1,2x\Delta u)\psi_{\alpha_2 0}(c_2,2y\Delta v)$, assuming that u and v are measured in wavelengths and that x and y are direction cosines on the plane of the sky.

Hodge [7] gives a method which may be used to calculate the eigenvalues of the DE for the p.s.w.f.'s. One can modify his procedure in order to yield the eigenvalues $b_{\alpha n}(c)$ of the more general DE for the spheroidal functions. $\psi_{\alpha 0}$ then is computed by evaluating the expansion in fractional order Bessel functions given by Rhodes in [10, Eq. 14]---the eigenvalues determine the expansion coefficients. This expansion, convergent for all real α, is an efficient means of calculating the functions for values of the parameters over the range of interest.

Given an accurate determination of $b_{\alpha 0}(c)$, a more straightforward method, if one only is interested in $|\eta| < 1$, is simply to numerically integrate the DE, starting at $\eta = 0$. This method, though, is guaranteed to blow-up near $\eta = \pm 1$, because of the pole in the DE.

Fortran subroutines implementing these two methods are given in a technical report [11]. In addition, I have computed rational approximations to selected 0-order spheroidal functions which are potentially useful g.c.f.'s. Table 1 lists the coefficients of the approximations for five choices of support size, m = 4, 5, 6, 7, and 8 cells, and five choices of the weight function exponent, α = 0 (the 0-order p.s.w.f.), 1/2, 1, 3/2, and 2. For $m \geq 6$, two approximations are used to cover the full range $0 \leq |\eta| \leq 1$. Each approximation is of the form

$$\psi_{\alpha 0}(c,\eta) \simeq \frac{\sum_{k=0}^{n} p_k(\eta^2 - \eta_2^2)^k}{\sum_{k=0}^{d} q_k(\eta^2 - \eta_2^2)^k}, \quad \eta_1 \leq |\eta| \leq \eta_2, \quad \text{where } c = \frac{\pi m}{2}.$$

Quotients of polynomials in (η^2 - constant), rather than η^2, are used in order to avoid cancellation error when the sums are evaluated by Horner's rule. On each line of the Table, the numerator coefficients, $p_0,\ldots,p_n$, are listed first, followed by the denominator coefficients, $q_0,\ldots,q_d$. The latter can be distinguished from the numerator coefficients by the fact that in each case $q_0 = 1$. Each approximation is an approximately best uniform rational approximation, in the sense that the maximum relative error

Table 1. Coefficients of rational approximations to selected 0-order spheroidal functions.

m	α	η_1	η_2									
4	0	0	1	1.584774−2	−1.269612−1	2.333851−1	−1.636744−1	5.014648−2	1	4.845581−1	7.457381−2	
	$\frac{1}{2}$			3.101855−2	−1.641253−1	2.385500−1	−1.417069−1	3.773226−2	1	4.514531−1	6.458640−2	
	1			5.007900−2	−1.971357−1	2.363775−1	−1.215569−1	2.853104−2	1	4.228767−1	5.655715−2	
	$\frac{3}{2}$			7.201260−2	−2.251580−1	2.293715−1	−1.038359−1	2.174211−2	1	3.978515−1	4.997164−2	
	2			9.585932−2	−2.481381−1	2.194469−1	−8.862132−2	1.672243−2	1	3.756999−1	4.448800−2	
5	0	0	1	3.722238−3	−4.991683−2	1.658905−1	−2.387240−1	1.877469−1	−8.159855−2	3.051959−2	1	2.418820−1
	$\frac{1}{2}$			8.182649−3	−7.325459−2	1.945697−1	−2.396387−1	1.667832−1	−6.620786−2	2.224041−2	1	2.291233−1
	1			1.466325−2	−9.858686−2	2.180684−1	−2.347118−1	1.464354−1	−5.350728−2	1.624782−2	1	2.177793−1
	$\frac{3}{2}$			2.314317−2	−1.246383−1	2.362036−1	−2.257366−1	1.275895−1	−4.317874−2	1.193168−2	1	2.075784−1
	2			3.346886−2	−1.503778−1	2.492826−1	−2.142055−1	1.106482−1	−3.486024−2	8.821107−3	1	1.983358−1
6	0	0	.75	5.613913−2	−3.019847−1	6.256387−1	−6.324887−1	3.303194−1	1	9.077644−1	2.535284−1	
	$\frac{1}{2}$			6.843713−2	−3.342119−1	6.302307−1	−5.829747−1	2.765700−1	1	8.626056−1	2.291400−1	
	1			8.203343−2	−3.644705−1	6.278660−1	−5.335581−1	2.312756−1	1	8.212018−1	2.078043−1	
	$\frac{3}{2}$			9.675562−2	−3.922489−1	6.197133−1	−4.857470−1	1.934013−1	1	7.831755−1	1.890848−1	
	2			1.124069−1	−4.172349−1	6.069622−1	−4.405326−1	1.618978−1	1	7.481828−1	1.726085−1	
	0	.75	1	8.531865−4	−1.616105−2	6.888533−2	−1.109391−1	7.747182−2	1	1.101270	3.858544−1	
	$\frac{1}{2}$			2.060760−3	−2.558954−2	8.595213−2	−1.170228−1	7.094106−2	1	1.025431	3.337648−1	
	1			4.028559−3	−3.697768−2	1.021332−1	−1.201436−1	6.412774−2	1	9.599102−1	2.918724−1	
	$\frac{3}{2}$			6.887946−3	−4.994202−2	1.168451−1	−1.207733−1	5.744210−2	1	9.025276−1	2.575337−1	
	2			1.071895−2	−6.404749−2	1.297386−1	−1.194208−1	5.112822−2	1	8.517470−1	2.289667−1	
7	0	0	.775	2.460495−2	−1.640964−1	4.340110−1	−5.705516−1	4.418614−1	1	1.124957	3.784976−1	
	$\frac{1}{2}$			3.070261−2	−1.879546−1	4.565902−1	−5.544891−1	3.892790−1	1	1.075420	3.466086−1	
	1			3.770526−2	−2.121608−1	4.746423−1	−5.338058−1	3.417026−1	1	1.029374	3.181219−1	
	$\frac{3}{2}$			4.559398−2	−2.362670−1	4.881998−1	−5.098448−1	2.991635−1	1	9.865496−1	2.926441−1	
	2			5.432500−2	−2.598752−1	4.974791−1	−4.837861−1	2.614838−1	1	9.466891−1	2.698218−1	
	0	.775	1	1.924318−4	−5.044864−3	2.979803−2	−6.660688−2	6.792268−2	1	1.450730	6.578684−1	
	$\frac{1}{2}$			5.030909−4	−8.639332−3	4.018472−2	−7.595456−2	6.696215−2	1	1.353872	5.724332−1	
	1			1.059406−3	−1.343605−2	5.135360−2	−8.386588−2	6.484517−2	1	1.269924	5.032139−1	
	$\frac{3}{2}$			1.941904−3	−1.943727−2	6.288221−2	−9.021607−2	6.193000−2	1	1.196177	4.460948−1	
	2			3.224785−3	−2.657664−2	7.438627−2	−9.500554−2	5.850884−2	1	1.130719	3.982785−1	
8	0	0	.775	1.378030−2	−1.097846−1	3.625283−1	−6.522477−1	6.684458−1	−4.703556−1	1	1.076975	3.394154−1
	$\frac{1}{2}$			1.721632−2	−1.274981−1	3.917226−1	−6.562264−1	6.305859−1	−4.067119−1	1	1.036132	3.145673−1
	1			2.121871−2	−1.461891−1	4.185427−1	−6.543539−1	5.904660−1	−3.507098−1	1	9.978025−1	2.920529−1
	$\frac{3}{2}$			2.580565−2	−1.656048−1	4.426283−1	−6.473472−1	5.494752−1	−3.018936−1	1	9.617584−1	2.715949−1
	2			3.098251−2	−1.854823−1	4.637398−1	−6.359482−1	5.086794−1	−2.595588−1	1	9.278774−1	2.530051−1
	0	.775	1	4.290460−5	−1.508077−3	1.233763−2	−4.091270−2	6.547454−2	−5.664203−2	1	1.379457	5.786953−1
	$\frac{1}{2}$			1.201008−4	−2.778372−3	1.797999−2	−5.055048−2	7.125083−2	−5.469912−2	1	1.300303	5.135748−1
	1			2.698511−4	−4.628815−3	2.470890−2	−6.017759−2	7.566434−2	−5.202678−2	1	1.230436	4.593779−1
	$\frac{3}{2}$			5.259595−4	−7.144198−3	3.238633−2	−6.946769−2	7.873067−2	−4.889490−2	1	1.168075	4.135871−1
	2			9.255826−4	−1.038126−2	4.083176−2	−7.815954−2	8.054087−2	−4.552077−2	1	1.111893	3.744076−1

$$\max_{\eta_1 \leq \eta \leq \eta_2} \left| \frac{\psi_{\alpha 0}^{\text{approx}}(c,\eta) - \psi_{\alpha 0}(c,\eta)}{\psi_{\alpha 0}(c,\eta)} \right|$$

is, to about three significant decimal digits, the smallest maximum relative error that can be achieved with the chosen approximating form. In each case, the maximum relative error is less than a few times 10^{-6}. Not all of the digits displayed in the higher-order coefficients are significant. (Note that the weight factor needed to obtain the g.c.f. is not included in these approximations.)

Use of a support width $m > 6$ is not recommended, except with extended precision arithmetic, because of the extreme correction which would be required at the map edges. $(1 - \eta^2)\psi_{10}$, with $m = 6$, is the default choice of g.c.f. used in one of the NRAO mapping programs. More details are given in [11, 12].

Plots of the map-plane tapering achieved with a few of the spheroidal functions, for the case $m = 6$, are shown in Figure 2, along with comparisons with two other common g.c.f.'s: a 1 cell x 1 cell "pillbox" g.c.f., and a Gaussian-tapered "sinc" function. The two characteristic width parameters of the latter g.c.f. have been adjusted to maximize $R_0(3\pi)$, the quantity which is maximized over all square-integrable g.c.f.'s by the 0-order p.s.w.f. Plots of a function $\dot{C}(\eta)$, defined as the ratio of the intensity of an alias from position η outside of the field of view, to the intensity the feature would have if it actually lay within the map field, at the position of its alias, are given in [11], for various choices of support size m. Space limitations do not permit including here this more useful form of display.

5 FURTHER APPLICATIONS

In addition to the obvious application to mapping, use of an optimal g.c.f. can aid, in another way, in wide-field mapping/deconvolution. Consider the case of a wide field consisting of N isolated source components (or isolated sources, if you prefer) and assume that the location and extent (i.e., the support) of each of the source components is known. (Often an approximation to the support of the brightest source components, at least, can be obtained from a low-resolution map.) Then it may be unnecessary to compute a map over a single grid large enough to encompass all of the source components. Instead, N small maps can be computed, over N small grids, each one encompassing a single source component. If a sufficiently good g.c.f. is used, then there are only negligibly small aliased responses in any of the N maps; that is, the main effect any source component has upon the $N - 1$ little maps other than its own is to contribute sidelobes which legitimately fall within the other regions.

Now, with a bit more bookkeeping than usual, the N maps can be cleaned in unison, by a modification of the Clark CLEAN algorithm [2]. The algorithm consists of two (nested) iteration loops. In each inner

Figure 2. Plots of the map-plane "tapering" $\log_{10}|\hat{C}(\eta)|$ achieved with various g.c.f.'s: *a*) spheroidal function, $m = 6$, $\alpha = 1$, along with the FT of a "pill-box" g.c.f. one cell in width; *b*) spheroidal function, $m = 6$, $\alpha = 0$; *c*) optimized Gaussian-tapered "sinc" function (see text); *d*) spheroidal function, $m = 6$, $\alpha = -1/2$. For the spheroidal functions, $\hat{C}(\eta) = \psi_{\alpha 0}(\pi m/2, \eta)$.

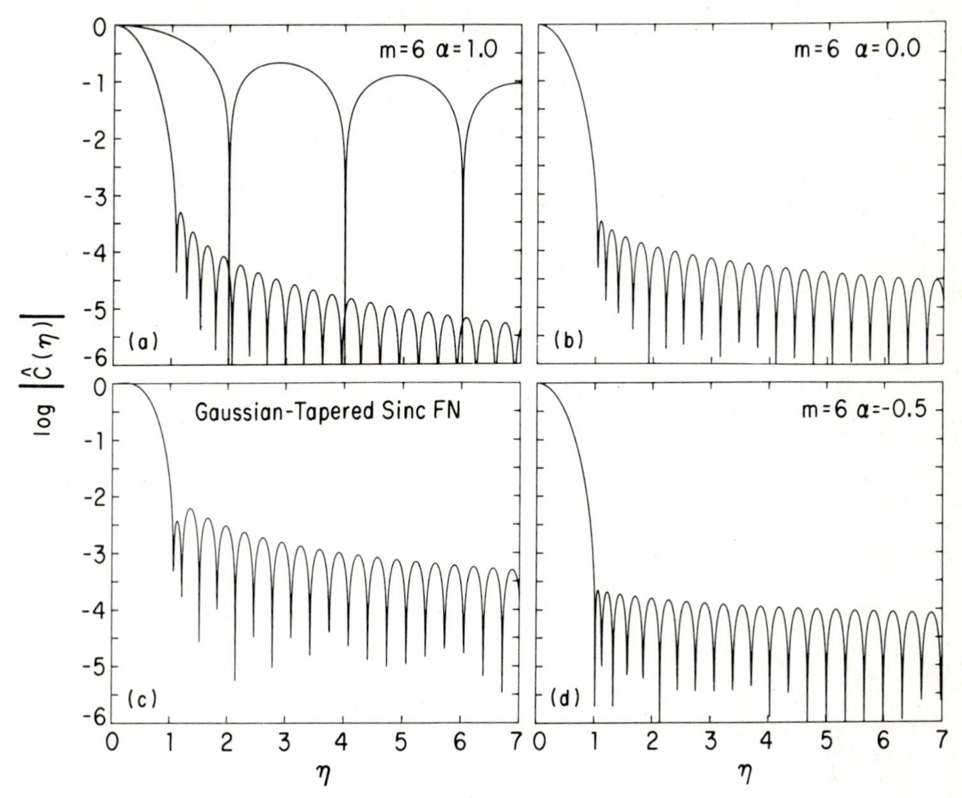

iteration the brightest residual is located, considering all of the residual maps together, and a component at that location is removed from that location's map alone (one hasn't computed the point source response out as far as the other map regions). After some (not too large) number of inner iterations, an outer iteration is performed. The outer iterations consist of computing the visibility residuals (not from gridded visibility data, but from the ungridded data), then re-gridding (N times over), again using one of the optimal g.c.f.'s, and then finally re-mapping, to obtain all N residual maps, properly computed. When the visibility residuals are computed, various corrections can be applied to the clean component model visibilities---e.g., corrections for non-coplanarity of the baselines, for non-monochromaticity or finite bandwidth, and for finite integration time; that is, albeit crudely, the algorithm, as a side benefit, can be made to perform a more general numerical inversion than deconvolution.

Many fields are too complex for this algorithm to be practical or effective. But, in some cases of a few widely separated source components, it may be very much more economical than the standard map/CLEAN combination, since the *N* small grids may contain many fewer "pixels" than would a single grid large enough to surround all of the source components. The algorithm is conceptually straightforward but, admittedly, tricky to program.

Finally, I shall mention a simple application to confusion-limited self-calibration. In self-calibration, given roughly calibrated data which are used to determine an initial source model (via mapping/deconvolution) satisfying certain constraints such as bounded support, positivity, only a believable level of roughness, etc., one solves for time-varying calibration parameters, or complex gains---one per antenna---which put the visibility data in best agreement with the model. (One is restrained from putting the data in too good agreement with the model, because with an *n* element array, at any given instant, there are $O(n^2)$ observations, but only $O(n)$ calibration parameters to be determined. The number of degrees of freedom can be held further in check by assuming that the complex gains are slowly-varying.) When the field is highly complex, with very many widely separated sources, the self-calibration procedure breaks down, because it is too difficult to model the entire field of view to which the interferometer pairs are sensitive.

One can alleviate this problem by pre-filtering the visibility data (*u-v* track by *u-v* track), before solving for the calibration parameters. Assume, for example, that one can find a rough model for the sources in a rectangular neighborhood *A* of the source of primary interest. Then, as in §2, for an arbitrary rectangle *B* in the *u-v* plane, one can find a function $C(u,v)$, confined to *B*, and such that $\hat{C}$ is highly concentrated in *A*. By convolving the data along each *u-v* track (separately) with *C*, and using a source model tapered by $\hat{C}$, one can filter out some of the "confusion." But since the tracks are approximately linear in the neighborhood of a given (u,v) location, the filtered visibility datum at that (u,v) still may have a contribution due to source components off in the direction $\pi/2 + \tan^{-1}(dv/dt \,/\, du/dt)$ from the field center, since the fringe rate due to such components is close to 0. Nevertheless, in many cases, enough of the data may be filtered well enough to aid considerably in the solution for the calibration parameters. And this technique is very inexpensive to apply.

Because of the problem of ionospheric phase corruption, such a scheme may be a necessity in the case of a low-frequency array, such as the 75 MHz array which has been proposed by R. A. Perley of the NRAO. The "directionality" problem mentioned at the end of the last paragraph can be overcome by the technique of bandwidth synthesis: in place of each single *u-v* track, one then has a family of closely spaced tracks from different frequency channels; and these 2-dimensional data can be filtered effectively, assuming that the frequency resolution and the time resolution are not too coarse.

ACKNOWLEDGEMENTS

I wish to acknowledge the benefit of discussions with Dr. E. W. Greisen, who, in particular, pointed up the importance of suppressing the aliasing due to a source's own sidelobes. I have made extensive use of his computer programs in this work. The ideas of §5 originated in discussions with Drs. W. D. Cotton and R. A. Perley.

The National Radio Astronomy Observatory (NRAO) is operated by Associated Universities, Inc., under contract with the National Science Foundation.

REFERENCES

1. Brouw, W.N. (1975). Aperture synthesis. In Alder, B., Fernbach, S. & Rotenberg, M. (eds.), Methods in Computational Physics, 14, pp. 131-175. New York: Academic Press.
2. Clark, B.G. (1980). An efficient implementation of the algorithm CLEAN. Astron. Astrophys., 89, 377-378.
3. Fedotowsky, A. & Boivin, G. (1972). Finite Fourier self-transforms. Quart. Appl. Math., 30, 235-254.
4. Greisen, E.W. (1979). The effects of various convolving functions on aliasing and relative signal-to-noise ratios. VLA Scientific Memorandum No. 131 (NRAO-VLA, P.O. Box O, Socorro, NM 87801, U.S.A.).
5. Grünbaum, F.A. (1981). Eigenvectors of a Toeplitz matrix: discrete versions of the prolate spheroidal wave functions. SIAM J. Alg. Discr. Meth., 2, 136-141.
6. ____________. (1983). The limited angle reconstruction problem. In Shepp, L.A. (ed.), Computed Tomography. Proc. Symposia in Appl. Math., 27, pp. 43-61. Providence: Amer. Math. Soc.
7. Hodge, D.B. (1970). Eigenvalues and eigenfunctions of the spheroidal wave equation. J. Math. Phys., 11, 2308-2312.
8. Jarem, J.M. (1980). Construction of doubly orthogonal functions. J. Math. Phys., 21, 28-31.
9. Landau, H.J. & Pollak, H.O. (1961). Prolate spheroidal wave functions, Fourier analysis and uncertainty--II. Bell Syst. Tech. J., 40, 65-84.
10. Rhodes, D.R. (1970). On the spheroidal functions. J. Res. Nat. Bur. Standards, 74B, 187-209.
11. Schwab, F.R. (1980). Optimal gridding. VLA Scientific Memorandum No. 132 (NRAO-VLA, P.O. Box O, Socorro, NM 87801, U.S.A.).
12. ___________. (1981). Rational approximations to selected 0-order spheroidal functions. VLA Computer Memorandum No. 156 (NRAO-VLA, P.O. Box O, Socorro, NM 87801, U.S.A.).
13. Slepian, D. & Pollak, H.O. (1961). Prolate spheroidal wave functions, Fourier analysis and uncertainty--I. Bell Syst. Tech. J., 40, 43-63.
14. Slepian, D. (1964). Prolate spheroidal wave functions, Fourier analysis and uncertainty--IV. Bell Syst. Tech. J., 43, 3009-3057.
15. ___________. (1983). Some comments on Fourier analysis, uncertainty and modeling. SIAM R., 25, 379-393.
16. Stratton, J.A. (1935). Spheroidal functions. Proc. Nat. Acad. Sci. U.S.A., 21, 51-56.

DISCUSSION

J. NOORDAM

'Prefiltering' of data before SELFCAL is an interferometer-based operation and thus destroys the SELFCAL-assumption of only telescope-based errors.

F.R. SCHWAB

Good point. I'm not advocating this 'pre-filtering' approach in all applications, but only in seemingly 'hopeless' cases for self-calibration - e.g. in the case of a low-frequency array where ionospheric problems may be devastating and in which the interferometer pairs may be sensitive to so much of the sky that it's impossible to model the unfiltered visibility function.

P.E. DEWDNEY

With a relatively small 'patch' of the gridding function used (say 6 x 6 u-v cells), the boundaries of the function will change slightly thereby changing the effective convolution from input point to input point. The correction in the map plane (untapering) will have to be an average correction which may not be a very good average for sparse samples in the u-v plane. In the limit the (untapering) correction depends upon the pattern of tracks in the u-v plane and will be different for each map. This can be calculated empirically by simulating point sources at all map points (or a representative sample of them), but this procedure is computationally expensive.

F.R. SCHWAB

With a large number of well-distributed antennas, the effect diminishes. I don't think it's much of a problem with the VLA. I don't know how to make better use of the available information when you want to take advantage of the computational economy of the FFT algorithm.

R.N. BRACEWELL

Gridding involves both averaging to reduce variance and interpolation to obtain values on a grid. It is curious that the procedures we adopt do not actually interpolate (in the sense of passing a surface through the data) even though interpolation is a mature subject contributed to by generations of mathematicians and long practised by applied mathematicians, for example in meteorology.

F.R. SCHWAB

There are some excellent papers on scattered-data interpolation in recent volumes of Mathematics of Computation. But there may be hundreds of <u>noisy</u> visibility samples in the neighbourhood of a given u-v grid cell, and there may be 'redundant' observations (same (u,v)) which disagree. So I think we need to have smoothing combined with the interpolation. In well-sampled regions of the u-v plane, the discrete convolution summations produce a fairly smooth surface, provided that the noise (and the visibility function) is well-behaved.

R.N. BRACEWELL

Another curious feature of gridding is that the 'gridding convolution function' $C(u,v)$ does not actually perform convolution. To see this, note that the operation performed is space-variant, depending on the local value of N, the number of data in the support of C (typically 6 x 6 pixels).

F.R. SCHWAB

Yes, but

$$\frac{m_1 \Delta n m_2 \Delta v}{N} \sum_{i=1}^{N} C(u-u_i, v-v_i)\ V(u_i,v_i) \rightarrow (C*V)(u,v) \text{ as } N \rightarrow \infty .$$

Some sort of filtering is essential if you want to take advantage of the FFT, and I don't know of any other way to use the available information to suppress aliasing.

The AIPS implementation and the VLA pipeline implementation of gridding convolution do use a slightly different form of normalization. The u-v plane is divided into cells (before gridding), the number of samples in each cell is recorded, and, in the convolution summations, each visibility sample is weighted by the reciprocal of the number of points in its own, fixed neighbourhood, rather than by the N I've used above. So the space-variant effect that you note is removed.

J.P. HAMAKER

In the plots you showed, the taper in the map plane resulting from your Fourier-plane convolution assumes very low values near the map edges, in the order of 1/1000. Consequently when you apply the inverse taper, you must enhance map noise by a large factor even in the central part of your map. This apparently is the price you pay for getting really good alias suppression.

F.R. SCHWAB

This point, which was raised in one of the early VLA Scientific Memoranda, is refuted in a memo by Barry Clark ['Gridding and signal-to-noise ratios', VLA Scientific Memo No. 124, April 1976]. For the spheroidal g.c.f. $(1-\eta^2)^{\alpha}\psi_{\alpha 0}(c,\eta)$ the relative S/N along the x-axis of the map is given by $SNR(x)=g(x)/g(0)$ where

$$g(x) \equiv \frac{\psi_{\alpha 0}(c, 2x\Delta n)}{\sqrt{\sum_{k=-\infty}^{\infty} \psi_{\alpha 0}{}^2(c, 2x\Delta n+2k)}} , \quad c = \pi m/2$$

In [11] it's shown that SNR is monotonic in the interval between the centre of the map and the map edge, stays near unity (for moderate m, say $m \geqslant 4$) over most of the interval, and drops abruptly to near $\sqrt{2}/2 \cong 0.7$ (its limiting value as $m \rightarrow \infty$) at the map edge.

However, round-off errors that occur in the FFT computation <u>do</u> get amplified. My experience, working in 32-bit floating-point arithmetic, is that m oughtn't to exceed about 6. Note that the correction is less extreme for larger α. With $m > 6$ and α near 0, you'll indeed notice craziness in the corner pixels of the map if you use 32-bit arithmetic. For details see [11].

You <u>must</u> have really good aliasing suppression for high dynamic range mapping, because of the ever-present problem of aliasing of a source's own sidelobes.

9.3 EFFECTS OF IMAGES, IMAGE CONSTRAINTS, AND ALGORITHMS ON VLBI IMAGE RECOVERY USING HYBRID MAPPING

D.G. Steer, M.R. Ito, P.E. Dewdney

Dept. of Electrical Engineering, U.B.C.,
Dominion Radio Astrophysical Observatory, H.I.A.,
P.O. Box 248, Penticton, British Columbia, V2A 6K3, CANADA.

ABSTRACT. Simulation experiments have been conducted to study the effects of images, image constraints, and recovery algorithms on the reconstruction of VLBI images. The hybrid mapping techniques of "Image-Constraint", "Phase-Closure", and "Self-Calibration" were compared for conditions of large antenna phase errors. It was found that the success of these algorithms, both in terms of the accuracy of the final image and the speed of convergence, depended on both the nature of the image and the algorithm. The most powerful algorithm, "Self-Calibration", was found to be the most successful, however its speed and accuracy were significantly reduced as the contrast in the test object was decreased.

INTRODUCTION. One of the difficulties with Radio Long Baseline Interferometry (LBI) is forming an image from the measured correlation coefficients. We have tested number of existing imaging algorithms to determine how they perform for different objects and amounts of "a-priori" information and we will briefly review here the results of our simulations. This work will not be wholly new to regular LBI data processors, but we believe it to be the first attempt to compare the performance of several algorithms with different types of images.

TEST OBJECTS AND EXPERIMENTS. Figure 1 shows a diagram of the generalized process for imaging with measured visibilities and an initial estimate of the image. Crimmins and Schafer et al. have indicated that this process will converge if the image is constrained to be confined and positive. Using this generalized technique, our experiments consisted of using various algorithms for each operation, and working with different data sets to see how the image was recovered.

We illustrate our experimental results with one of the objects we have tested which is shown in figure 2. This object has been formed from a portion of an image made with the DRAO synthesis telescope. The object field was Fourier transformed and sampled to provide a set of simulated LBI visibility records to which antenna phase errors in the range $\pm 2\pi$ were added. Figure 3 shows the distorted "image" made by Fourier transforming this simulated data.

The data reduction is complicated by the detail that the correlation coefficients are measured in the aperture plane at sample points determined by the geometry of the array. These "sampling tracks" are curved and the data must be converted to a rectangular grid before the efficient FFT algorithm can be used to transform between the data and image domains. The gridding is not an easy task and in our experiments we have avoided the problem by defining our sampling tracks to be linear, and aligned to match a grid. In our case, with images containing 4096 pixels and an 8 antenna array, 28 of the 33 possible grid lines are sampled. While this pattern is artificial, it does remove from our tests the effects due to the gridding algorithm. These linear aperture sampling tracks are similar to those produced by an LBI array with a large North-South baseline component which is observing a source at 0° declination.

THE IMPORTANCE OF BEING POSITIVE. Perhaps the simplest process is the "image-constraint" algorithm. This is a technique which was first applied to LBI astronomy by Fort & Yee. In this case the data constraint is the simple combining of the measured amplitudes with the model visibility phases. These "hybrid" visibilities are then transformed to the image domain to form a "hybrid" image. This hybrid image is then CLEANed to apply the positivity and confinement constraints and to deconvolve the beam. The resulting modified image, formed from the components detected by CLEAN, is then transformed to obtain a new set of model visibility phases for the next iteration.

Figure 4 shows the image recovered using this simple algorithm. Although the recovered image is not perfect, it does correctly reproduce all the major features. Convergence is asymptotic and the accuracy of the reconstruction is essentially limited by the number of iterations performed and by the ability of CLEAN to remove the beam.

This reconstruction was begun with an initial model composed of only the positive components of figure 3. We were able to successfully recover images using this algorithm as long as they contained a bright feature. Images without such a feature would converge to the zero-phase image when started from an arbitrary initial model. The correct image could be obtained in these cases, however, by using an initial model sufficiently biased towards the correct image to break the symmetry. It is apparent from these experiments, that positivity, and confinement and the CLEAN algorithm are the driving forces behind the image reconstruction. More complex algorithms involving phase data provide speedier convergence and more detailed images, but they are not fundamental to the correction of the phase errors.

THE IMPORTANCE OF PHASE CONSTRAINTS. To extend the range of images recoverable, the standard procedure is to include the additional constraint in the data domain that the phase errors originate at each antenna and only the differences are measured by the correlation coefficients. For the Phase-Closure (Readhead 1978) and the Self-

Calibration (Cornwell 1981) algorithms, the data constraint involves the combination of the measured and model phases in such a way that the hybrid visibilities maintain the same phase-closure sums as the data. Although the algorithms are quite similar, we found the Self-Cal to be superior. This was principally because the phase-closure algorithm requires the selection of a "reference antenna" and the best choice was found to depend on the array configuration and the object being imaged. Although there appeared to be a relation between the effectiveness of the selected reference antenna and the aperture coverage of the baseline set involving that antenna, we were unable to find an algorithm for selecting the optimum reference antenna.

Figure 5 shows the image recovered from the distorted data using the Self-Cal/CLEAN process. In this case the reconstruction is quite accurate and has more detail than the result from the image-constraint algorithm. The convergence is quite fast, this example was generated in 40 iterations, and the accuracy of the image is essentially limited by the ability of CLEAN to remove the beam.

Although we found Self-Cal/CLEAN to be the most powerful algorithm, we were able to make test objects that were un-recoverable. These consisted of only broad featureless areas of brightness, and these would not converge.

CLEANING AND MAXIMUM ENTROPY. The standard phase recovery methods use the CLEAN algorithm to deconvolve the telescope beam from the hybrid image in order that the image constraints can be applied. The CLEAN algorithm also plays an important role in the early iterations of the phase recovery by separating the bright features from the confusion in the hybrid image.

The Maximum Entropy Method (MEM) is another algorithm for deconvolving the beam. In this case, the effect is to produce the smoothest positive map that is consistent with the data and the noise. When the MEM algorithm was simply substituted for CLEAN in the phase recovery process, we found that the zero-phase image would be formed when starting from an arbitrary initial model. For phase errors, the entropy function has several local maxima. The largest entropy occurs for the zero-phase image and when starting from an arbitrary model, the algorithm will most often converge to the global maximum at zero-phase.

We were able, in many cases, to recover images with the MEM algorithm in combination with a modified Self-Cal phase algorithm. We modified the process by clipping the image produced by the MEM at each iteration to leave only the bright features. As the iterations proceed, and the features become correct, the clipping level is reduced to allow the lower level details to be filled in. By this technique, we are effectively making use of the positivity and confinement constraints enforced by the MEM algorithm, but are discarding the "entropy" feature of the reconstructed image. Again, as with CLEAN, the final reconstruction accuracy is limited by the accuracy with which the MEM is able to deconvolve the beam.

Figure 6 shows the image recovered by this combination of Self-Cal and MEM. The reconstruction of figure 6 should be compared with figure 7 which shows the image reconstructed by the MEM from data with no phase errors.

CONCLUSIONS. In addition to the fact that imaging is possible in the presence of large visibility phase errors, we draw the following conclusions from these experiments:

1) The constraints of positivity and confinement are the main forces behind the success of the image recovery from large phase errors. The phase-closure information allows faster convergence and more detail, but the image constraints are the main forces behind the phase recovery.

2) Self-Cal and CLEAN proved to be the most powerful combination of algorithms. This is principally because the Self-Cal algorithm, by concentrating on the antenna based nature of the phase errors, is independent of the array configuration and the need for a reference antenna. The CLEAN algorithm is also important both for its ability to successfully deconvolve the beam, and its ability to filter out the main features in a hybrid image. This is especially important during the initial iterations when the model hybrid image is largely incorrect.

3) It helps to have bright features in object. While a bright feature will clearly act as a "calibrator" for the Self-Cal algorithm, it is also effective for the image constraint algorithm as it provides a dominant feature for CLEAN to work with. Note that the feature need not be separated from other areas of brightness, merely much brighter.

4) The maximum entropy method for deconvolving the beam cannot be simply substituted for the CLEAN algorithm. During the early iterations, the MEM feature of providing a smooth reconstruction drives the process to the symmetric zero-phase solution. If the MEM algorithm is combined with a filtering operation to select only the major features of the hybrid image, then a MEM image can be recovered.

ACKNOWLEDGEMENTS. Through the cooperation of the University of British Columbia and the Herzberg Institute of Astrophysics, this study was performed at the Dominion Radio Astrophysical Observatory near Penticton. One of the authors, D.G. Steer, gratefully acknowledges a scholarship from the Natural Sciences and Engineering Research Council of Canada and the H.R. MacMillan Family Fellowship. Special thanks to Dr. L.A. Higgs for making available the data used for the object shown in figure 2 and for providing a speedy CLEAN program.

REFERENCES.

Cornwell, T.J. & Wilkinson, P.N. (1981). "A New Method for Making Maps with Unstable Radio Interferometers". Mon. Not. R. Astr. Soc., Vol. 196, pp. 1067-1086.

Crimmins, T.R. & Fienup, J.R. (Feb. 1983). "Uniqueness of Phase Retrieval for Functions with Sufficiently Disconnected Support". J. Opt. Soc. Am., Vol. 73 #2, pp. 218-221.
Fort, D.N. & Yee, H.K.C. (1976). "A Method of Obtaining Brightness Distributions from Long Baseline Interferometry". Astron. Astrophys. Vol. 50, pp. 19-22.
Readhead, A.C.S. & Wilkinson, P.N. (July 1978). "The Mapping of Compact Radio Sources from VLBI Data". Astrophys. J. Vol. 223, pp. 25-36.

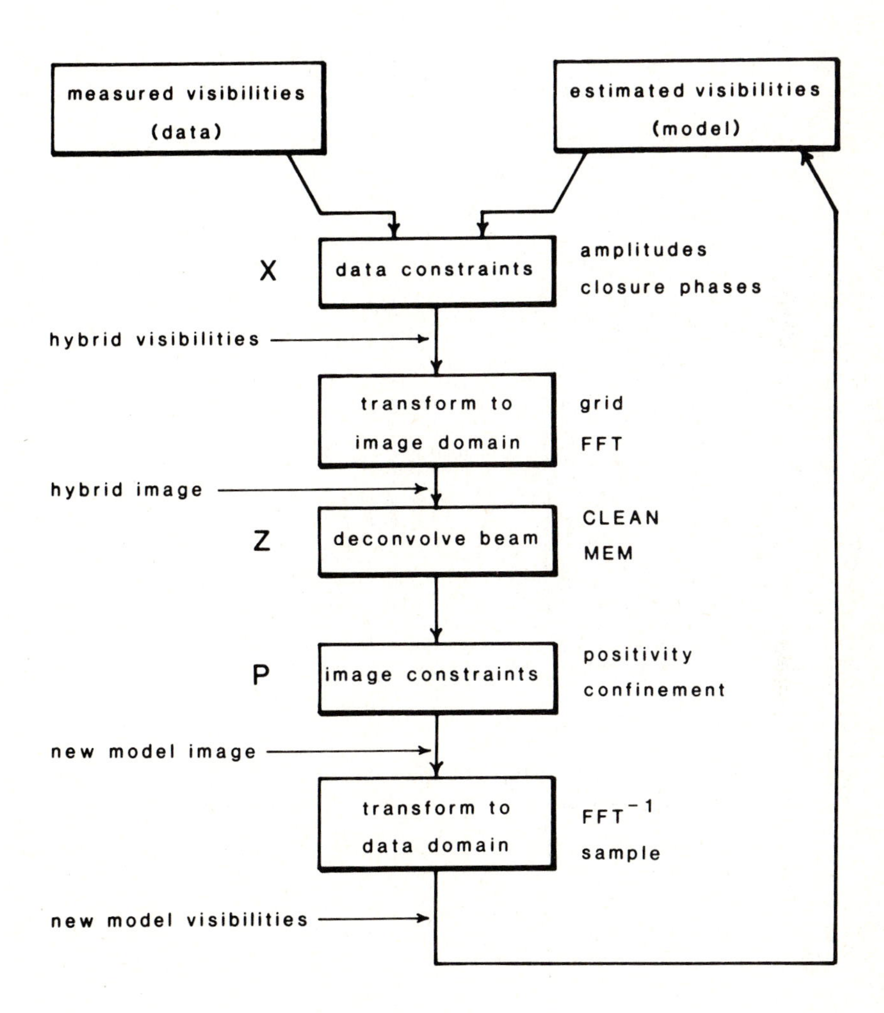

FIGURE 1 General Image Recovery Algorithm

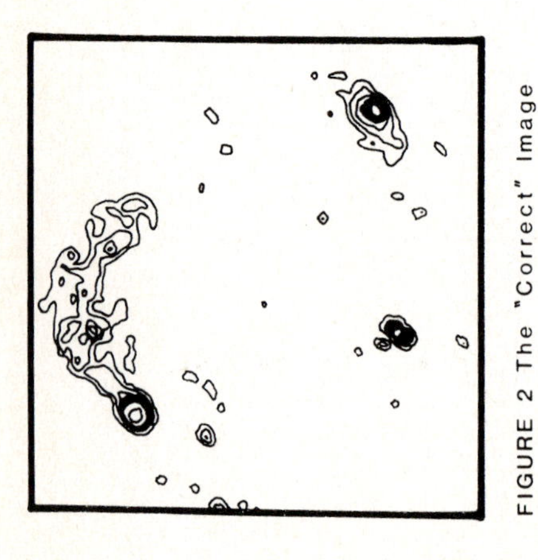

FIGURE 2 The "Correct" Image Reconstructed by CLEAN from undistorted data

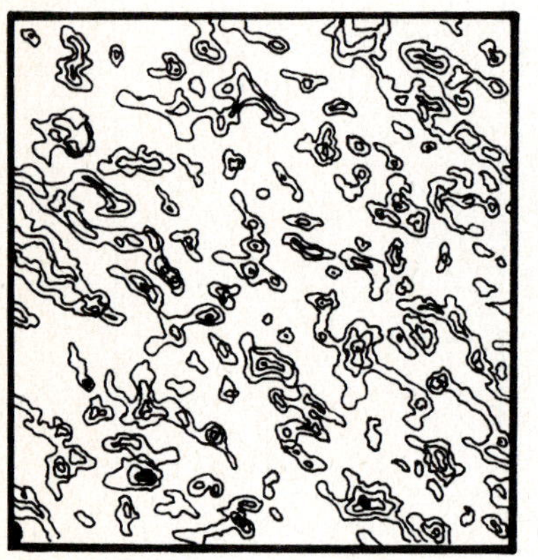

FIGURE 3 The "Distorted" Image 360° antenna phase errors

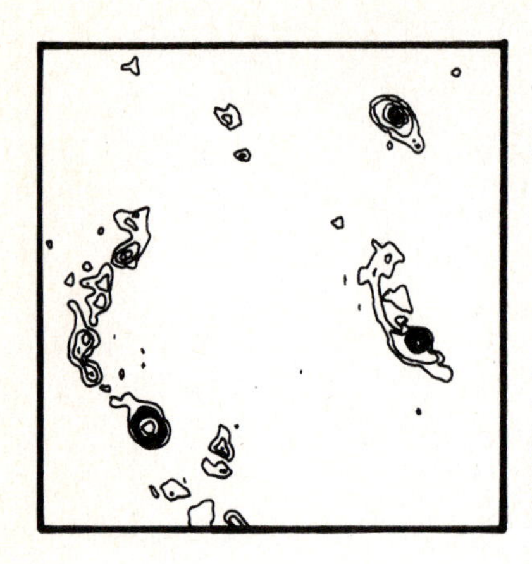

FIGURE 4 Image recovered by "Image-Constraint" (NO phase information)

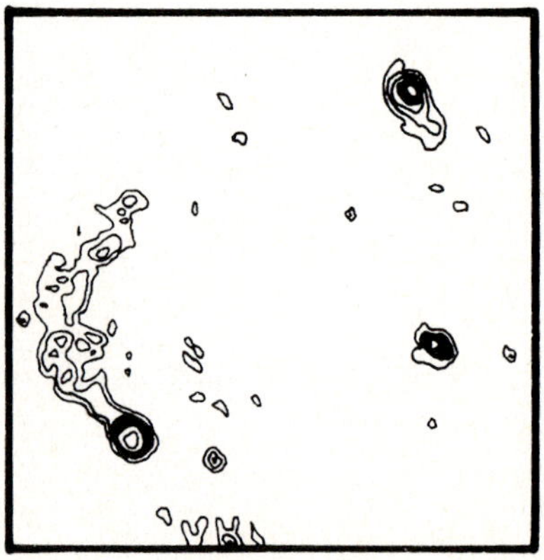

FIGURE 5 Image recovered by "Self-Cal/CLEAN" (closure phase information)

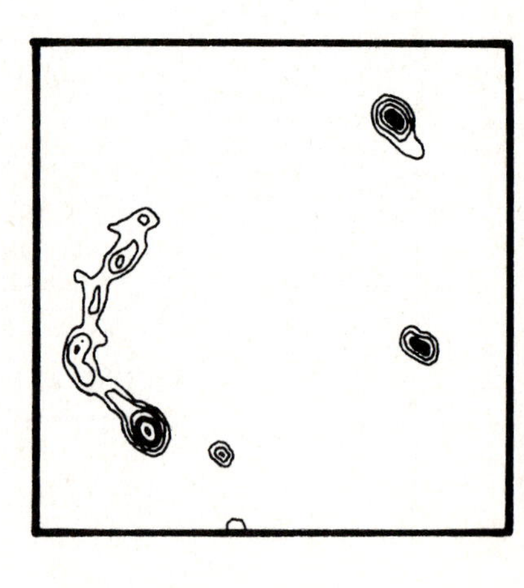

FIGURE 6 Image recovered by "Self-Cal/MEM" (closure phase information)

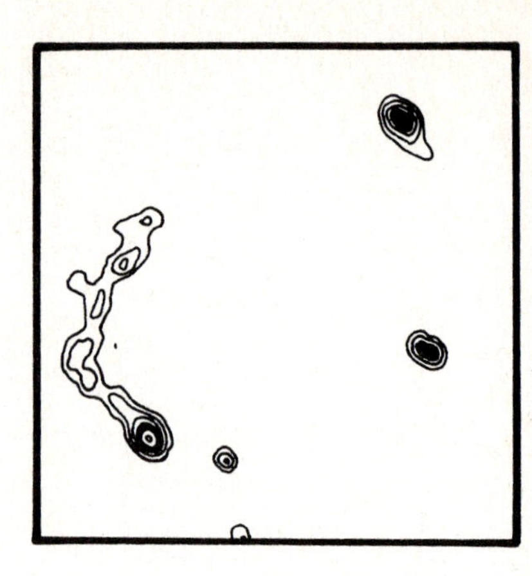

FIGURE 7 Image Reconstructed by the MEM from undistorted data

DISCUSSION

J.E. BALDWIN

Is it possible to make a quantitative statement about the fraction of the total flux which should be in compact components or how bright those components should be in relation to the noise for the final image to be good.

D.G. STEER

Yes, it is possible to make such a statement. However I am not in a position to answer. One limit which I suspect is more severe than necessary is that there must be enough flux in the main compact component for the major features to remain separated in the auto-correlation image.

J.D. O'SULLIVAN

CLEAN has just been described by Cornwell & Wilkinson as a 'sharpness' indication in drawing an analogy with adaptive optics schemes. MEM has been described as a maximally smooth estimate. Does this imply that sharper images are better for the purposes of phase-less reconstruction methods?

D.G. STEER

For the phase-less reconstruction algorithm the image is built up one component at a time. The components detected by the CLEAN algorithm are ideal for this process. The final restored image however may be smooth or sharp at the choice of the observer during the final restoration operation.

9.4 IMAGE RESTORATION TECHNIQUES - HOW GOOD ARE THEY?

G.A. Dulk,* D.J. McLean, R.N. Manchester, D.I. Ostry and P.G. Rogers
Division of Radiophysics, CSIRO, Sydney, Australia

1 INTRODUCTION

Maps of the radio sky made with synthesis telescopes are imperfect. The most prominent defects usually arise from incomplete sampling of the complex visibilities by the antenna array. In regular arrays, aliased responses (grating rings) from sources which may or may not be in the mapped area are generally present. If the map-making procedure involves sampling of the visibility data on to a regular grid and transform by the fast Fourier transform (FFT) algorithm, then further aliasing is introduced. Receiving systems contribute noise to a map and, to a greater or lesser extent, introduce gain and phase errors. Fluctuations in the atmosphere and ionosphere also contribute to phase errors in the measured visibilities. Finally, antenna pointing errors and, in altitude-azimuth mounted antennas, beam non-circularities result in modulation of the detected signal and hence errors in the map.

Great ingenuity has been shown by astronomers and others in devising methods to remove these defects from maps, leaving behind an image of the 'true' sky, albeit convolved with some smoothing function. Image restoration procedures such as CLEAN (Högbom 1974; Schwarz 1978) and the maximum entropy method (MEM) (e.g. Gull & Daniell 1978) reduce the effects of incomplete sampling of visibilities and aliased responses in regular arrays and are commonly used. Aliased responses resulting from the FFT can be largely eliminated by use of suitable convolving functions in the visibility plane (e.g. Schwab 1978, 1983). Self-calibration techniques (e.g. Cornwell & Wilkinson 1981, 1983) can be used to remove antenna-based complex gain errors - e.g. errors contributed by the atmosphere or the receiver system.

In this paper we have the limited objective of exploring how well image-restoration techniques reproduce the actual sky brightness distribution from observations made with an incomplete array in the presence of random receiver noise. Gain errors are assumed to be zero and self-calibration techniques are not considered. We use an array simulation program developed as part of the Australia Telescope design studies. Three image restoration techniques are considered: CLEAN, MEM and a process we call the constrained positivity method (CPM). This last

---

*Also Department of Astro-Geophysics, University of Colorado, Boulder, U.S.A.

method was originally suggested by Högbom (1974), but as far as we are aware it has not been widely used, although variants have been employed in reconstructing limited phase data (e.g. Fienup 1978; Rogers 1980).

2 DYNAMIC RANGE AND FIDELITY

The dynamic range of a map is often used as a measure of its quality. We define this quantity as

$$D = -10 \log(3\sigma/M) ,$$

where σ is the r.m.s. fluctuation in source-free regions of the map and M is the maximum intensity in the map. The dynamic range measures the ability to detect weak sources in the neighbourhood of a strong source.

In real life it is seldom possible to measure how well an image corresponds to the actual source distribution. However, in a simulation we do have this ability, since the sky distribution is known. We define a new quantity, the map fidelity, which measures how well the image of a source corresponds to the actual source distribution. Fidelity is defined by the equation above with σ equal to the r.m.s. fluctuation in the difference between the map and sky distributions computed over the region for which the sky brightness exceeds 1% of its peak value. Before the difference is taken, the map is effectively scaled so as to minimize σ.

3 THE DIRTY MAP

A simulated observation has been generated using a 5×12-h synthesis with the 3-km section of the planned Australia Telescope compact array. This section contains five antennas 22 m in diameter, giving a total of 50 baselines for the five-day synthesis. These are arranged in a non-redundant way on a regular east-west array of 20-m increment. The shortest baseline is 40 m, the longest is 3 km and the distribution is roughly uniform. An observing frequency of 1.4 GHz and bandwidth of 1 MHz are assumed and system noise is computed assuming a system temperature of 35 K and a dual polarization system.

The brightness distribution for the test source, known as SPIRAL, is shown in Figure 1(a). The source consists of a sequence of 24 circular Gaussian components, each of full width at half-power 38" arc, and has a total flux density of approximately 3.4 Jy. The intensity of each component is 1.5 dB less than that of the preceding one, so the component intensities cover a total range of 34.5 dB. Since we later wish to subtract the restored maps from this sky distribution, the angular size of the components is such that they are resolved by the synthesized beam. The assumed source declination is -50°.

The dirty map which would result from a synthesis observation of this source with the array described above is shown in Figure 1(b). Observed visibilities have been convolved on to a u-v (spatial frequency) grid of

Figure 1. (a) Sky brightness distribution for the test source (SPIRAL) and (b) the corresponding dirty map simulation for a 5 × 12-h synthesis with the 3-km section of the proposed Australia Telescope array. Contour intervals are -10%, -5%, -4%, -3%, -2% and -1% (short dashes), -0.5% and 0.5% (long dashes), 1%, 2%, 3%, 4%, 5%, 10%, 20%, 30%, 40%, 50%, 60%, 70%, 80% and 90% (full line) of the map maximum. The pixel size is 5" arc in each coordinate; only the central quarter of the mapped region is shown.

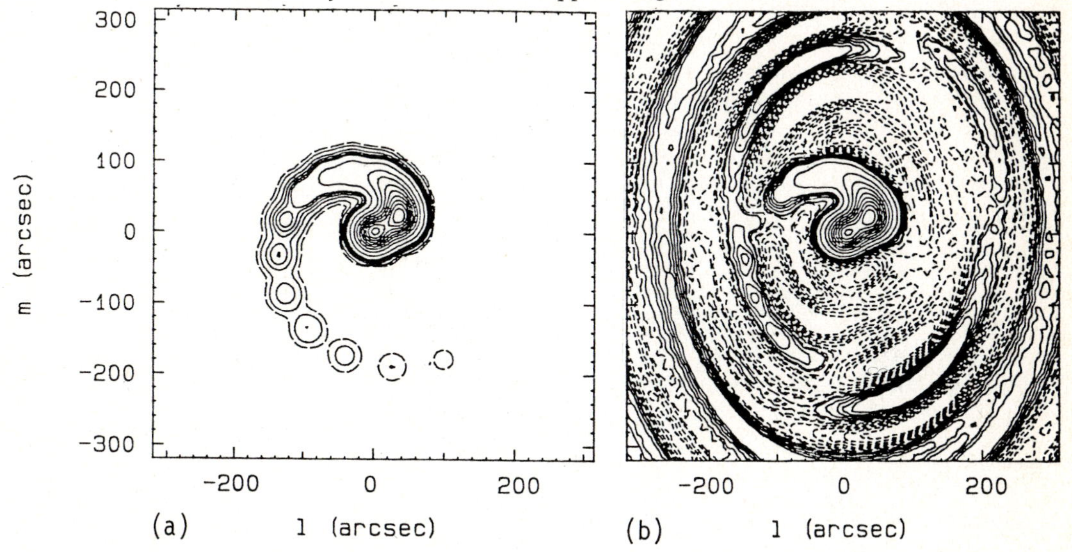

size 256 × 256 points using a Kaiser-Bessel convolving function (Schwab 1978) and weighted uniformly. Dynamic ranges and fidelities for the dirty map are given in Table 1.

4 *CLEAN* RESTORATION

The dirty map shown in Figure 1(b) was cleaned using the 'smoothness stabilized' CLEAN of Cornwell (1983); an impulse of 25% of

TABLE 1
Dynamic ranges and fidelities

Image	SPIRAL Dynamic range (dB)	SPIRAL Fidelity range (dB)
Dirty map	10.6	11.2
CLEAN map	22.6	14.6
MEM map	25.1	19.3
CPM map	20.6	19.9

the main beam height was used. A total of 3000 components were removed from the dirty map and restored using a Gaussian beam of half-power widths equal to those of the dirty beam to produce the cleaned map shown in Figure 2(a). Figure 2(b) illustrates the difference between the cleaned map and the sky distribution. Dynamic range and fidelity parameters for the cleaned map are given in Table 1. The dynamic range is about 22 dB, corresponding to an improvement of 12 dB over the dirty map. This dynamic range is just that expected from the ratio of source intensity to random noise in the map. However, the fidelity is considerably less than this figure, and the reason for this is clearly shown in Figure 2(b); these contours are negative at component maxima and positive on the wings of components, indicating that the map components are broader than those of the sky distribution. This arises because the restoring beam low-pass filters the clean components and also because the clean components are always centred on grid points. Use of a better restoring function or a finer grid in the map plane would, at least to some extent, overcome these problems.

5 *MEM* RESTORATION

An MEM restoration is shown in Figure 3(a). The Maximum Entropy Data Consultants Ltd. implementation of the MEM algorithm (cf. Gull & Daniell 1978) was used. The data file was formed by averaging visibilities into the nearest grid point; the number of samples per

Figure 2. (a) Map of SPIRAL restored with the smoothness stabilized CLEAN algorithm and (b) the corresponding (map - sky) difference distribution. Contour intervals for the map are as for Figure 1 and for the difference are at intervals of 20% of the difference maximum with negative contours dashed and zero omitted.

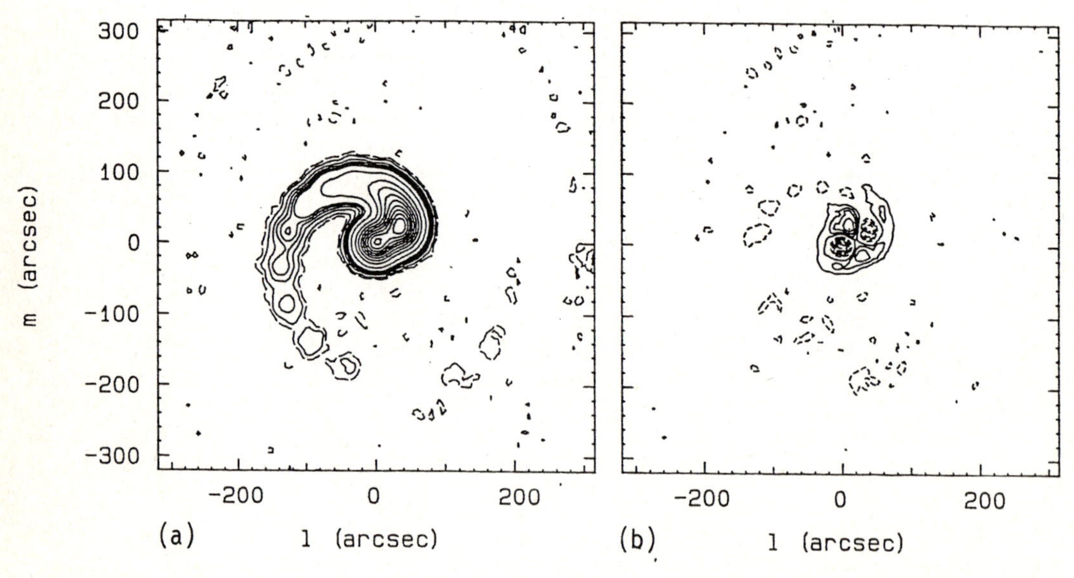

Figure 3. (a) Map of SPIRAL restored using the MEM algorithm and (b) the corresponding (map - sky) difference distribution. Contour levels are as in Figure 2.

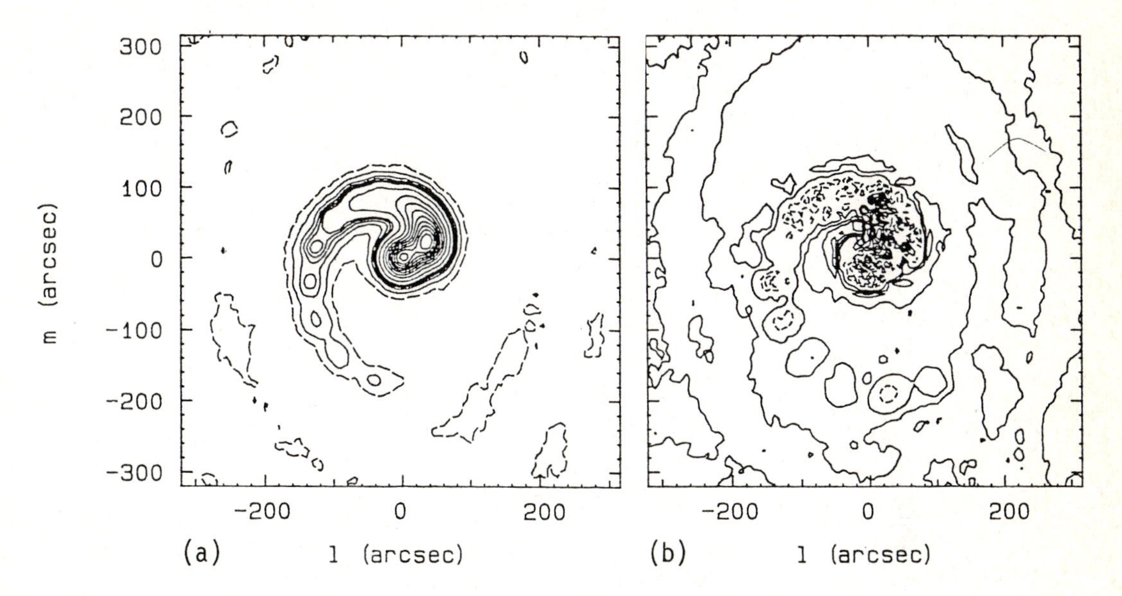

grid point was used to provide a measure of the accuracy of the data. This box-car convolution was necessary to provide independent data points for MEM. A uniform image was used as the starting point for the restoration. The iteration was stopped after 20 cycles when the rate of improvement of map fidelity was low.

The difference map is shown in Figure 3(b) and dynamic range and fidelity parameters are listed in Table 1. The dynamic range exceeds the 22 dB level, illustrating the noise suppression property of MEM, and the map fidelity is considerably better than for the CLEAN restoration. Unlike CLEAN, MEM tends to over-resolve the bright components of the sources. As well as suppressing noise, MEM suppresses weak components of the source. This is illustrated in the difference map (Fig. 3b), where the weaker components appear negative with respect to the background. Further iterations do not significantly reduce this effect.

6 *CPM* RESTORATION

The constrained positivity method (CPM) of image restoration is an alternating projection technique which exploits the fact that real images (at least of Stokes parameter I) are always positive. The algorithm uses the dirty map as the initial estimate and operates as follows.

(1) An intensity histogram of the map is calculated and the nominal zero level is taken as the lowest level at which a population

peak occurs. Intensities in the map below this level are multiplied by a constant factor less than one.

2. The resulting image is then Fourier-transformed into the u-v plane.

3. At grid points corresponding to measured visibilities, the map visibility is replaced by a weighted average of the measurement and the current prediction.

4. All visibilities outside an elliptical region defined by the measured visibilities plus a guard zone are multiplied by a constant factor less than one.

5. A new map is then formed by transforming these visibilities.

6. Steps 1 to 5 are repeated until convergence is obtained.

Like MEM, this algorithm requires independent u,v data points, that is, box-car convolution, in the u-v plane.

The restored map after 50 iterations is shown in Figure 4(a), the difference map in Figure 4(b), and dynamic ranges and fidelities in Table 1. Although dynamic ranges have not quite reached the level corresponding to system noise, the map is of high fidelity. This is illustrated in Figure 4(b), which shows that there is little or no increase in error

Figure 4. (a) Map of SPIRAL restored using the CPM algorithm and (b) the corresponding (map - sky) difference distribution. Contour levels are as in Figure 2.

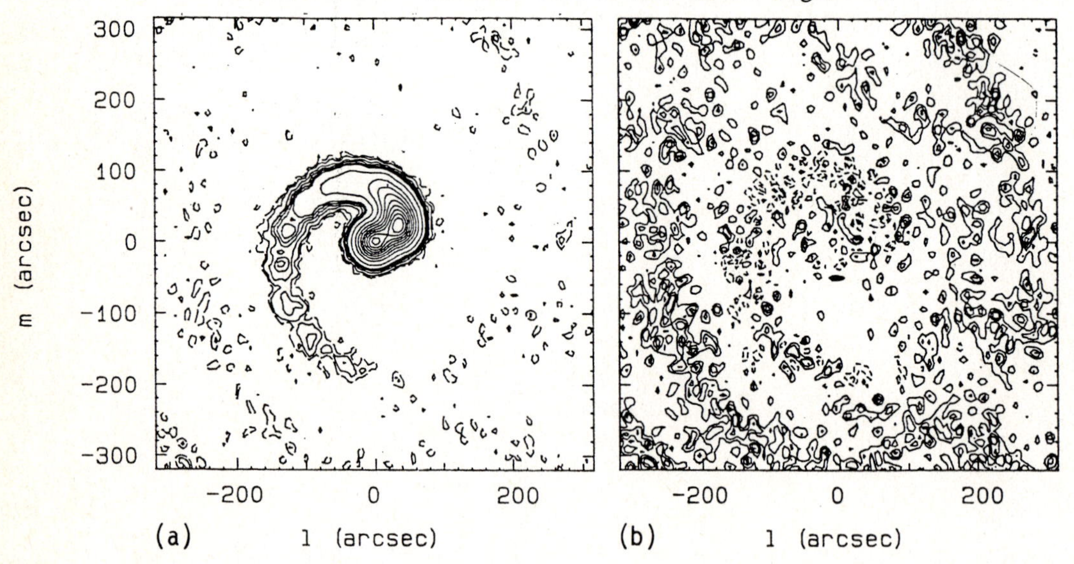

magnitude at the source location. Much of the error in this map appears to be concentrated in the sidelobe structure of the dirty map. Further development of the algorithm may eliminate this effect.

7 CONCLUSIONS

An investigation has been made of the degree to which three different image restoration techniques, namely CLEAN, MEM and CPM can restore a 'dirty' image generated by an earth-rotation synthesis array. Of the three methods, MEM produces the best dynamic range because of its noise suppression property; however, this is rather misleading, since the routine also suppresses weak sources. The best map fidelity (or reproduction of source structure) is achieved by the CPM method. CLEAN tends to under-resolve source components whereas MEM tends to over-resolve them. The processing times required by the three techniques were comparable.

8 ACKNOWLEDGMENTS

We thank J. Skilling for providing us with the MEM routine and J.A. Roberts for assistance with the MEM computations.

References

Cornwell, T.J. (1983). A method of stabilizing the CLEAN algorithm. Astron. Astrophys., 121, 281.

Cornwell, T.J. & Wilkinson, P.N. (1981). A new method for making maps with unstable interferometers. Mon. Not. R. Astron. Soc., 196, 1067.

Cornwell, T.J. & Wilkinson, P.N. (1983). Self-calibration (this volume).

Fienup, J.R. (1978). Reconstruction of an object from the modulus of its Fourier transform. Opt. Lett., 3, 27.

Gull, S.F. & Daniell, G.J. (1978). Image reconstruction from incomplete and noisy data. Nature, 272, 686.

Högbom, J.A. (1974). Aperture synthesis with a non-regular distribution of interferometer baselines. Astron. Astrophys. Suppl., 15, 417.

Rogers, A.E.E. (1980). Methods of using closure phases in radio aperture synthesis. Proc. Soc. Photo-Opt. Instrum. Eng., 231, 10.

Schwab, F. (1978). Suppression of aliasing by convolutional gridding schemes. VLA Sci. Mem. No. 129.

Schwab, F. (1983). Optimal gridding of visibility data in radio interferometry (this volume).

Schwarz, U.J. (1978). Mathematical-statistical description of the iterative beam removing technique (Method CLEAN). Astron. Astrophys., 65, 345.

9.5 RADIO ASTRONOMICAL SYNTHESIS PROCESSING: SOME REMARKS.

W.N. Brouw
Netherlands Foundation for Radio Astronomy, P.O. Box 2,
7990 AA Dwingeloo, The Netherlands.

Introducing "Processing" after all the talks at this Symposium on the details of the processing of radio astronomical data and image processing, is a superfluous task. I will, therefore, restrict myself to two general remarks. Both remarks, one on software and one on hardware, will be directed to two groups of people who have been neglected at the Symposium.

The first group consists of the poor observational radio astronomer, who has requested time on a telescope, whose proposal was accepted and who did his observations. He wants to reduce his data and is not interested to prove that his final map is unique, because it can not be factorized. The second group consists of the people who have to provide the computing power for the first group.

Since the Groningen Symposium in 1978, the change, in my view, in processing can be summarized by:

- Software : FITS/AIPS
- Hardware : Display/AP

Software

Data processing programmers all agree that it is a waste of time to reinvent the wheel many times over, but:

- their software is always:
  - not usable on our machine
  - inefficient
  - maybe good for their observations, but not for ours
- their data format
  - does not take into account our beautiful data format
  - has too low (or high) a precision

Hence, pre 1978, we did reinvent the wheel. This has changed drastically in the last few years. Although the above arguments are still true, we use this software anyway. The main reasons for this are:

- the use of FITS (Flexible Image Transport System) at all major observatories
- the availability of AIPS (the NRAO Astronomical Image Processing System); which, once accepted by an installation, made the acceptance of other outside software packages much easier
- the availability of a, more or less, standard 32-bit machine
- the ever increasing software cost, vis a vis diminishing funds

- the disappearance of overall system knowledge
- the increasing number of digital imagers in other wavelength regions
- the amount of telescope data made "distributed" processing essential

This change in altitude, and acceptance of a few standards, makes the exchange of manpower and data easier, and also allows concentration on the really important software issues.

Hardware

Colour image displays have become much cheaper in the last few years. Therefore, individual institutes can afford to buy them, increasing the possibilities of distributed processing. Although the features of the different displays still vary widely, an effort is already under way to define a set of standard minimum requirements.

The AP (Array Processor) has become a standard addition to many radio astronomical processing systems. The main reason used in acquiring an AP, is that it gives you 12 Mflops for 25% of the price you paid for your 1 Mflop mini. In general, however, this is not true. If we assume an AP on a 1 Mbyte bus, with an 12 Mflops maximum rate, a 1024 point Fast Fourier Transform will take:

- load data	10 ms
- unload data	10
- start AP	5
- do FFT	15
	40 ms

In our 1 Mflops mini it will take about 100 ms. As soon as we have to load the data from disk, the time needed for both AP and mini processing will increase with 30-200 ms for a disk load and the same for a disk unload. From the above it is clear that an array processor is only worthwhile in those cases where we have more than 1000 datapoints, and more than 10-20 flops/datapoint. In radio astronomical data processing this is only true if the AP has a very large memory, which, in general, cannot be afforded by most institutes. Timing in practical situations, e.g. the VLA spectral line processor, show indeed the limitations as given above.

I believe that (radio-)astronomical dataprocessing will be better off with a DP (Data Processor) than with an AP. In this context I would like to define a DP as:

- 100 Mbyte of 1 Mbyte/sec core
- controller that can:
  - read/write a "line" of data
  - read/write a "tile" of data
  - read/write with bit-reversed addressing
  - write destructive or additive

At present the cost of such a device is higher than that of an AP, however, with the advent of 256 bit chips it should become a viable proposal.

DISCUSSION

P.T. RAYNER

Display devices are becoming cheaper and more popular but there are few standards. This hinders the exchange of software between institutions with different image display/processors.

W.M. BROUW

This has been true. However, a list of minimum standard features has now been more or less agreed upon, making transitions relatively easy.

S. ANDERSON

The times quoted for array processor FFT performance are somewhat pessimistic. We have developed an array processor with performance illustrated by the following example: 1024 point (complex) FFT in 2.1 msec including all times (loading from multiport memory of 10 Mbytes, execution, *etc.*, unscrambling...)

W.M. BROUW

Loading from disk is necessary in radio astronomy owing to the large number of pixels per image. This time greatly exceeds the actual processing time.

9.6 FCP - AN IMAGE PROCESSING SOFTWARE SYSTEM

R.J. Sault
School of Electrical Engineering,
University of Sydney, NSW 2006, Australia

Abstract. An interactive image processing software system is described. Using a co-ordinated and 'user-friendly' approach in the design stage, the system provides users with simple, yet powerful access to images, processing programs and routines. The reduction of data from the Fleurs Synthesis Telescope is used as an example. The system has been running since 1981, and has been used for processing various classes of images, such as medical and radio-astronomy images.

INTRODUCTION

The increasing sophistication in indirect measurement hardware has, in general, led to a more distant relationship between the data actually measured and the quantity desired. The sophistication and complexity of the algorithms used in data reduction has, in turn, increased. A co-ordinated approach in the development of reduction software is now becoming a necessity. This is required both to minimize the difficulty in developing software for new algorithms, and to maximize the usefulness of software already available.

This paper describes a software system which has been designed to meet these needs. The system, known as the Fleurs Control Program (FCP) has been developed at the University of Sydney's School of Electrical Engineering, and runs on the Faculty of Engineering VAX 11/780 computer.

This paper is divided into two main sections. Firstly, an overview of the system is given, and some of the considerations in the systems design are discussed. A more detailed description can be found in Sault (1981). Secondly, the use of this system in the reduction of data from the Fleurs Synthesis Telescope (FST) is used as an example.

AN OVERVIEW OF THE SYSTEM

There are two, overlapping, ways of viewing the software system, depending on whether one is a user (a non-programmer, who is only interested in the reduction of the data), or a programmer (who is also interested in developing new algorithms).

To the user, the system appears as a set of application

programs, each of which performs a single, well defined task, such as data calibration or display. He will generally use many of these application programs, some repeatedly, in the reduction of the data. All the programs behave in a similar manner, and use the same conventions and file structures.

To the programmer, the system consists of a set of standard routines to aid him in developing programs either for his own personal use, or for general use. The routines can perform either 'housekeeping' or 'number-crunching' functions.

A few of the more significant features of the system are described below.

User interaction

Some of the major considerations in determining the way the user interacts with programs are:

1 Interactive processing is very desirable; it is simple to use, and decisions can be made as the processing proceeds.
2 The system must have a good batch capability. Batch processing is essential in performing large amounts of repetitive processing, or for performing lengthy tasks (e.g. iterative and slowly converging processes such as CLEAN).
3 Inputting parameters to a program must be simple, yet potentially concise, flexible and powerful. To a surprisingly large degree, the acceptance of a processing package by users is determined by the ease with which the processing can be specified.
4 While the user must be capable of changing every conceivable parameter which determines the processing task, it would be ridiculous for him to have to specify the more obscure parameters which rarely vary from their default value.
5 Gaining parameters from the user must be reasonably simple for the programmer.
6 The program, as a whole, must be reasonably tolerant of the inevitable mistakes made by users.

In our system all interaction with the user is handled by one subroutine. Whenever an application program requires a parameter, which the user is allowed to change, it calls this subroutine, with a description of the parameter, an optional default value, an optional informative message, and a hint as to how important the parameter is. By concentrating all user interaction into one subroutine, the application program can relieve itself of much error checking, and also allow very flexible ways of entering the parameters.

As far as the user is concerned, he can enter parameters at either of two stages. Firstly, the user can enter parameters, on the command line, when he initiates the processing program. This is the method used for batch processing, and for entering obscure parameters. Secondly, the user will be prompted for the more important parameters, if they were not entered on the command line. This is the normal interactive mode. This is also the simplest and most verbose method of entering parameters, with the user usually being given an informative message about the parameter required.

By using a command procedure, the user can redefine which parameters he is prompted for. He can also indicate that user-defined or system-defined defaults are to be used. Command procedures can be used to simplify running commonly used sequences of programs. Symbol definition and do-loop structures, for repeated execution of a program, are also available to the user.

File and data handling

A key feature of any processing system is, of course, the file structure and data I/O. As our school has wide ranging processing interests, a very flexible file structure was required. The file structure had to allow for irregularly sampled data (e.g., aperture synthesis correlation data) as well as regularly sampled data (of one, two, three and potentially higher dimensions). Several number representations of the data, ranging from 8-bit integers to complex floating point numbers, also had to be supported.

Another requirement was to be able to store an essentially infinite amount of ancillary information (such as processing history, or the relation of the data to sky co-ordinates) in the file header. This was to allow a processing program to save this information, so that it could be retrieved later by a co-operating program, or by the user.

The structure chosen is essentially the FITS (Flexible Image Transport System) (Wells *et al.*, 1981) structure, which is a standard format for image transport on magnetic tape. Some modifications, however, were needed to make it more applicable to an in-house disc system. In a FITS file, the file header consists of ACSII card images. Each piece of ancillary information concerning the file, and the data, is specified by a card image. The number of card images that can be placed in the header is unlimited.

Standard routines are available in the system to handle the header, as well as to open, close, read and write data files. The programmer can be ignorant of the file structure actually used, and (to a large degree) the FITS standard. System routines handle all of this for him.

Processing different classes of data

Again because of the wide variety of data handled at our school, a requirement was that it be easy to write a program which operates on different classes of data. For example it should be relatively easy to write a program which will work equally well on 32 x 32 medical images (stored as 8-bit integers), a 512 x 512 radio-astronomy maps (stored in a floating point format) or possibly even one dimensional speech data, when the processing to be performed on them is otherwise the same.

In our system, three features aid in doing this. Firstly, dynamic memory allocation is usually used for array storage. This allows processing of data sets of sizes up to the computer's memory limit. Secondly, the I/O routines automatically convert data from one number representation (e.g., integer) to another (e.g., real) during the I/O process. The number representation produced by the I/O operations can be varied at any time, so the processing routine can work in the most natural representation for its problem. Finally, the I/O routines tend to be independent of the data's dimensionality. The same routines are used regardless.

Standard routines

The software system also includes a set of commonly used subroutines, such as FFT, mathematics and graphics routines, to aid the programmer.

THE FST REDUCTION PROGRAMS

This second part of the paper describes the use of the software system in the reduction of data from the FST.

The FST produces radio-maps of the sky brightness using Earth rotation aperture synthesis. It consists of 64 6 m antennae and 6 14 m antennae, normally operating at 1415 MHz. Christiansen (1973) gives a comprehensive description of the telescope.

Figure 1 gives a simplified flowchart of some FCP commands that would be used in the data reduction. The data can either be new data from the telescope, or old archived data (the archived data is in a very different format to the new data). At present the software for the reduction of the correlation data to the map stage will handle Fleurs data only. The map-handling routines will work for any data. (We can process, in a straight forward manner, map data from any institution which can write a FITS tape). The display routines work on any image (not just radio-astronomy maps).

CONCLUSION

A processing software package has been described. Its aim has been to relieve the researcher from the mundane side of new algorithm development, while at the same time producing highly flexible and user-friendly software. Since coming into general use in October 1981, the system has steadily continued to develop, expand and attract more programmers and users.

ACKNOWLEDGEMENTS

The author is grateful to the computing staff of the Mt. Stromlo Observatory for making their software available to us, to Mr. D.J. Skellern for assistance in the systems design, to Professor T.W. Cole, and to the numerous people who have made constructive criticisms of the system since it came into general use. The author also acknowledges the support of a Commonwealth Postgraduate Research Award.

Figure 1: FST reduction programs

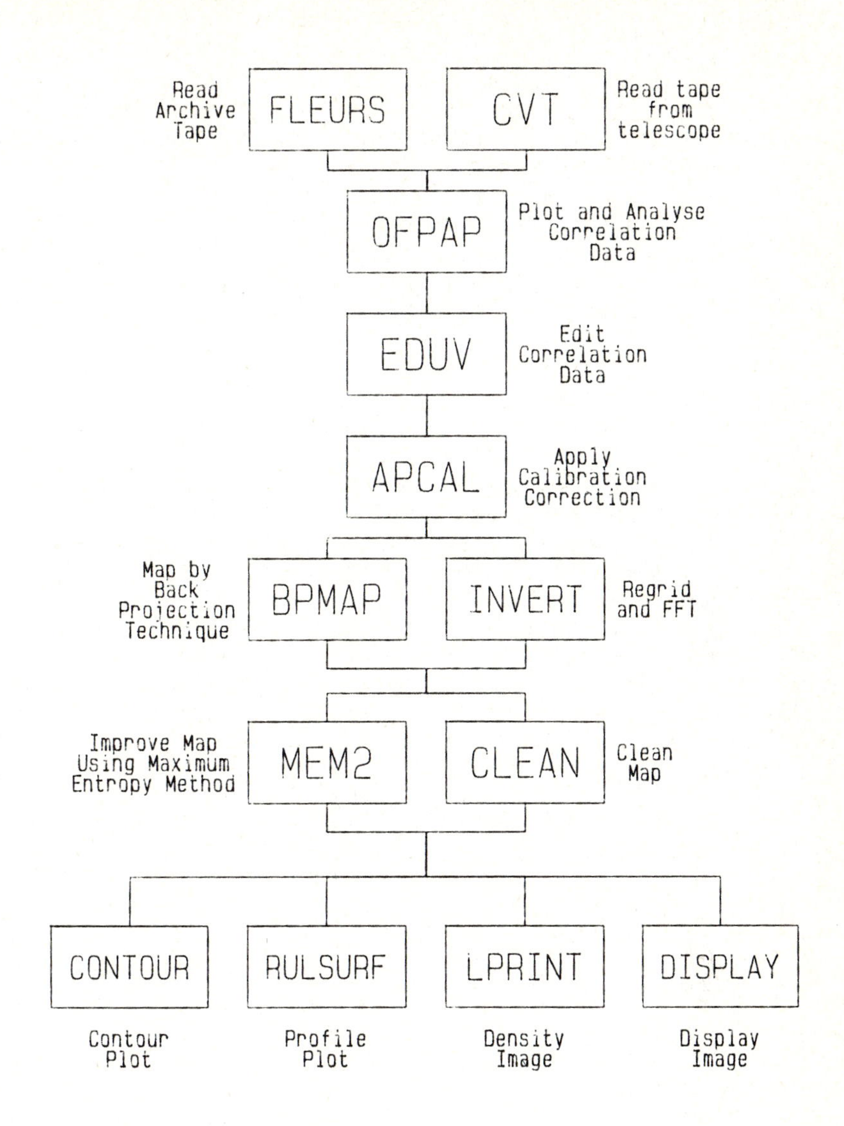

References

Christiansen, W.N. (1973) ed. The Fleurs Synthesis Telescope, Proc. IREE, No. 8, Vol. 34.

Sault, R.J. (1981). FCP Users Manual.

Wells, D.C., Greisen, E.W. and Harten, R.H. (1981). FITS: A Flexible Image Transport System, Astron.Astrophys., Suppl. Ser., p. 363-371, Vol. 44.

DISCUSSION

P. STEFFEN

You didn't mention a very important thing for an interactive data processing system: that is to have multiprocessing from a single terminal. Normally after starting a process that is very time consuming your terminal is blocked for further input and the system is no longer an interactive one. A multiprocessing system (like GIPSY) is able to start some processes, allow for interaction with the user during execution etc. while your terminal serves as the output device for the process running but is also available for the input of other work, e.g. editing.

R.J. SAULT

I'm not sure that the extra complication to the system, and extra confusion to the user is worth it. We use 'batch' for long jobs.

P.T. RAYNER

Any image/data processing system must support the astronomer/programmer. This means that interfaces to the system must be simple and well defined.

R.J. SAULT

I strongly agree. This is the short-coming of some other systems.

9.7 MAP CONSTRUCTION WITH THE MOLONGLO OBSERVATORY SYNTHESIS TELESCOPE

D.F. Crawford
School of Physics, University of Sydney, NSW 2006, Australia.

ABSTRACT
The Molonglo Observatory Radio Telescope uses earth rotation to synthesise a complete two dimensional aperture. The method of direct fan beam synthesis is used rather than the more usual summation in spatial frequency space followed by Fourier transformation. This method is used because of variations in the primary beam shape and because real time fan beams are provided by the hardware. Synthesis methods, calibration correction, point source finding, and parameter fitting are described.

INTRODUCTION

The Molonglo Observatory Telescope is a simple two antenna interferometer in which each antenna is a 778 m by 11.6 m trough reflector operating at 843 MHz. Both antennas are aligned in an E-W direction and can be steered in the N-S direction by mechanical rotation. Steering in the E-W direction is by phasing each of the 3872 feed elements in each antenna, and then combining the signals so that each antenna is divided into 44 separate sections. The outputs from each section are further processed to provide the required phase and delay gradient. The 44 signals are added together in a resistance array which provides incremental phase gradients to produce 64 real time fan beams. Signals from corresponding beams from each antenna are multiplied to produce 64 real time interferometer beams. By switching the phase gradient by a small amount every second these 64 beams are time multiplexed to produce either 128, 256 or 384 beams in each 24 s sample. Each beam has an E-W width of 43 arcsec (HPBW) and at meridian passage a N-S width of 2.3 degrees. The hardware beams have a separation of 22 arcsec and the time multipled beams have a separation of 11 arcsec which is just under half the Nyquist sampling requirement. A more complete description of the instrument can be found in Mills (1981) and Durdin, Little & Large (1983).

In a typical observation the same point in the sky is tracked for 12 hours (or less for declinations north of -30°) so that the fan beams rotate and all spatial frequencies are observed. The computational problem is to produce a map of the sky, including gain, pointing and phase corrections; to improve the map by cleaning; to locate sources;

and to measure source flux densities and positions. In this paper the processing undertaken at Sydney University is discussed; processing at the Observatory is described in Durdin, Little & Large (1983).

MAP GENERATION

The usual rotational synthesis procedure accumulates data as points in the spatial frequency (u,v) plane, and then interpolates them onto a rectangular grid. The map in the (x,y) plane is produced by a Fast Fourier Transformation. An important requirement of this method is that the primary beam shape must not vary throughout the observation. This makes it unsuitable for the Molonglo telescope where the primary beam derives from a rectangular aperture. Because of direct interactions between the feed elements and other interactions within the waveguides, together with the foreshortening of the effective apperture, the gain of the telescope can vary by over a factor of five as the pointing moves from the meridian to 60 degrees from the meridian. Fortunately this gain variation is easily removed by scaling the sampled data. Unfortunately the interactions produce a change in the beam widths during the observation. The E-W width remains unaltered and all the change takes place in the N-S direction, with the beam width (HPBW) varying by about 30% over the full range of meridian distance. With 384 beams the map width is 1.17 cosec(δ) degrees and it is apparent that this variation in beam width will make a large variation in the relative gain between the centre of the map and points near the edges.

The problem of non circularity and variability of the primary beam may be overcome by fan beam synthesis. An elementary form of fan beam synthesis was first used by Christiansen and Warburton (1955) and the method was developed theoretically by Bracewell (1956). A practical system suitable for the Molonglo Synthesis Telescope was devised and briefly described by Mills, Little & Joss (1976). More recently Perley (1979) has given a detailed theoretical analysis of a similar method and described its implementation for the University of Maryland's TPT array. The method works directly in the (x,y) plane, adding the response for each sample to all points in the map and is equivalent to the method of direct Fourier transform. In practice the algorithm takes each map pixel in turn and estimates by interpolation within the observed beams the response for that meridian distance which passes through the pixel. This reponse is then added to the value for that pixel. The operation is repeated for every data sample, with the only operation needed at the end of the observation being a simple normalisation. It is immediately obvious that the major disadvantage of this method is the large number of operations required. If the map has NxN pixels and there are M samples then MN^2 operations are required. Compare this with the $O(N^2\log(N))$ operations needed for the fast Fourier method. If the primary beam does not vary, both methods give identical results. The difference is that in the fan beam summation the Fourier transform is done (in hardware at Molonglo) before summing the data, whereas in the Fourier method the transform is done after combining the data.

The variations in beam width at Molonglo are counteracted by using a gain factor which depends on the N-S distance of the map point from the centre of the beams. This method of treating the variation in primary beam shape is an approximation and should only be used where the corrections are small. One consequence is that not only does the noise amplitude vary radially but it also has a correlation structure for sources away from the field centre which depends on position angle.

There are other important advantages of fan beam synthesis. (i) If an on line computer can be used, the map is produced in real time and is available as soon as the observation is completed. In this case the only computational requirement is that the calculation be done fast enough. Real time map construction is done at Molonglo on a Hewlett Packard A700 computer using a similar algorithm to that in Sydney but with a coarser interpolation (Durdin, Little & Large 1983). A NOVA 2/10 computer is used to record the raw data on magnetic tape for later analysis either at the telescope or back in Sydney on a CDC CYBER 720 computer. (ii) The basic calculation is simple and could be done in a microprocessor with a large memory. (iii) There is no restriction on the position of the map grid points. For example they can be in a 1950.0 right ascension grid with a separation that suits the final requirement, independent of original beam spacing and orientation. All that is required during the computation is the ability to compute the meridian distance to each map point for every sample. Related to this independence is complete freedom as to the position and size of the map with respect to the observed field. (iv) The final advantage is that all the problems associated with aliasing and with interpolation in the (u,v) plane are avoided.

An important factor in choosing the hardware production of real time beams was the ability to study transient, or time varying phenomena. It is easy to switch the telescope from an interferometer to a total power instrument with an improvement in the signal to noise ratio. A typical application is the observation of pulsars where only several of the central beams are needed.

CLEANING

When a map is made which includes point sources, the shape of the point response function (typically called the dirty beam) depends on what spatial frequencies are present, their relative weights and the hour angle coverage. The purpose of cleaning is to replace this dirty beam by a clean beam, a Gaussian or similarly smooth shape, that facilitates the astronomical interpretation. The ideal cleaning operation would change the map from one of the sky convolved with the dirty beam to one convolved with the clean beam. Although cleaning is not well defined (different algorithms can give different results) it does work successfully for the majority of maps. One of the advantages of the Molonglo telescope is that for a 12 hour synthesis (possible for declinations south of -30°) all spatial frequencies between its lower and upper limits have been observed, so that its dirty beam is already very good (Mills 1981). The first negative side lobe has an amplitude of 8% and the first positive sidelobe has an amplitude of 0.7%. Furthermore the configuration remains fixed so that the dirty beam shape is always known. The cleaning program takes advantage of this to use an analytic dirty beam rather than requiring a separate dirty beam map. One result of this is that memory is saved and it is easy to have point sources at any position - they are not restricted to grid points. Another advantage of this type of dirty beam is that a much greater proportion of the source flux density can be subtracted at each iteration. If there is less than 12 hours coverage the dirty beam is still calculated but now using an hour angle coverage table derived from the observation.

The cleaning program uses a method similar to the Clark algorithm (Clark 1980) in that as many components are removed at each iteration as possible. The choice is made by selecting components in decreasing magnitude, omitting those too close to an existing component, stopping at the first one less than 10% of the largest component. All these components are removed at once. The procedure terminates when either a component with the given flux density is reached or the standard deviation of all the pixels fails to decrease in three consecutive iterations. There is no limitation to the number of components that can be removed or subsequently restored. Restoration can be either with a Gaussian, the positive lobe of the fully synthesised map beam or other simple analytic beamshapes.

An alternative method to cleaning, which is close to the traditional spatial frequency reweighting technique is under development. It modifies the taper along the two antennas in order to alter the transit beamshape in such a way that when the data are summed into a map, the resultant map's point response function has the required shape. Although there are restrictions on the possible beamshapes (lower side lobes can only be achieved at the expense of degraded resolution) the method is feasible since all spatial frequencies, within the range of the telescope, are available. The taper modification could be done on line in the microcomputer which controls the gain of each section, or it could be done subsequently by convolution of the data with an appropriate filter. This latter method is the most flexible and can be applied to old observations. In practise the convolution would be done with Fast Fourier Transforms. The major advantage of this method is the avoidance of the computationally expensive cleaning operation. This is especially true for large extended sources where the degradation in beamwidth can be tolerated and the normal cleaning operation is laborious.

CALIBRATION

Although maps are produced in real time at the telescope, the raw data are also recorded for permanent storage and later analysis. A significant improvement in the map can be achieved if calibration errors can be measured and applied before re-mapping. The simplest method is to measure the errors for calibration sources (with short observations) before and after the observation and assume that a linear correction is sufficient. Because of the stability and redundancy of the telescope this method is adequate for many observations. However a further improvement can be made by the method of self calibration if there is a large point source within the field. The errors that can be measured for each sample are the gain, pointing (an offset in meridian distance measured in the plane of the tilted telescope), and the average phase between the two antennas. The gain is determined by fitting the transit beam for a point source to the data and measuring its amplitude. The pointing error comes from the best position fit, while the phase angle is obtained from the amplitude of a transit beam fitted with quadrature phase. These errors are averaged for several minutes to give a calibration file which is used in the synthesis of a new map, or for other operations such as source fitting. For a strong source typical uncertainties in these have a one standard deviation of 0.6% in gain, 0.7 arcsec in pointing and 1 in phase angle. As well, a convolution with a low pass filter can be used to remove high frequency noise components. This reduces the noise by 25%.

SOURCE FITTING

Another important requirement is the ability to find and determine the parameters of sources. So far this has been done only for point sources, for which a variety of operations exist. The simplest is to search and fit to pixels in a map, using the least squares method, and the dirty map beam. This is fast and can be used for the weakest sources but does require a map to have been synthesised. An alternative is to locate and fit sources in the raw data. First consider the estimation of flux density for a known point source. For each sample the exact meridian distance can be calculated and the flux density estimated by fitting the transit beam shape. These estimates are then averaged over all the samples to get the source flux density (and its uncertainty). There is no restriction on the size of the source or on the number of samples used. This estimate is fast and gives the same result as that made from the map to within small variations due to the treatment of local base line levels. It can also be used to measure sources outside the map area but still within the primary beam.

Source searching in the raw data is a more difficult task, and can only be applied to strong sources. This is because we cannot take advantage of the correlations between samples that enable a weak source to be successfully mapped. The method depends on the equation for the meridian distance $\sin(M.D.) = A\sin(l) + B\cos(l) + C$, where M.D. is the meridian distance, l is the sidereal time and A, B and C are constants depending on the location and orientation of the telescope as well as the source position. The method is to obtain estimates of A, B and C and from them calculate the source position. There are two search algorithms available, the first is used when there is no knowledge of the positions, and the second improves known positions. The first locates peaks in the data for each hour angle sampled and by matching these peaks to source positions builds up a list of possible sources. The only requirement is that the source must be within the primary beam and have a flux density greater than about one or two standard deviations of the noise in each sample. This search is usually followed by another pass fitting for flux density and removing spurious effects. The second algorithm does a full least squares fit using the derivatives of the transit beam shape to obtain better estimates of the source position. In all these methods of fitting to the raw data an ordered procedure is followed in that the strongest source is fitted first. Its effect is then removed by subtracting its expected response from the data before the next source is fitted. This procedure is repeated in sequence for the following sources. It is seen that the methods of source finding are complementary, the map method finds and fits all sources within the map, whereas the raw data method is restricted to stronger sources but can find sources outside the map and does not need a map to have been synthesised.

ACKNOWLEDGEMENTS

The observatory is supported by a grant from the Australian Research Grants Committee. Contributions have also been received from the Sydney University Research Grants Committee and the Science Foundation for Physics within the University of Sydney. Computing facilities at Sydney are provided by the University of Sydney.

REFERENCES

Bracewell, R.N. (1956). Australian J. Phys., 9, 198.
Christiansen, W.N. & Warburton, J.A. (1955). Australian J. Phys., 8, 474.
Clark, B.G. (1980). Astron. Astrophys., 89, 377.
Durdin, J., Little, A.G. & Large, M. (1983). This conference.
Mills, B.Y., Little, A.G. & Joss, G.H. (1976). Proc. Astron. Soc. Aust., 3, 33.
Mills, B.Y. (1981). Proc. Astron. Soc. Aust., 4, 156.
Perley, R.A. (1979). Astron. J., 84, 1443.

DISCUSSION

U.J. SCHWARZ

How is the position of a component determined in the CLEANing process described?

D.F. CRAWFORD

By finding the best position from a bi-quadratic fitted to pixels near a maximum.

9.8 TAURUS Data Reduction Software: The Handling of 3D Data Sets in Optical Astronomy.

Keith Taylor
Anglo-Australian Observatory, Epping, NSW.2121, Australia.

Richard N.Hook
Sussex University, Brighton, E.Sussex, U.K.

Paul D.Atherton
Kapteyn Laboratory, Groningen, The Netherlands.

Introduction. The details of the TAURUS technique for obtaining emission-line velocity information over large fields, in the optical, has been covered in depth elsewhere, (Taylor 1978, Taylor and Atherton 1980, Atherton et al. 1982) ; the purpose of this paper is to describe the techniques used to reduce the 20 Mbyte 3D data sets created by TAURUS into understandable astrophysics. Without going into the details of the optical techniques employed, we shall take it that TAURUS is an imaging Fabry-Perot (FP) interferometer which creates 3D data arrays, I(x,y,z), where x and y are the spatial co-ordinates and z maps to wavelength (or velocity) linearly. However, the effect of the FP is to create surfaces of constant wavelength within the 3D array which are not simple x,y planes, but are paraboloids periodic in z, where each surface corresponds to a different order of interference.

This feature of the TAURUS data cubes requires us to devise specialized software in order to flatten the iso-velocity surface and it is the theory and practice of these techniques which I wish to expound in the first half of the discussion. Later on I shall be describing the algorithms which are used to create 2D velocity and line width maps from the corrected data.

The Phase-Map. The periodic, paraboloidal surfaces of constant wavelength, shown schematically in Figure 1, are due to the familiar Fabry-Perot fringe pattern which, for a given wavelength, λ_0, is represented by a 3D Airy-function ;

$$\Phi(x,y,z) = \{ 1 + (2N/\pi)^2.\sin^2[2\pi t(z)\cos\theta_{xy}/\lambda_0]\}^{-1} \qquad (1)$$

where θ_{xy} is the incident angle at the FP corresponding to position (x,y) in the detected image plane, N is the finesse of the FP etalon and t(z) is the etalon optical gap, linearly dependant on z, with a constant of linearity 'k'. The z-period of the Airy-function is usually referred to as the free-spectral-range (FSR), which, from inspection with equation (1), is given by ;

$$\Delta z_{xy} = (\lambda_0/2k)\sec\theta_{xy} \qquad (2)$$

Clearly, if we were to put an x-y plane through the paraboloid of constant λ_0, we would obtain the 2D Airy-function, $\Phi_z(x,y)$, shown in Figure 2, which is the familiar FP ring pattern. Now if the 2D variation of surface brightness at any wavelength, λ, is given by $B_\lambda(x,y)$ then the distribution of observed intensity, $I_z(x,y)$, is ;

$$I_z(x,y) = \Phi_z(x,y) \cdot B_\lambda(x,y) \qquad \text{for any value of } z. \qquad (3)$$

Indeed TAURUS data is acquired by detecting a sequence of 2D arrays, $I_z(x,y)$, sampled at an interval δz over a full FSR, Δz_0.

Note that the operator in equation (3) is simply a product,(.), not a convolution,(*), and hence this is NOT analogous to interference rings of a radio-interferometric point-spread function. The spatial point-spread function for TAURUS is simply the gaussian seeing profile introduced by atmospheric effects.

The raw 3D data cube can, more usefully, be viewed as a 2D array of 1D spectra, where the observed profile, $I_{xy}(z)$, is given by the convolution of the incident spectra, $B_{xy}(\lambda)$, at any point, (x,y), and the periodic instrumental Airy-profile, $\Phi_{xy}(z)$, such that ;

$$I_{xy}(z) = \Phi_{xy}(z) * B_{xy}(\lambda) \qquad (4)$$

Clearly, for a monochromatic source, $B_{xy}(\lambda_0)$, the observed periodic profiles, $I_{xy}(z)$, will be identical except for a 'phase-shift', p_{xy}, introduced by the $\cos\theta$ term in equation (1). In fact ;

$$p_{xy} = n.\Delta z_0(\sec\theta_{xy} - 1) \qquad (5)$$

where n is the on-axis interference order for wavelength, λ_0, and Δz_0 is the on-axis FSR.

The effect of the Phase-shift can be seen most graphically by considering a spectrum as a 1D cut in the z-direction of Figure 1. Depending on the particular (x,y) pixel chosen, the 1D spectral profile will cross the paraboloidal surface at different z-values and these differences, with respect to the on-axis pixel (x_0,y_0) are the phase-shifts referred to in equation (5). Phase-correcting a 1D profile, $I_{xy}(z)$, simply involves, therefore, resampling the profile from a new origin corresponding to the value of the phase-shift at the co-ordinate (x,y).

In order to map this phase-shift in 2D we illuminate the focal-plane of TAURUS with a spatially uniform monochromatic source, and by scanning the etalon (which corresponds to varying z in δz steps through one FSR) we create our 'calibration-cube' which is subsequently used to flatten the surface of constant velocity in the real data. An example of the type of data obtained from a calibration-cube is shown in Figure 2.

From the calibration-cube we can readily determine the on-axis image co-ordinate, (x_0,y_0), the FSR and the finesse, N ; knowing the etalon optical gap, t(z), the pixel-size and the calibration wavelength, λ_0, we can predict the theoretical phase map, as given in equation (5). This theoretical surface is a very close approximation to the locus of intensity maxima within the calibration-cube. However towards the edges of the data, spatial distortions within the detector and the optics begin to become important so that the theoretical surface is no longer a good approximation.

To cope with this effect, the calibration-cube is phase-corrected with its own, theoretically generated, phase-map. This operates on the paraboloidal surface to give an approximately flat, constant wavelength surface with a residual curvature resulting from the optical and detector induced destortions. A smoothed version of this 'residual' map is then added to the theoretical phase-map to give an 'imperical' phase-map which, when applied to both the calibration-cube (from which it is derived) and subsequent data obtained on astronomical sources, gives the data cubes an iso-velocity surface which is flat to better than 0.1δz.

Extracting Kinematic Information. Once the data has been phase-corrected, all that remains is to extract from the phase-corrected data-cubes, velocity, line-width and intensity information by fitting

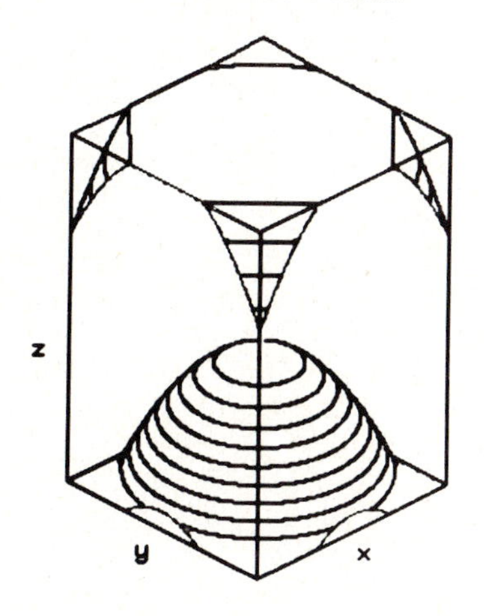

Figure 1. A 3D representation of the 'periodic paraboloids' which map the surface of constant wavelength within the data-cube.

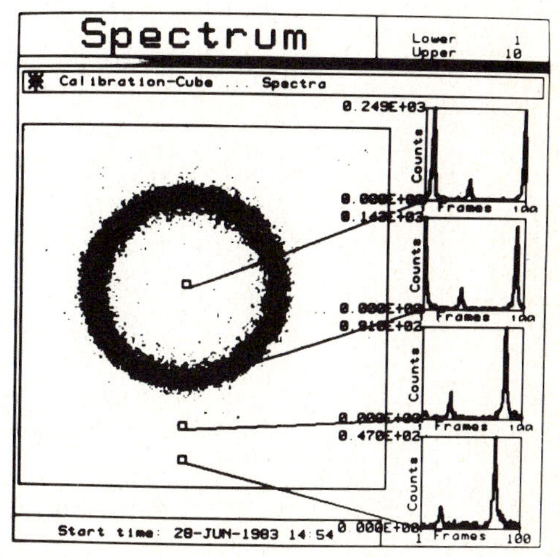

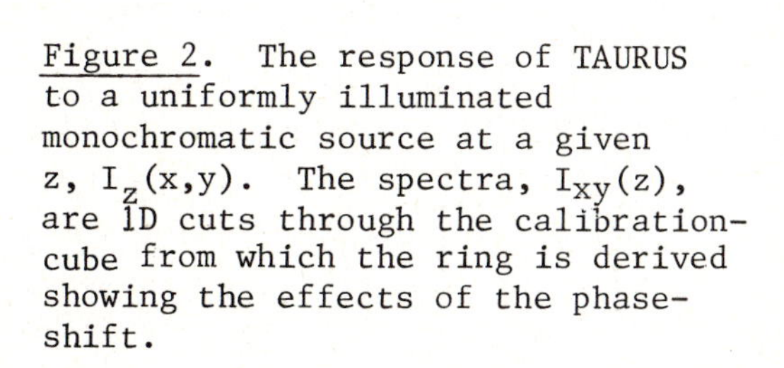
Figure 2. The response of TAURUS to a uniformly illuminated monochromatic source at a given z, $I_z(x,y)$. The spectra, $I_{xy}(z)$, are 1D cuts through the calibration-cube from which the ring is derived showing the effects of the phase-shift.

gaussians (for example) to the 1D profiles, in order to map the kinematics of a particular line-emitting object. Clearly, not all profiles contain good S/N information and so the process of extracting the kinematics from a data-cube cannot be entirely automated, however the data-cubes are typically dimensioned, 256x256x128 and hence there is a clear need to maximize the amount of non-interactive processing involved in the data reduction.

After a long process of refinement we are now converging towards a fairly standard approach to this section of the data reduction, which to my knowledge, varies significantly from the techniques employed, at present, in radio astronomy.

The first step is to create an 'emission-map', which is simply a 2D array representing the intensity in the emission-line component of each profile, $I_{xy}(z)$. This is acheived by taking the integral of the profile after high-pass filtration and background subtraction. The image so formed dramatically improves the visibilty of the spatial structure of the line emitting regions when compared with a simple profile summation algorithm, (a compacted array, in TAURUS jargon). This distinction is clearly demonstrated in Figure 3, where the two techniques have been applied to the same data-cube.

The emission-map is then used as a logical mask, whereby all spatial pixels containing values greater than a predetermined threshold in their emission-map have their spectral profiles fitted automatically by a gaussian ; all pixels below this threshold are ignored. In this manner a set of 2D maps, called the gaussian-parameter-block, (GPB), containing, respectively, the height, velocity, width and background for each single (or double) gaussian fit, as a function of position, (x,y), is created.

This automatic process, of course, cannot extract the kinematic information in its entirety and an interactive process is required to clean-up badly fitted profiles and to extend the fits into regions of low S/N. The problem here is to view the 3D information in a way which allows the user to decide whether a particular spatial or spectral feature is real, and to do this he has to be able to look for spatial and/or spectral coherence. The 'GIPSY' technique, which we, initially adapted for our own purposes, is to view the data as a series of x-z slices, allowing the user to assess the spatial coherence of faint features at a given y value ; by displaying more than one slice at any one time, spectral coherence can also be checked. The technique then involves interactively drawing polygons around the good S/N regions in each x-z slice, and hence forming a complex 3D volume within the data-cube, within which a moments anaysis can be acheived.

The problems with this approach are that the process of creating the 3D polygon surface within the data-cube is extremely expensive in interactive time at the computer terminal, but more seriously the technique is guaranteed to produce sensible-looking numbers for the 1st and 2nd moments within the volume defined, without a simple means of comparing the results with the raw spectral data from which they are derived. As a result of the need, both to assess spatial and spectral coherence and also to view the moments solution, or gaussian fits, to the spectral data, we have developed a software package which plots a spatial subset of the data-cube as a page of profiles, with their respective gaussian fits, on an interactive TV display ; a cursor is then used to select the pixel-profile of interest. The following menu is then presented to the user, which allows him to interact with the data and as a consequence up-date and edit the GPB.

Options:

C ... Display Pixel Co-ordinates	S ... Set another position
N ... Next screen	G ... Fit a 4 parameter Gaussian
R ... Fit Gaussians over a region	D ... Fit a Double Gaussian
U ... Unfit a Gaussian	E ... Enlarge plot
T ... Enter 2D display options	L ... Leave routine

This technique covers our needs since it allows the user to assess the reality of spectral features while giving him a direct means of appraising the fits to each profile spectrum. However as we become more familiar with using these techniques it is becoming obvious that a great deal more can be done towards automating the process since the majority of decisions that are made interactively are done on the basis of spatial continuity. We now have software that allows a large degree of automatic checking of the GPB but the package needs a little more development work before we can reduce the process to one of almost total automation.

A Special Example. Of course not all data-cubes contain simple gaussian pixel-profiles and as an example of a non-standard use of TAURUS, I present here in pictorial form data taken on the [OI],λ6300 line in the region surrounding the Becklin-Neugebauer (BN) infrared complex in the Orion molecular cloud. The [OI] line from the background HII/HI interface in Orion appears as a gaussian of roughly constant width and velocity, however, super-imposed on this is a family of high negative velocity (< 500km/s) Herbig-Haro objects associated with outflow from the BN-IR complex, (Axon and Taylor 1983). The montage of pictures, shown in Figure 4, is derived from the same data set ; the left hand images being created by simply summing the data-cube over restricted region in z, or velocity space, while the right hand images are taken from the same data-cube which has had the gaussian, representing the background nebula, fitted and then subtracted from the data cube. In this way we are able to increase the visibility of the high velocity HH objects against the background emission and view, for the first time, their clustering in a cone whose apex is centred near the BN-complex.

References.

Atherton,P.D.,Taylor,K.,Pike,C.D.,Harmer,C.F.W.,Parker,N.M. and Hook,R.N. 1982. Mon.Not.R.astr.Soc.,201,661.

Axon,D.J. and Taylor,K. 1983. Mon.Not.R.astr.Soc., In Press.

Taylor,K. 1978. Proc. 4th. Intern.Coll. on Astrophys., (Ed. M.Hack), p469.

Taylor,K., and Atherton,P.D. 1980. Mon.Not.R.astr.Soc.,191,675.

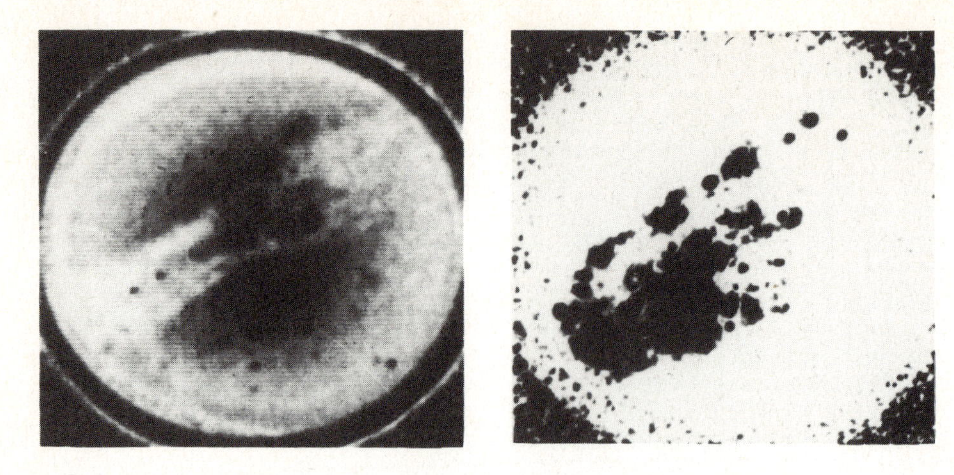

Figure 3. TAURUS images of the central 4 arcmin. of Centaurus A, taken from the same data set. The left-hand image is the compacted array, equivalent to viewing the object through a 20Å band-pass H_α filter. The right-hand image is the emission-line map showing the distribution of H_α in the disk.

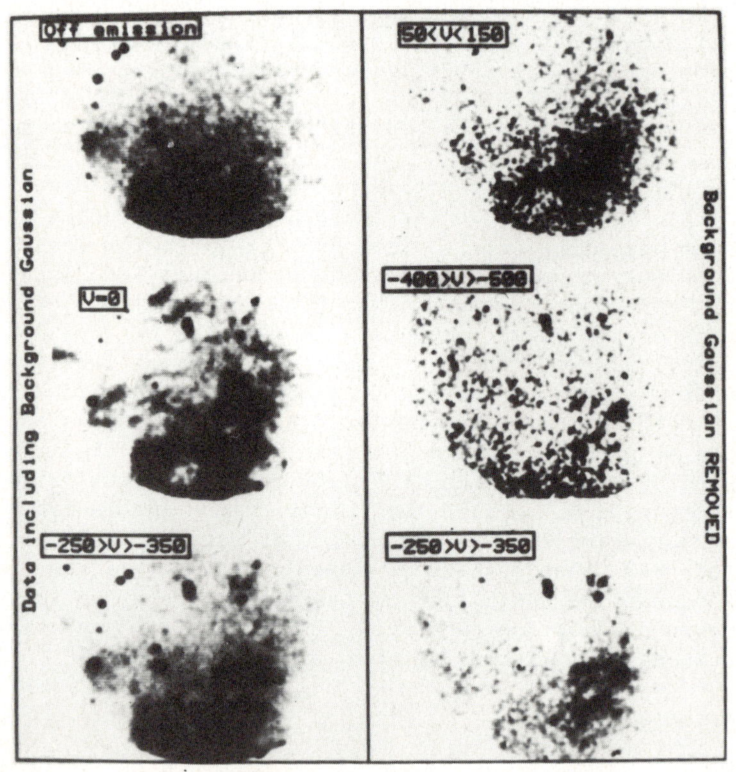

Figure 4. Images of the Orion Nebula taken from the same [OI], λ6300, data cube. The left-hand images are taken from the original, phase-corrected data, while the right-hand images are from the background subtracted version. The bright sources seen in the bottom right image are all high velocity HH objects. The cross marks the position of the primary embedded IR source, IRc2.

10

Specialized hardware

10.1 A DIGITAL FFT SPECTRO-CORRELATOR FOR RADIO ASTRONOMY

Yoshihiro Chikada, Masato Ishiguro, Hisashi Hirabayashi, Masaki Morimoto, Koh-Ichiro Morita, Keisuke Miyazawa, Kiyoshi Nagane, Kazuo Murata, Arata Tojo, Shizuyo Inoue, Tomio Kanzawa, and Hiroyuki Iwashita

Nobeyama Radio Observatory, Tokyo Astronomical Observatory, University of Tokyo.
Nobeyama, Minamimaki, Minamisaku, Nagano-ken, 384-13 Japan

Abstract. A new digital FFT spectro-correlator has been successfully built for the five-element synthesis telescope and the 45 meter telescope of the Nobeyama Radio Observatory, and is under performance tests. It has 320 MHz maximum bandwidth with fifteen correlations divided into 1024 frequency channels each. The advantages over the conventional direct correlation receivers and the implementation techniques are discussed.

1 *INTRODUCTION*

High spatial resolution spectroscopy with a radio interferometer gives us a large amount of information about many interesting aspects of astronomy. However, when applying this technique at millimeter wavelength, there are some difficulties in implementing a spectro-correlator of sufficient total bandwidth and frequency resolution. The difficulties are as follow. (A). The instantaneous bandwidth should be large enough to cover the galactic rotation at millimeter wavelength, e.g., larger than several hundreds of megaherz, though digital components do not have such processing speed. The limits on the speed can be overcome by introducing parallel architecture to increase total throughput, though the size of the hardware will be prohibitively large if one adopt the conventional algorithms. (B). There are analog techniques with sufficient bandwidth and frequency resolution such as acousto-optics or SAW (surface acoustic wave) techniques, but they are not matured to be used in spectro-correlators which need to have more complex structure than in spectrometers of single dishes (Chikada 1981).

Yen (1974) pointed out the important role of the FFT (Fast Fourier Transform) (Cooley & Tukey 1965) in astronomy, and several attempts (Morris & Wilck 1978) have been made to implement dedicated FFT processors for single-dish spectroscopy. Chikada (1979,1981) proposed a method to realize a spectro-correlator of wide bandwidth and high resolution where size is drastically reduced by the application of the FFT algorithm. Here the signals are directly converted into digital form, Fourier transformed into frequency domain by the parallel pipeline FFT processors, and then correlated to give (cross) power spectra by term-by-term multiplications.

For the Nobeyama five-element interferometer (Ishiguro 1981), a digital FFT spectro-correlator, which is called FX, was assembled in the Kawasaki Factory of Fujitsu Limited and is under performance tests. The preliminary results show that the digital part cleared the aimed bandwidth of 320 MHz and the system works as expected. The tests on the system performance will be completed in October 1983.

The FX spectro-correlator (Figure 1, Table 1) processes five inputs of 320 MHz instantaneous bandwidth, which is presently limited to 80 MHz by the input bandwidth and the sampling rate of analog-to-digital converters. It has 1024 complex frequency channels for each of fifteen correlations. Its frequency resolution can be changed by tuning the clock rate of the analog-to-digital converters, and the highest resolution is 4.9 kHz. A similar spectrometer is also being made for the Nobeyama 45 meter telescope (Morimoto 1981). One FFT processor is time-shared to process four inputs of 10 MHz maximum bandwidth. Four power spectrums are output, each of which have 1024 frequency channels. This processor can be used as the sixth FFT processor of the spectro-correlator when the 45 meter telescope is connected to the interferometer.

The FX is made up of about 4500 LSI (large scale integrated circuits) chips of CMOS (complementary metal-oxide-silicon) gate arrays. The gate array has 3900 or 2000 gates in it, operates at 10 MHz clock rate,

Table 1. Specification summary of the Nobeyama FX.

Maximum bandwidth		
Digital processor	320 MHz	
A to D converter	80 MHz	On the present stage.
No. of complex frequency channels per correlation	1024	
No. of FFT processors		
Interferometer	5	
45 meter telescope	1	This can be used as the sixth processor of the interferometer.
No. of correlations		
Interferometer	15	
45 meter telescope	4	
Highest resolution		
Interferometer	4.9 kHz	
45 meter telescope	0.153 kHz	
Noise-to-signal ratio	< 10 %	The ratio of the noise power to the input power.

and consumes 100 mW per chip or less. Four kinds of LSI were implemented, which are butterfly, multiplier, accumulator, and corner-turner (See Section 4). They are arranged to form parallel pipelines of 6-8 bit precision arithmetics.

The signal processing in a spectro-correlator is briefly reviewed in Sections 2 and 3. The principle of operation and the implementation of the Nobeyama FX is reported in Section 4 and also its preliminary test results in Section 5. In Section 6, we will discuss the selection of algorithms for spectro-correlators in general.

Figure 1. Block schematics of the interferometer and 45 meter telescope of the Nobeyama Radio Observatory (NRO). Backends other than the digital FFT spectro-correlator and spectrometer (thick lines) are not shown in this figure.

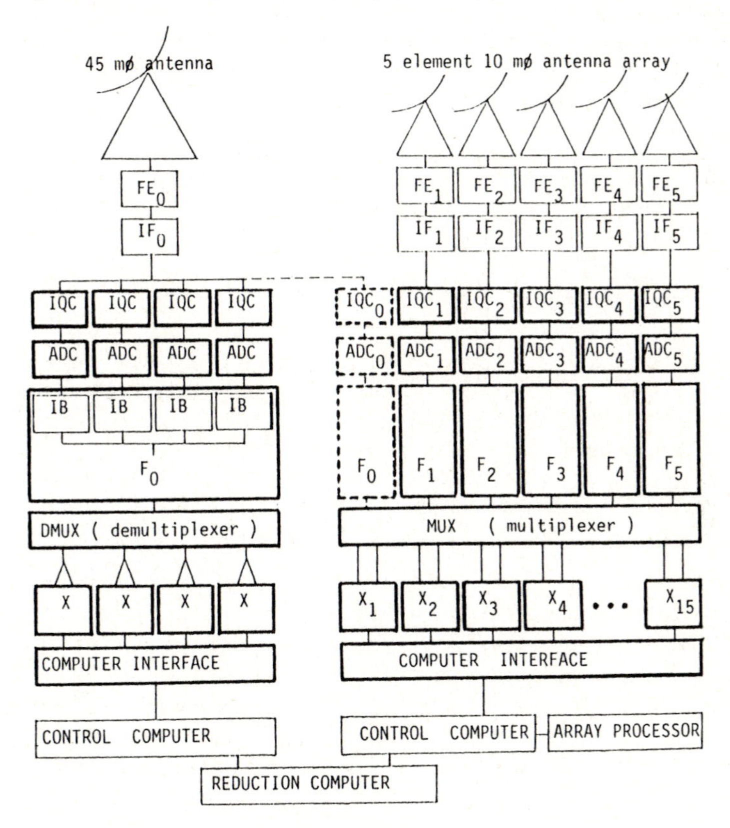

FE : Receiver front-end.
IF : Intermediate frequency amplifier.
IQC : In- and quadrature-phase frequency converter.
ADC : Complex analog-to-digital converter.
F : FFT processor.
IB : Input signal buffer (the first corner turner).
DMUX : Demultiplexer.
MUX : Multiplexer.
X : Correlator in frequency domain.

2 *STATEMENT OF THE PROBLEM*

The notations used in this paper are summarized below.

- u : Projection of a position vector of an element antenna onto a plane perpendicular to the line of sight to the center of the object source.
- x : Vector to the point on the map from the map center.
- t : Time.
- f : Frequency.
- U : Difference between two u vectors.
- T : Time difference.
- s : Real time signal.
- p : Detected power per unit time (n/B).
- P : Averaged power.
- X**n : X^n
- log x : $\log_2 x$, base 2 logarithm of x.
- O(n) : Order of n.
- B : (Maximum) bandwidth of the backend.
- N : Number of element antennas.
- n : Number of frequency channels (at the maximum instantaneous bandwidth).
- M : Number of pixels in a map taken simultaneously.
- K : max(N,M).
- O : Number of operations.
- G : Number of gates.
- C : Complexity of an operation in unit of gate $\text{operation}^{-1}\text{MHz}^{-1}$ or J operation^{-1}.

In Figure 2, the signal flow in a spectro-correlator is shown. The signal s(u,t) is (after the transform(s), if necessary,) detected via convolution in u and/or t domain or via term-by-term multiplication in x and/or f domain to give p, which is then averaged to give the spectral map P(x,f) (after the transform(s), if necessary,). On each of the three stages, i.e., s, p, and P, there are four different forms of representation which are Fourier transforms of the others on the same stage. We can distinguish six different paths from s(u,t) to P(x,f) (Figure 3). Two pairs, (a) and (b), and (e) and (f), only have a difference in the order of Fourier transforms on the same stage, so there are four different types of path. (For simplicity, we confine our discussion in the cases where the vectors, u, U, or x, are one-dimensional vectors. There are 24 paths and 8 types for the two-dimensional case.) What we will clarify in this paper is which is the best path for a spectro-correlator or a spectrometer, for a given n, N, and M.

3 *ALGORITHMS*

The necessary number of operations depends on which path is taken from s(u,t) to P(x,f). For the direct correlation receiver (Weinreb 1963; Weinreb et al. 1963), one needs O(N**2 n**2) operations from the signal s(u,t) to the convolution p(U,T), and O(Mn) operations for accumulation on the path from p(U,T) to P(U,T). Synthesized maps for all frequency channels are obtained after the post-detection Fourier transforms on the path from P(U,T) to P(x,f) where the processing costs are negligible compared with those of the real-time operations. Therefore on the path (a) or (b) in Figure 3, s(u,t) - p(U,T) - P(U,T) - P(x,f), one needs O(N**2 n**2) and O(Mn) operations.

On the other hand, on the path where the Fourier transform to frequency domain take place first (the path (c) in Figure 3), the N signal flows from N antennas are transformed into amplitude spectrums by O(N n log n) operations of N n-point FFT's, and then (cross) correlated by O(N**2 n) term-by-term multiplications where the signal amplitude in a frequency channel of the i-th antenna is multiplied by the one in the same frequency channel of the j-th antenna and the products are summed (resampled) if there are any baseline redundancies. The resultant p(U,f) is then accumulated to give cross power spectrums at each point on the u-v plane (P(U,f)) by O(Mn) operations. P(U,f) is then transformed to the map P(x,f) by the post-detection two dimensional Fourier transform where the processing costs are negligible compared with those of the real-time operations. Therefore on the path,

Figure 2. Data flow diagram in a spectro-correlator. Double lines indicate Fourier transforms, and figures on the lines indicate the necessary number of operations. See Section 2 for the notations.

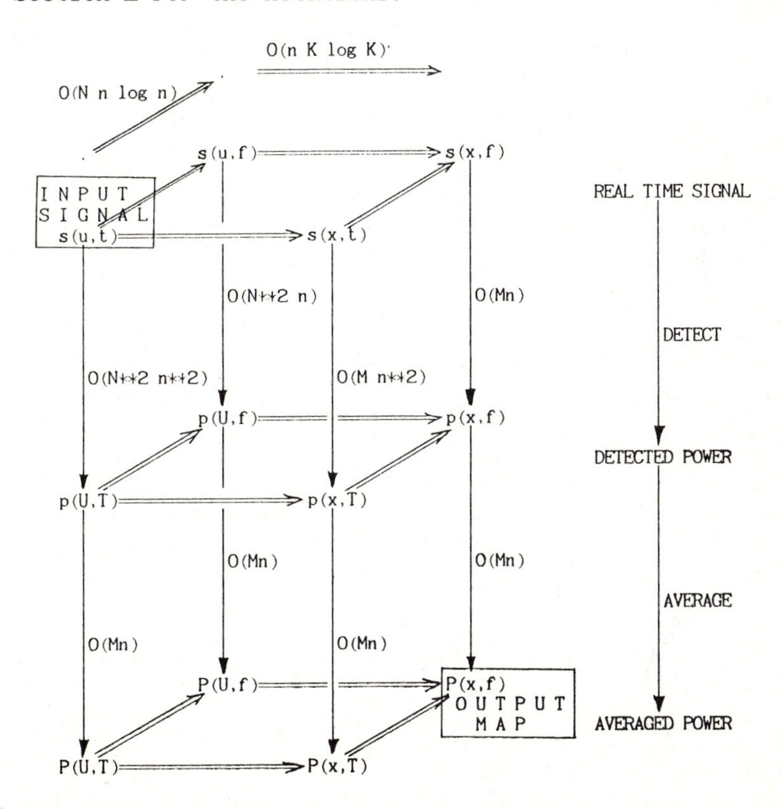

s(u,t) - s(u,f) - p(U,f) - P(x,f), one needs O(N n log n), O(N**2 n), and O(Mn) operations. This is the path of the Nobeyama FX. On this path (FX), one needs much less operations than on the path of the direct correlation receiver in the cases where the number of frequency channels n or the number of element antennas N is sufficiently larger than 1 (Chikada 1981). On the paths, (d), (e), and (f) in Figure 3, it can be easily seen, one needs less operations if N**2 is sufficiently larger than M, i.e., antenna array has redundant baselines.

The hardware size depends not only on the number of operations per seconds but also on the complexity of a unit operation and on the necessary signal-to-noise ratio (SNR). The complexity is approximately proportional to the number of bits per word for addition and for storage or to its square for multiplication. In cases of most FFT spectro-correlators, as in the Nobeyama FX, the real-time FFT needs to be calculated with 6-8 bit precision arithmetic for each of the real and imaginary parts of a complex word. The necessary number of bits per word has weak dependence on the number of points to be transformed as long as the number of points is sufficiently large. The correlation receiver has to perform multiplication and accumulation to complete convolution, so that the complexity is not as small as one expects counting only multiplications. The actual gate count and complexity of the Nobeyama FX and those of a conceptual correlation receiver will be given in Section 6.

Figure 3. Six paths from the input signal s(u,t) to the output map P(x,f). The necessary number of operations are summarized in the rows, (1) for the real-time transforms, (2) for the detection, and (3) for the averaging.

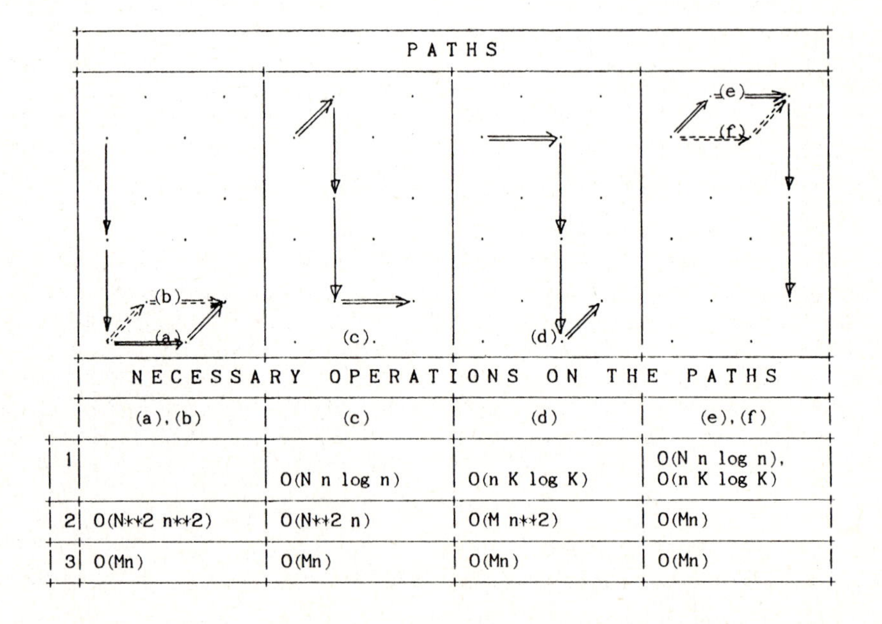

	NECESSARY OPERATIONS ON THE PATHS			
	(a), (b)	(c)	(d)	(e), (f)
1		O(N n log n)	O(n K log K)	O(N n log n), O(n K log K)
2	O(N**2 n**2)	O(N**2 n)	O(M n**2)	O(Mn)
3	O(Mn)	O(Mn)	O(Mn)	O(Mn)

In the case of the digital FFT spectro-correlator or spectrometer, the backend noise arises in the analog-to-digital converter and in the digital processor. The latter of the Nobeyama FX is less than five per cent of the input power for n up to 1024 and the former is approximately five per cent for a three bit analog-to-digital converter. In correlation receivers, the noise arises mainly in the analog-to-digital converter and its ratio to the input power is 57 per cent and 14 per cent for one bit and two bit analog-to-digital converter, respectively. In the SNR range tolerable for astronomical use, the size of the correlation receiver is more sensitive to the necessary SNR than that of the digital FFT spectro-correlator.

4 *IMPLEMENTATION*

4.1 *Architecture*

The Nobeyama FX consists of four sections (Figure 1), which are the IQC (in- and quadrature-phase frequency converter), ADC (complex analog-to-digital converter), F (FFT processor), and X(correlator). The spectro-correlator has five sets of IQC, ADC, and F, and fifteen X's. There are empty card slots reserved for the sixth ADC and F to be used in the future six element operations. The signal connections between F's and X's can be arranged through multiplexers for such operations as well as the normal five element operations. The spectrometer of the 45 meter telescope has four IQC's, four ADC's, one F, and four X's. F is time-shared to process four input signal streams.

4.2 *In- and quadrature phase frequency converter (IQC)*

IQC has two functions. One is to extract the signal in the bandwidth to be observed and convert it down to the frequency range acceptable for ADC. The other function is to provide in-phase signal and quadrature-phase signal to a pair of analog-to-digital converters as the real part and the imaginary part of the original signal. A pair of mixers are used to realize the above functions (Figure 4). They are fed by the local oscillator signals which have a phase difference of 90 degrees. The output frequency range is not base-band, except for the case of the maximum bandwidth of ADC, and is selected to be from B to 2B where B is the bandwidth to be observed.

4.3 *Complex analog-to-digital converter section (ADC)*

The bandwidth of ADC is 80 MHz at the first implementation step and is to be 320 MHz at the further steps. The input signal is sampled at the clock rate equal to the signal bandwidth and converted into a complex digital word having three bits for each of the real and imaginary parts (Figure 4). The digitized signal is sent to F after the signal clock rate is reduced by 1/32 through a 1-to-32 serial-to-parallel converter, while the total signal throughput rate remains the same as before.

4.4 *FFT processor (F)*

F is a 1024 point complex FFT processor whose input, at maximum bandwidth, is 320 mega complex samples per second (Figure 4). It outputs a 1024 channel amplitude spectrum every 3.2 micro seconds. The input signal is shifted into a buffer register called the corner turner, which has 32 wide x 32 word input shift registers and similar output shift registers. The shift direction of the latter is rotated by 90 degrees from that of the former, so that the j-th data on the i-th input shift register is "snap-shotted" onto the j-th output shift register and then shifted out at i-th clock cycle. Similar permutation is known as the bit-reversal operation in the radix-2 FFT (Cooley & Tukey 1965). The clock frequency fi of the input shift register is,

Figure 4. Block schematics of the in- and quadrature-phase frequency converter (IQC)(upper left), the complex analog-to-digital converter section (ADC)(lower left), and the F (FFT processor) (right). BPF, LPF, and LO are a band pass filter, a low pass filter, and a local oscillator, respectively. The frequency range of the ADC input is selected to be from m fADC to (m+1) fADC where m is an integer and fADC is the clock frequency of the A/D converters. The F (FFT processor) transforms 1024 point complex data into amplitude spectrum in every 3.2 micro seconds. The 1024 point data is processed in the 32-parallel-input pipelines in 32 clock cycles.

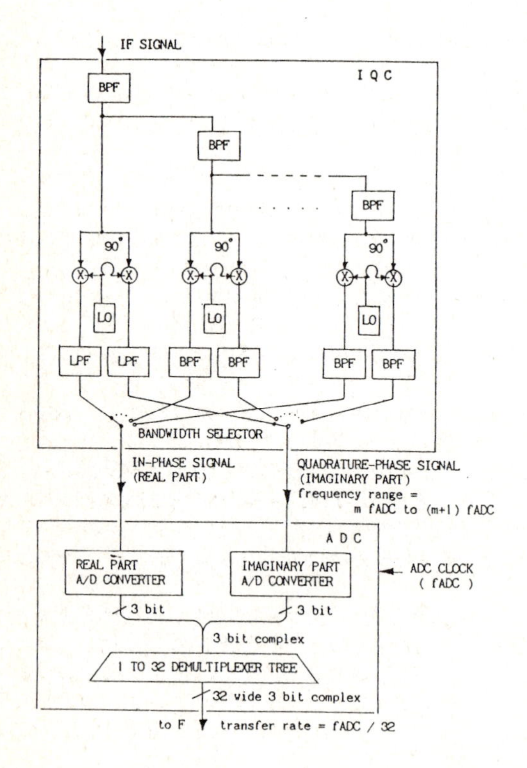

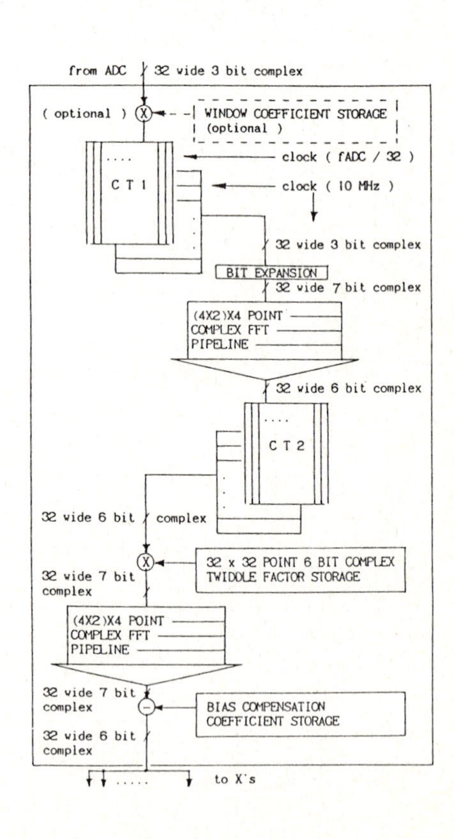

$$fi = fADC / 32 = B / 32 , \qquad (1)$$

where fADC is the clock frequency of ADC and B is the bandwidth to be observed. The clock fo of the output shift register is fixed to 10 MHz and the "snap-shot" (parallel load onto output shift registers) takes place every 3.2 micro seconds. When B is smaller than 320 MHz (maximum bandwidth of F), there are overlaps among the input data segments of successive Fourier transforms.

The X output pk(f) for k-th segment consists of three parts, which are the contribution Sk(f) of the k-th input segment from ADC, the quantization noise contribution Qk(f) of F and X, and the cross-term Ck(f) between the input and the quantization noise. In the case of no overlaps, assuming that the input signal is white noise, there is no dependence between two successive input data segments, and, therefore, there is no dependence between the successive output pk(f) (k=1,2,3,...) from X in a frequency channel f. In the case of finite overlaps, Sk(f) has dependence on that of the successive data segment, Sk'(f) (k'=k+1, etc.,), while Qk(f) and Ck(f) are less dependent on Qk'(f) and Ck'(f), because the data are shuffled in F so that the quantization noise for the k-th segment produced in latter stages has no dependence on that of the k'-th. Therefore, after averaging over finite time, the estimation errors of the power spectrum are smaller in the case of finite overlaps than in the case of no overlaps.

The spectrometer of the 45 m telescope has four corner turners as the temporary data storage. Each retains its data while the data of others are processed.

The signal from the corner turner(s) is then fed to a 32 point FFT pipeline, which has a 32 wide parallel input port. It outputs a 32 point Fourier transform every clock cycle of 100 nsec. It has the structure of full implementation of the signal flow graph of the 4-2-4 radix FFT. Then it passes through the second 32 x 32 corner turner, the 32 x 32 point twiddle factor multiplier, and the second 32 point FFT pipeline, so that the complete 1024 point FFT is performed.

On the final stage of F, there is a bias compensator, which subtracts the expected value of the bias in the output of F to avoid unnecessary SNR degradation (Chikada 1979,1981).

4.5 *Correlator (X)*

The amplitude spectrum output from F is then (after demultiplexing in the 45m spectrometer) fed to the X's (Figure 5). Here the amplitude spectrum from the i-th F is multiplied by the complex conjugate of the one from the j-th. After the demodulation of the switching, the resultant (demodulated) power spectrum is accumulated in the FIFO's (first-in first-out registers) and in the final accumulation RAM's (random access memories).

The representation of values is offset binary throughout the accumulation stages. The carry rate slows down in the successive stages, so that the sampling rate on latter stages can be less. Therefore, the memories of these stages can be more cost-effective "deep" memories, i.e., more bits and less I/O lines. The read-outs of the final accumulation storage are multiplexed and sent to the host control computer at request.

4.6 *Semi-customized LSI's*

As stated above, the FX has regular and repetitive architecture and has no feed- back loops, therefore we can easily divide it into several basic functions that can be integrated in state-of-art LSI's (large scale integrated circuits). Four types of semi-custom LSI's are implemented, which are CT (16x16 bit corner turner), MPY (6 bit complex x 6 bit complex multiplier) whose output is rounded to an 8 bit complex word, BTF (7 bit complex butterfly) which calculates the sum and difference of the two inputs, and ACC (accumulator) which accumulates 9 bit complex word into a 32 stage FIFO). They are 2000-3900 gate CMOS

Figure 5. Block schematics of a X (correlator), which multiplies 1024 channel complex amplitude spectrums channel-by-channel and accumulates these products into the storage.

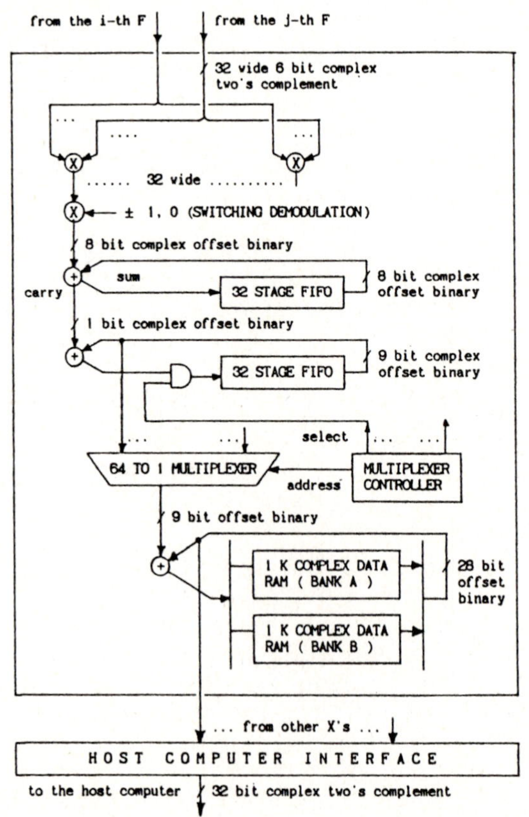

Figure 6. Functional diagrams of (a) 16 x 16 bit corner turner LSI, (b) 7 bit complex butterfly LSI, (c) 6 bit complex x 6 bit complex multiplier LSI, and (d) 9 bit complex 32 word accumulator LSI.

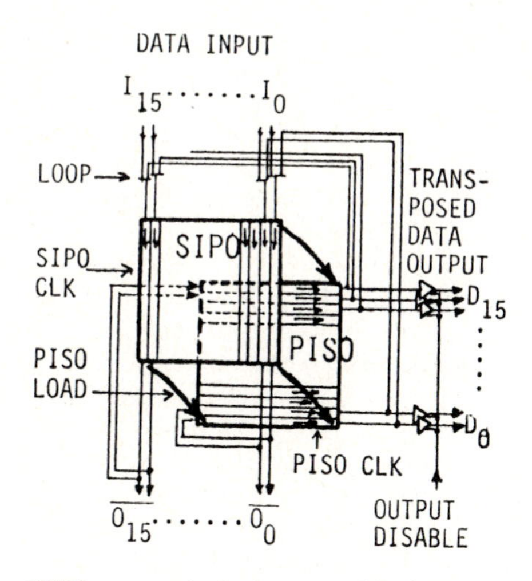

SIPO : serial-in parallel-out register.
PISO : parallel-in serial-out register.
CLK : clock.

(a)

A
7c
B
7c
OVFL CLIP
S
7c
OVFL CLIP
D
7c
SCALING
CLOCK

OVFL CLIP : overflow clipper.

(b)

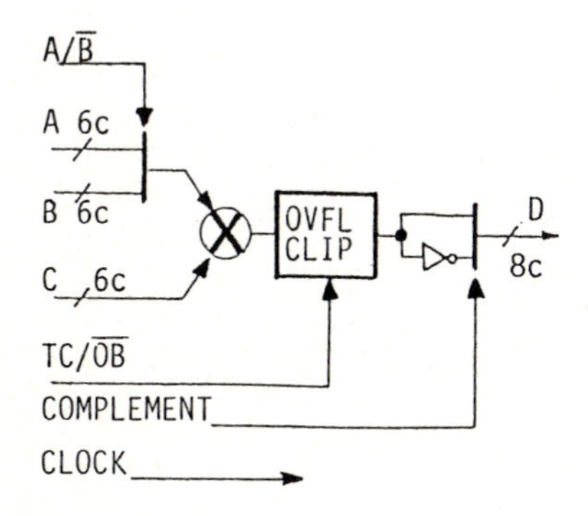

TC : two's complement.
OB : offset binary.
OVFL CLIP : overflow clipper with TC/OB selector.

(c)

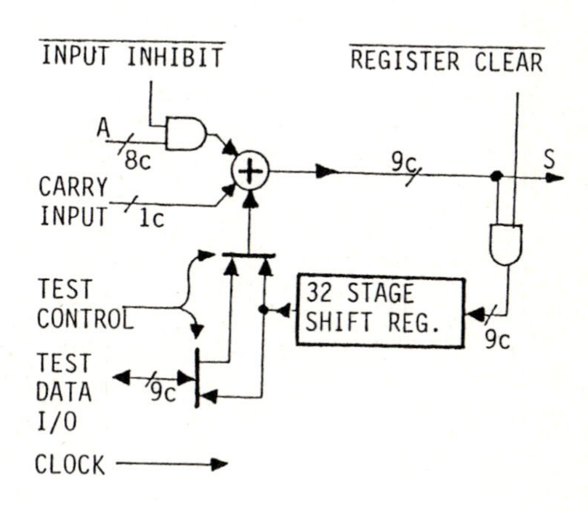

(d)

gate arrays which operate at 10 MHz clock rate and consume less than 100 mW per a chip. They are themselves pipelines to assure the above processing speed. The overhead for the pipeline registers in them is approximately one half of their total gate count. Almost all of the necessary arithmetic operations are implemented in them , while the interface circuits, i.e., drivers and receivers, the sequence controllers, etc., are implemented with standard LSTTL's. The number of gates for these do not exceed 10 per cent of the total, though their power consumption is a major part of the total power consumption. The functions of the LSI's are shown in Figures 6.

5 *PRELIMINARY RESULTS OF PERFORMANCE TESTS*

Although performance tests and the tuning have not yet been completed, some preliminary results can be shown. The digital FFT processors (F's) and the correlators (X's) were tested for their processing speed and they proved to function correctly at the maximum bandwidth of 320 MHz. They were also tested for their logical validity and they gave the identical output with that of a simulator in a computer. The NSR (noise-to-signal ratio) of the FFT processor was measured on the output of the simulator to be less than 3 per cent. Although total NSR, including IQC, ADC, and X, has not been measured, the above result shows that it will be less than 10 per cent.

Table 2. Summary of the size of The Nobeyama FX.

		C T	B U T	M P Y	A C C	total
Gate count / chip (*)		3367	1897	3593	3671	—
Chip count	F	106	176	160	0	442
	X	0	0	32	64	96
	5 F + 15 X	530	880	1280	960	3650
	F+4X (45m)	178	176	192	256	802
	Total	708	1056	1472	1216	4452
Total gate count (in unit of mega gates)	F	.36	.33	.58	0	1.27
	X	0	0	.12	.24	.35
	5 F + 15 X	1.79	1.67	4.60	3.52	11.58
	F+4X (45m)	.60	.33	.69	.94	2.56
	Total	2.38	2.00	5.29	4.46	14.14

(*). Equivalent number of 2 input NAND gates.

6 *DISCUSSION*

In this Section, we will make comparison between the FX method and the direct correlation method. As we stated in Section 3, the comparison should be made, considering (a) necessary operations per unit time (Of, Ox, and Oc for F, X, and correlation receiver, respectively), (b) complexity of the operation (Cf, Cx, and Cc) and (c) the signal-to-noise ratio (SNR). The necessary number of operations per unit time (which is the reciprocal of the frequency resolution) is, for the FX method,

$$Of = N\ n \log n \quad \text{for F's and} \quad Ox = N(N-1)\ n/2 \text{ for X's.} \qquad (2)$$

That of the correlation receiver is,

$$Oc = N(N-1)\ n{**}2 \qquad (3)$$

Given the algorithm and the backend bandwidth B, one can calculate necessary number of gates G,

Figure 7. Approximate gate count (lower) for the unit cell (upper) of k bit correlation receivers. The circuits are assumed to be implemented with the same technologies as the FX LSI's. To make a whole system of a correlation receiver, N(N-1)nB/10 MHz unit cells are necessary. The complexities are approximately 11 and 30 gate operation^{-1}MHz^{-1} for one-bit and two-bit correlation receivers, respectively.

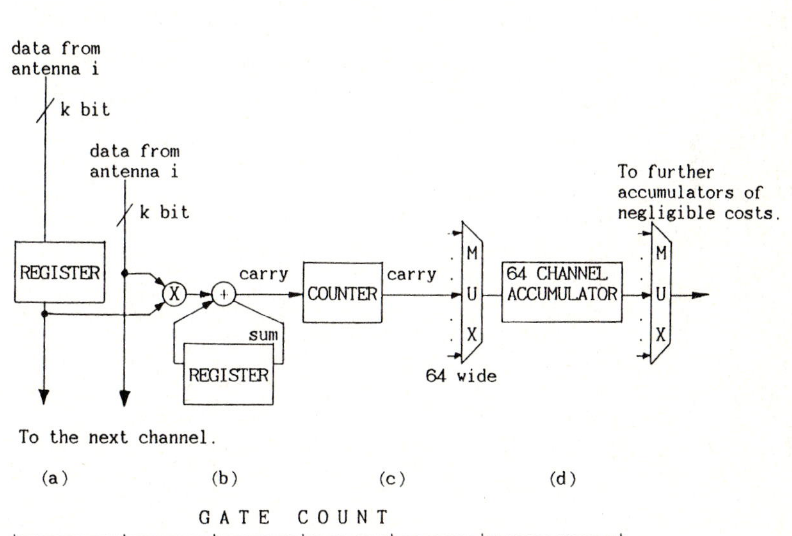

GATE COUNT

Bit/word	(a)	(b)	(c)	(d) (*)	Approx. total
k = 1	7	5	42	3671/64	110
k = 2	14	150-250	42	3671/64	250-350

(*). Assumed that ACC of the FX LSI's is shared among 64 channels.

$$G = Cg\ O\ B/n, \quad (4)$$

where Cg is the complexity of a operation in unit of gate operation^{-1}Hz^{-1}. Though Cg has strong dependence on the algorithm, it has weak dependence on the architecture, e.g., parallel processing, pipelining, etc. Cg, and hence G, depends also on the particular type of devices in use, TTL, ECL, CMOS, etc. If one needs a cost comparison, and hence a performance comparison, between receivers using different types of devices, one should use the complexity Cp in unit of J operation^{-1} instead of Cg,

$$P = Cp\ O\ B/n, \quad (5)$$

where P is the total power consumption of the arithmetic units in the system. Cp depends only on the speed-power product of the process technologies of the gate. Cp is derived from Cg by the relation,

$$Cp = Pg\ Cg, \quad (6)$$

where Pg is the power consumption per gate.

In the FX method, the complexity has a weak dependence on n on the path (c), on M on the path (d), or on n and M on the paths (e) and (f), through the word length of the arithmetic which should be larger for

Table 3. Complexity of the operation for n = 1024, N = 6, and M = 30.

	F (*1)	X (*2)	FX (*3)	1 bit cor.RX (*4)	2 bit cor.RX (*4)	unit
No. of gates	1.27	0.35	12.8	108	295	mega gates
No. of operations	10.24	1.024	–	31457	31457	kilo operations
Complexity (Cg)	400	1100	–	11	30	(*5)
(Cp) (*6)	10	27	–	0.28	0.75	nJ operation^{-1}

(*1). 1024 log 1024 operations per 3.2 micro sec.
(*2). 1024 operations per 3.2 micro sec. 1024 complex frequency channels.
(*3). The number of gates is for six F's and fifteen X's which have 320 MHz bandwidth and 1024 complex frequency channels for each correlation.
(*4). The number of gates given for a correlation receiver which consists of thirty two identical sub-systems having 10 MHz bandwidth. Each sub-system has thirty 1024 channel correlators. The unit cell of the correlator is shown in Figure 7.
(*5). The unit is gate operation^{-1}MHz^{-1}.
(*6). The FX LSI consumes 25 micro W gate^{-1} at 10 MHz clock rate.

Figure 8. Gate count ratio of the FX method over the conventional correlation receivers : (a) for minimum redundancy antenna arrays and (b) for redundant antenna arrays, corresponding to the paths (c) and (e) in Figure 3, respectively. The Fx method has the advantage under the line Gfx/Gc1 = 1 over the one-bit correlation receiver (right), or under the line Gfx/Gc2 = 1 over the two-bit correlation receiver (left).

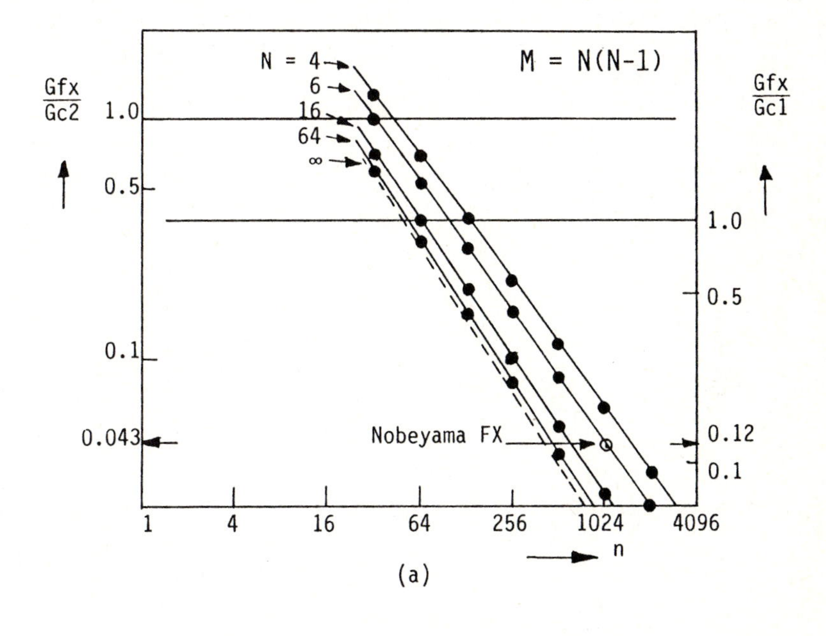

(a)

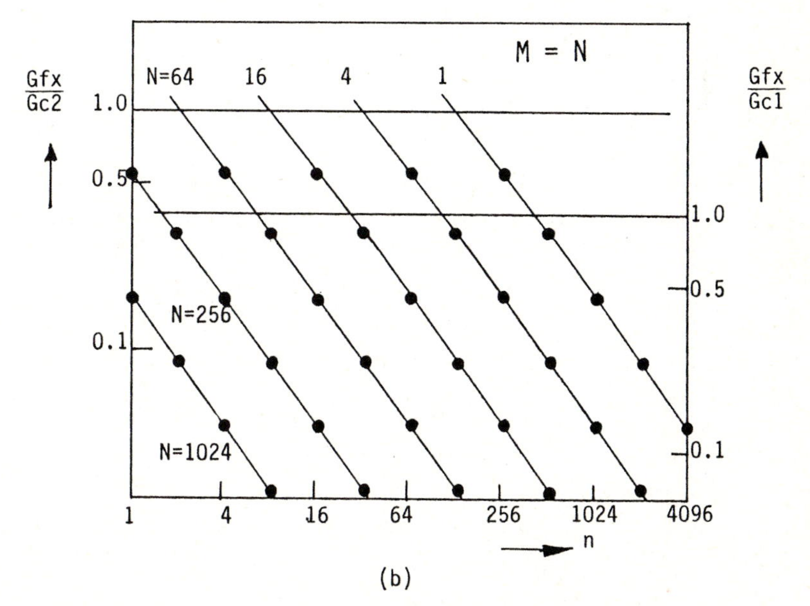

(b)

greater n or M. Because this dependence is very weak one can compare the necessary gate count of the FX method with those of the correlation receiver using the complexity calculated from the Nobeyama FX as long as n, M, and nM is not very much larger than the Nobeyama FX.

We make a comparison between the FX and a conceptual correlation receiver of which the basic cell is shown in the Figure 7. In the Table 3, the gate count and the performance (speed) are given for the case where it is implemented with the same type of devices, i.e., the same Pg as the FX. Therefore one can interprete the gate count ratio calculated from the figures in the Table 3 as the cost ratio which does not depend on the type or the scale of integration of devices, even if they are different between the two receivers, as long as the speed-power product of a gate is the same.

The ratio of the gate count is, in the case where N**2 is not very much larger than M, from the figures in Table 3,

$$\frac{Gfx}{Gc2} = \frac{400\ N\ B\ \log n + 1100\ N(N-1)/2\ B}{30\ N(N-1)\ n\ B}$$

$$= \frac{400 \log n + 550\ (N-1)}{30\ (N-1)\ n}, \qquad (7)$$

where Gfx is the gate count of the FX method and Gc2 is that of the two-bit direct correlation method. The Nobeyama FX has the advantage over the correlation receiver of the same n(=1024) and N(=6),

$$Gfx\ /\ Gc2 = 1\ /\ 23\ . \qquad (8)$$

Even if the SNR degradation in one-bit correlation receiver is tolerable, the ratio of gate count is,

$$Gfx\ /\ Gc1 = 1\ /\ 8.4\ . \qquad (9)$$

In the cases where N**2 is sufficiently larger than M, i.e., the cases of redundant antenna arrays, the path (e) or (f) is preferable, where the ratio over the correlation receiver is, approximately,

$$\frac{Gfx}{Gc2} = \frac{400\ N\ B\ \log n + 400\ M\ \log M\ B\ + 1100\ M\ B}{30\ N^{**}2\ n\ B}$$

$$= \frac{400\ (\ N \log n + M \log M\ ) + 1100\ M}{30\ N^{**}2\ n}. \qquad (10)$$

The ratios for various n, N, and M are shown in the Figure 8.

7 *SUMMARY*

The digital FFT spectro-correlator of the Nobeyama Radio Observatory has been successfully implemented. It has been shown that, particularly for its wide bandwidth and its high spectral resolution, the FX method has an advantage over the correlation method, and is therefore well suited as a spectro-correlator or spectrometer for millimeter wavelength astronomy.

Acknowledgement

We wish to express our thanks to the people of Fujitsu Limited who are T. Miyazawa and T. Nakazuru of the Computer Engineering Department, K. Miura of the Supercomputer Planning Department, H. Ando of the General System Development Department of the Advanced System Product Division, H. Shiratori of the Mainframes Division, T. Hayashi, M. Yamasawa, and H. Gambe of the Transmission Systems Development Department, and M. Hiraoka of the Transmission Systems Engineering Department for the detailed discussions on the design of the FX, and also, for their important role in the production of the FX, S. Nagasawa and T. Miyano of the Computer Engineering Department, and T. Yamanaka of the Transmission Systems Engineering Department. Thanks are also due to the people of Nihon Tsushinki Company Limited for the discussions on the design of IQC, who are Y. Abe, K. Ougi, and T. Hoshino of the Research and Development Division.

References

Chikada, Y. (1979). Progress Report on the Development of the Backends. Proc. Symp. at Joukouji, held by Uchuu Denpa Kondankai (in Japanese).

Chikada, Y. (1981). Techniques for Spectral Measurements. Nobeyama Radio Obs. Tech. Rep., no.8; read at the URSI General Assembly 1981.

Cooley, J.W. and Tukey, J.W. (1965). An Algorithm for the Machine Calculation of Complex Fourier Series. Math. of Comput.,*19*, 297-301.

Ishiguro, M. (1981). The Nobeyama mm-Wave 5-Element Synthesis Telescope. Nobeyama Radio Obs. Tech. Rep., no. 7.

Morimoto, M. (1981). The 45-meter Telescope at the Nobeyama Radio Observatory. Nobeyama Radio Obs. Tech. Rep., no.6.

Morris, G.A.Jr., and Wilck, H.C. (1978). JPL 2**20 Channel 300 MHz Bandwidth Digital Spectrum Analyzer. The Deep Space Network Progress Rep. 42-46, pp.57-59, Pasadena, Calif., U.S.A.: Jet Propulsion Lab.

Weinreb, S. (1963). A Digital Spectral Analysis Technique and Its Application to Radio Astronomy. Ph. D. Thesis and R.L.E. Tech. Rep., no.412, Cambridge, Mass.: M.I.T.

Weinreb, S., Barrett, A.H., Meeks, M.L., and Henry, J.C. (1963). Radio Observations of OH in the Interstellar Medium. Nature,*200*, 829-31.

Yen, J.L. (1974). The Role of Fast Fourier Transform Computer in Astronomy. Astron. Astrophys. Suppl.,*15*, 483-4.

DISCUSSION

R.H. FRATER

What would it cost to reproduce the Nobeyama correlator for use at other observatories.

Y. CHIKADA

The dead copy will cost one million dollars (A$) or less.

J.D. O'SULLIVAN

How many bits of data and coefficients are carried through the transform?

Y. CHIKADA

The word length of the data is 6-8 bits for each of the real or imaginary parts. The coefficients are six bit complex words.

P.C. EGAU

In your slide of the IQC you had BPFs in some of your bands defining filters and LPFs in others. Why is this?

Y. CHIKADA

The output of the IQC is base-band signal if the maximum bandwidth of ADC is fully used. Otherwise, it is not base-band signal. In the latter cases, the signal is shifted down to base-band by the aliasing. The non-base-band converter is simpler and easier to make than a base-band converter.

10.2 THE WESTERBORK BROADBAND CONTINUUM CORRELATOR SYSTEM

J.D. O'Sullivan*
Division of Radiophysics, CSIRO, Sydney, Australia

Abstract. The new continuum correlator system for the Westerbork synthesis telescope provides an improvement of $\sqrt{8}$ in continuum sensitivity and provides the extra measurements necessary to employ phase closure and other methods without sacrificing sensitivity. The system is described and various factors influencing the design of such backends are discussed.

1 INTRODUCTION

A 5120-channel digital spectrometer has been in operation with the Westerbork synthesis radio telescope (WSRT) since 1977. This system (Bos et al. 1981) provided up to 10 MHz bandwidth, which was adequate for spectral line observations even at the maximum observing frequency of 5 GHz. Continuum observations requiring great sensitivity provided the motivation to build a new correlator capable of processing a greater bandwidth. A bandwidth of 80 MHz yields a $\sqrt{8}$ sensitivity improvement over the 10 MHz spectrometer system.

Other specifications were of great importance to the final form of the backend. The desire to provide full primary beam mapping capability dictated a multichannel backend. In addition, the highly successful redundancy method (Noordam & de Bruyn 1982) and other processing methods (Readhead & Wilkinson 1978) using the closure phase properties of the synthesis data required that correlations not previously required for WSRT operation be now measured.

The resultant correlator consists of 8192 correlator channels each 3 level x 3 level (two-bit) and operating at 20 MHz sampling rate.

2 REQUIREMENTS

2.1 Wide field mapping

Given a synthesized beam of approximately $\lambda/2L$ rad, where L is the maximum baseline and λ is the observing wavelength, and a field radius of approximately λ/D rad, where D is the antenna element diameter, then a maximum radial smearing of the synthesized beam by a factor,

---

*Formerly at Netherlands Foundation for Radioastronomy.

$$F = \frac{\Delta\nu}{\nu} \cdot \frac{\lambda}{D} \cdot \frac{2L}{\lambda} = \frac{\Delta\nu}{\nu} \cdot \frac{2L}{D} \text{ beamwidths,}$$

results from a fractional bandwidth $\Delta\nu/\nu$.

At 5 GHz an 80 MHz bandwidth will yield F = 3.8 beamwidths for the WSRT. The result is unacceptable decorrelation and radial smearing of features at the field edge.

Two solutions are feasible.

(1) The total bandwidth may be split into eight or more frequency channels. Suitable account of the centre frequency of each channel in the mapping software allows the scale change with observing frequency to be corrected.

(2) The problem may be viewed as a decorrelation or delay beam effect. Separate delay channels may be provided appropriate to various regions in the field.

The first solution was adopted primarily for reasons outlined in Section 3. The 80 MHz band is split into eight bands of 10 MHz.

2.2 Closure phase and redundancy

The WSRT consists of 14 antennas in an east-west line (Bos 1981). The 14 antennas provide a possible total of 91 interferometers. For image-forming purposes only 40 interferometers are necessary. Measurements of the outputs of all interferometers would involve a doubling of the correlator size. In fact, only a subset of all possible interferometers and polarizations were chosen to yield 256 complex correlator channels per band (O'Sullivan 1980a) instead of the possible 364 if all four polarization combinations are measured for all interferometers. Methods such as closure phase and redundancy utilize the extra measurements to estimate errors in the receiver system.

Of great importance to the success of such methods is the extent to which the receiving system agrees with the model assumed. For phase the model is,

$$m_{ij} = t_{ij} + e_i - e_j$$

where m_{ij} is the measured interferometer phase, t_{ij} is the true interferometer phase to be estimated, e_i, e_j are the telescope-based phase errors. A similar formula (with plus sign) holds for the logarithm of the amplitude parameters.

With methods such as redundancy, self-cal, etc. and the improved sensitivity of the continuum backend, these relations should hold to an accuracy of 30-40 dB per interferometer in order to attain the better than 40 dB (10^4:1) dynamic ranges reported (Noordam & de Bruyn 1982). In particular, the resolution of amplitude errors into telescope-based errors only is affected by a number of receiver or correlator imperfections (Thompson 1980). Possible causes of errors in the model include:

(1) delay-bandwidth decorrelation (tracking or offsets);
(2) decorrelation due to fringe stopping steps;
(3) decorrelation resulting from phase variations within the continuum band;
(4) antenna polarization defects;
(5) antenna pointing errors.

Both amplitude and phase 'closure' relations can also be degraded by such effects as:

(1) D.C. offsets (in analogue-to-digital converter or correlator channels);
(2) cross-talk between adjacent telescope inputs;
(3) interfering signals injected into telescope inputs (but not equally so as to correspond to celestial sources).

A state-of-the-art synthesis telescope backend is therefore required to meet a number of stringent demands.

3 CORRELATOR SYSTEM

Before describing the correlator system in detail, some factors relevant to the overall structure will be considered.

3.1 Band splitting or lag splitting

The original Weinreb lagged correlator scheme could be used to provide sufficient spectral or delay resolution for large field mapping at full bandwidth. Given an $m \times n$ bit system of bandwidth B and N channels, the correlator processing power is closely related to the total number of bit operations per second (BOPS) required to perform the correlation

$$C_W = 2NBmn \quad \text{BOPS} .$$

The processing power is a measure of the number of gates times their speed required to build the correlator. With the improvement in memory technology, the requirements of later counter accumulator sections of the multiplier-accumulator have become dominated by the requirements of the first stages and the multiplier itself.

A considerable reduction in required correlator capacity can be obtained by splitting the input signal into M bands with analogue filters before sampling at a rate lower by a factor M (Fig. 1). We then obtain a processing power C_A:

$$C_A = 2N\frac{B}{M}mn \quad \text{BOPS} .$$

Analogue band splitting was used for this reason. The lagged correlators of Figure 1 are in fact reduced to one lag only. Incidental advantages were interference rejection and tunability.

Figure 1. Band-splitting with analogue filters to reduce required correlator capacity for a given bandwidth and total number of channels.

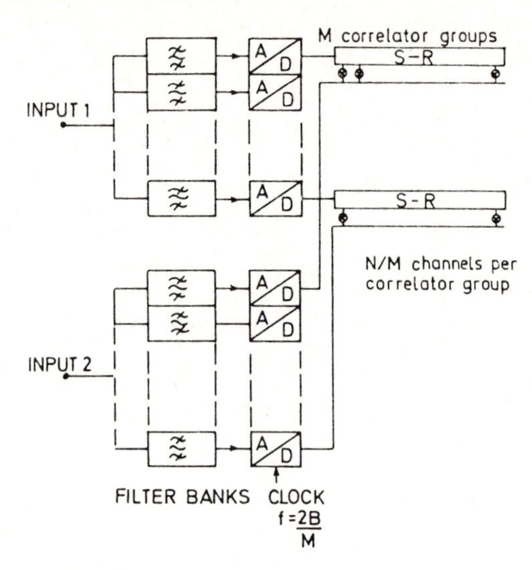

Future systems might better utilize the advantages of digital VLSI circuitry. One approach is to use an FFT processor (see e.g. Chikada et al. 1983). Another would be to use a 2M point FFT to lower the required correlator processing power to C_{FFT}:

$$C_{FFT} = 6N\frac{B}{M}mn \quad \text{BOPS} ,$$

where it is assumed that a complex correlation can be performed with three real correlations.

Some care must be taken in view of the fact that both the input and output of the transform must be quantized. Given that the transform or filter used allows the input correlations to be determined (e.g. a convolution theorem exists for the transform), then the data may be van Vleck corrected in two stages and many other possibilities open up. Such possibilities as FFT, Walsh-Hadamard transforms, rectangular transforms and number theoretic transforms have been considered in O'Sullivan (1982). Of these, FFTs or rectangular transforms perhaps performed with look-up table methods show great promise for significant reductions in required total correlator processing power or increase in the effective capacity of an existing correlator system.

3.2 Double- or single-sideband conversion to video

Selection of the eight frequency bands should ideally allow tunability in order to avoid interference and to allow total band compaction when lower bandwidths are selected for lower observing frequency operation. The phased single-sideband mixer commonly used suffers from sideband feed-through proportional to the magnitude of amplitude or phase errors in the local oscillator. The presence of unwanted sideband can compromise the phase closure properties and lead to 'ghosting' of strong sources (Bos 1983).

Since the cosine-sine splitting is done in the local oscillator before the double-sideband mixing, the correlated fringe direction for both sidebands is identical and there is no need to suppress one sideband. A side benefit arises from the necessity to measure all four (sine, cosine) x (sine, cosine) combinations to recover the 1.5 dB loss in signal-to-noise resulting from double-sideband spectrum folding (3 dB in correlator voltage outputs). Sine-cosine amplitude or phase errors cancel to first order provided they are common to all inputs.

The double-sideband system is simpler and meets the need of the continuum correlator system. For this reason it was selected.

3.3 Signal paths

The entire continuum correlator system is shown in Figure 2.

Figure 2. Block diagram of the broadband continuum backend.

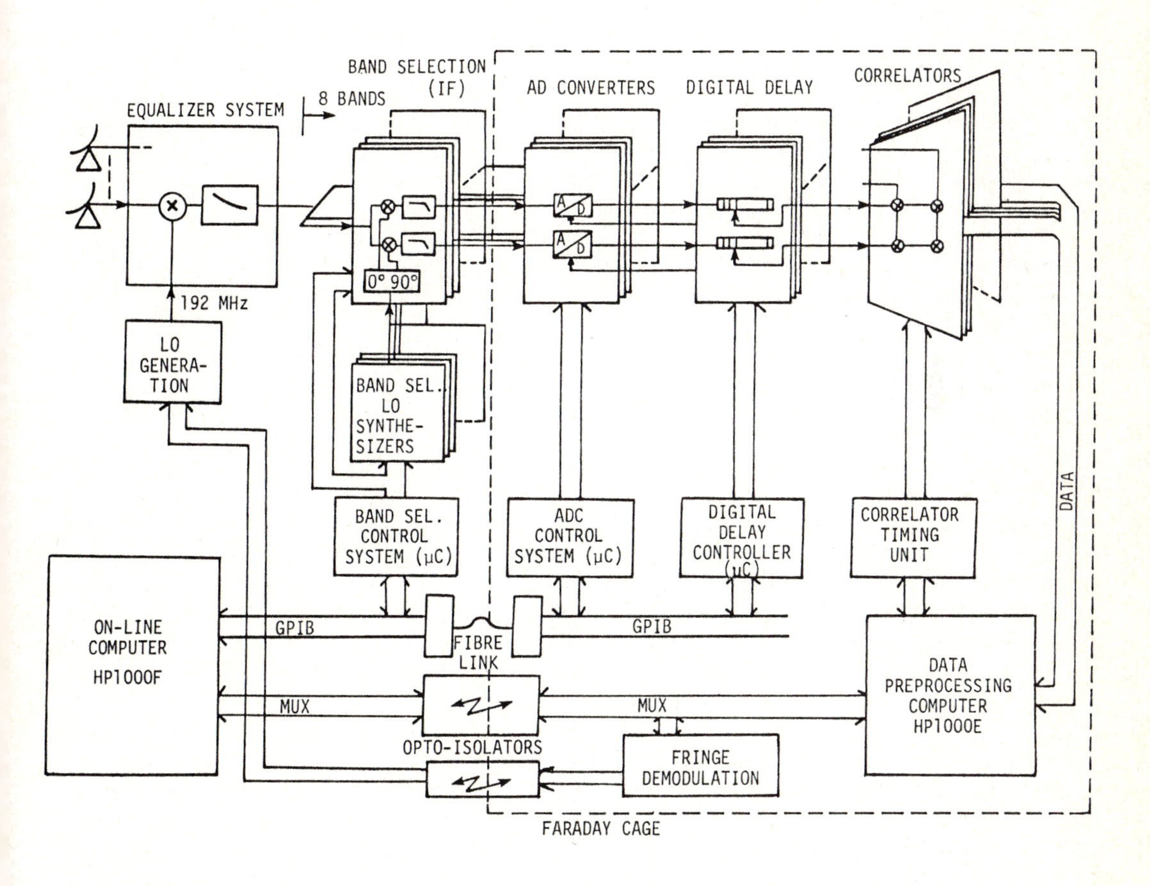

Broadband i.f. signals for both polarizations from each telescope are returned at 132 MHz centre frequency via coaxial cables to the central backend location. The signals are converted to the band 5-100 MHz with a fringe-rotated LO. Fringe rotation is in steps of 22.5° and accelerated beyond the natural rate to provide effectively higher resolution within one backend readout period of 10 s. The fringe rotation is correct for a frequency at the total band centre.

The equalization of the coaxial i.f. return cables is adequately performed by a single RC section to provide a single real pole-zero response.

Separation into eight distinct bands is performed by a double-sideband mixing down to a video band of 5 or 2.5 MHz (selectable). The centre frequency of each band is defined by a synthesizer which provides a local oscillator tunable anywhere in the 5-100 MHz equalized band. Twenty-eight i.f. modules are required for each frequency band, yielding a total of 224 modules. Each module provides a sine and cosine output.

The 448 sine-cosine video outputs are sampled and quantized in a three-level analogue-to-digital converter (ADC) operating at 20 MHz rate. This rate corresponds to a factor of 2 oversampling and results in 12% sensitivity degradation with respect to an analogue system. Each ADC digital output level is continuously controlled by feedback such that the fraction of counts of the three digital levels is constant relative to an external reference frequency (5 or 10 MHz). A voltage-frequency comparator chip allows high accuracy to be attained (O'Sullivan 1980b) and the analogue threshold level of one comparator in each ADC is monitored centrally to provide r.m.s. input voltage information. The voltage measurements are necessary to measure gain and system temperature via a switched noise source in each input. The system temperature measurement allows denormalization of the digitally correlated outputs.

The sample clock for the ADCs is shifted in steps of 3.125 ns (16 steps) to provide fine path delay corrections. The coarse corrections are provided by digital delay lines based on random access memories. All eight frequency bands and both polarizations for a given telescope are provided with the same digital delay, but fine delay and small shift register delay lines are providbd to allow possible offsets of up to 250 ns for the 16 inputs for any telescope.

Delay tracking is accelerated to roughly 4 ns s^{-1} above the natural rate and updated at 100 ms intervals. The effective delay resolution measured over a 10-s integration period is reduced to better than 0.6 ns corresponding to a maximum phase slope of 2° over 10 MHz.

The correlator system consists of 8192 correlators capable of 3 level x 3 level correlations at a 20 MHz rate. The correlations provided are hardwired.

The correlator is made compact with the aid of an octal 8-9 bit integrator chip (Millenaar 1980) using the Ferranti uncommitted logic array process.

Correlator outputs are read into a dedicated HP1000-E series processor at 156-ms intervals. At this stage phase switch demodulation, residual fringe demodulation and integration to 10 s are performed.

3.4 Control paths

Distributed processing plays an important role in the continuum correlator system. All interaction with the on-line computer is at the 10-s basic cycle (Bos et al. 1981) or slower to avoid burdening the on-line computer.

Within the correlator system, two methods are used to distribute control information and to gather monitoring information or correlated data:

(i) A general-purpose multiplexer or MUX system is used for high data rates >10-20 kbyte s^{-1}. The MUX is a 16-, 24- or 32-bit parallel bus with an 8-bit address which has been used extensively for the digital line correlator system.

(ii) The general purpose interface bus (GPIB) or IEEE 488 was used for relatively low data rate paths primarily because of ease of interfacing to microcomputer controllers.

Control of the band selection, the ADC system and the digital delay is performed by dedicated 6800 or 6809 microcomputers. Communication with the on-line computer is in each case via the GPIB. The controllers consist of a processor, optional memory board and a display/GPIB interface/ 12-bit ADC board. Manual operation is in each case via a front panel with keypads and a video display unit with alterable fields. No computer terminal operation is provided.

Communication with the GPIB is modelled on typical instrumentation usage and generally consists of a letter specifying command type followed by relevant parameters if any.

Hardwired controllers have also been used for fringe demodulation and correlator timing and readout control.

4 CONCLUSIONS

The continuum correlator system described has recently been used for the first astronomical observations. The design goals regarding sensitivity and closure phase accuracy appear to have been adequately achieved.

5 SUMMARY OF SPECIFICATIONS

Total bandwidth	80 MHz
Number of bands	8 x 10 MHz, 5 MHz (selectable)
Complex correlators/band	256
	160 fixed-movable
	12 movable-movable
	84 fixed-fixed
Total physical channels	8192
Quantization levels	3 x 3, 2 x 2 (selectable)
Oversampling factor	2
Degradation	12%

6 ACKNOWLEDGMENTS

Many people within the Netherlands Foundation for Radio Astronomy have contributed to this project. Those deserving special mention are R.P. Millenaar, who performed much of the detailed design and debugging, A.G. Poot, S. Zwier, Y. Koopman and A. Koster, who performed most of the construction and production work.

References

Bos, A. et al. (1981). A digital spectrometer for the Westerbork synthesis telescope. Astron. Astrophys., 98, 251-9.

Bos, A. (1983). On ghost source mechanisms in spectral line synthesis observations with digital spectrometers (this volume).

Chikada, Y. et al. (1983). A digital FFT spectro-correlator for radio astronomy (this volume).

Millenaar, R.P. (1980). An integrator circuit in ULA technology. NFRA Note 323.

Noordam, J.E. & de Bruyn, A.G. (1982). High dynamic range mapping of strong radio sources using the WSRT, with application to 3C84. Nature, 299, 597-600.

O'Sullivan, J.D. (1980a). On the choice of interferometers to be measured by the DCB. NFRA Note 322.

O'Sullivan, J.D. (1980b). Level regulation in the DCB analogue to digital converters. NFRA Note 329.

O'Sullivan, J.D. (1982). Efficient digital spectrometers - a survey of possibilities. NFRA Note 375.

Readhead, A.C.S. & Wilkinson, P.N. (1978). The mapping of compact radio sources from VLBI data. Astrophys. J., 223, 29-36.

Thompson, A.R. (1980). Closure accuracy and related instrumental tolerances. VLA Electronics Memorandum 192.

DISCUSSION

P.J. NAPIER

Do you use feedback to keep the threshold levels constant in the quantizers? Is it important to do this?

J.D. O'SULLIVAN

Yes. The threshold levels are servoed to provide a constant fraction of the total samples in each level. This allows further measurement of actual fractions to be ignored for further processing.

J. NOORDAM

Performance of the DCB can be measured in terms of the residual (interferometer-based) errors, which seem to be better than 0°.01 rms phase or 0.005% rms gain. This is quite satisfactory for attaining dynamic range in excess of 40 dB.

10.3 A CORRELATION SYSTEM FOR HIGH-QUALITY, WIDE-FIELD APERTURE SYNTHESIS

J.G. Ables
Division of Radiophysics, CSIRO, Sydney, Australia

ABSTRACT

Aperture synthesis radio telescopes suffer from a simple form of chromatic aberration that an optical lens designer would call a chromatic variation of magnification. This results from the scale of the synthesized image being inversely proportional to received frequency. There is no distortion at the image centre. However, for any finite received bandwidth there is a radial smearing which increases with distance from the centre, thus limiting the area over which good imaging is obtained. When large bandwidths are used to increase sensitivity for continuum mapping (sensitivity being proportional to the square root of bandwidth), it can readily be the case that good imaging is obtained over only a tiny portion of the field of view of the primary antenna elements. A major goal of the Australia Telescope design is to produce excellent images over the full extent of the primary field of view while utilizing the largest practical bandwidths that can be communicated from the individual antennas. This will require that the received bandwidth be subdivided into bands small enough to limit the smear at the field edge to an acceptable value, with the images from each band being reduced to a common scale before final combination. For individual antenna elements of diameter d, maximum distance between elements D, observing frequency f and subdivision bandwidth b the condition for full-field imaging is simply that $(D/d)(b/f) \ll 1$. For the Australia Telescope, a full bandwidth of 160 MHz is proposed. But for the 6-km array operating at 5 GHz, for example, b cannot exceed a few megahertz if the full field is to be mapped. Thus several tens of sub-bands must be used to cover the full bandwidth. The situation is much like that required for spectral line observations at millimetre wavelengths. (At longer wavelengths, spectral line work usually requires even finer frequency resolution over small total bandwidths.)

For the Australia Telescope a very large (by present standards) digital cross-correlator system is planned that will have the capability of satisfying the requirements of wide-angle, high-sensitivity continuum imaging as well as the spectroscopic needs ranging from 21 cm hydrogen line work to millimetre-wave molecular line studies. The correlator is based on an entirely new architecture which achieves the high computational speed required for total bandwidths well in excess of 100 MHz through a highly-parallel, systolic array of custom-designed VLSI circuits. This architecture has the inherent flexibility to provide also for high-resolution spectroscopy. This correlation system is briefly described and some results of an early attempt at the VLSI design are presented. Finally, some lesser-known advantages of the multi-channel cross-correlation approach advocated for the Australia Telescope design are discussed. These include a powerful means of discriminating against centre-of-field artefacts and improved sensitivity and image quality through better compensation for frequency-dependent instrumental effects and perfect quadrature between the measured real and imaginary components of the complex visibility function.

DISCUSSION

D.J. SKELLERN

What fabrication process do you expect to use for the correlator chips?

How high would you rate the 'risk' aspects of VLSI implementation of the correlator chip for the AT?

J.G. ABLES

At present we plan to use the 'standard' NMOS process with minimum line width of 5 μm, reducing to 3 μm if possible. CMOS looks attractive, but we have not investigated it in any detail yet.

The 'risk' is, I think, very high if adequate CAD tools are not available for simulation. Both logic level and analog simulation are important.

10.4 VLSI HARDWARE FOR PROCESSING ROTATIONAL SYNTHESIS DATA - A PRACTICAL ALTERNATIVE TO MASSIVE COMPUTER SYSTEMS

D.J. Skellern
School of Electrical Engineering, University of Sydney, N.S.W. 2006, Australia

Abstract
The use of special-purpose VLSI processors to carry out demanding computing tasks in rotational synthesis data processing is proposed. A systolic computer system which uses the back-projection method of radio image formation is described. The back-projection method of image formation is attractive from both the imaging and VLSI design viewpoints. High concurrency and simple data flow favour a linear processor array for the system, which will use a custom nMOS back-projection chip to implement each processor. The chip design, operation, size and packaging are discussed.

1 *INTRODUCTION*

The computer processing facilities of most observatories are currently based upon large minicomputer systems with attached array processor(s) to achieve high computation rates. The increased hardware capability of rotational synthesis radiotelescopes, especially for multi-frequency operation, has run ahead of the capacity of these economical computing systems to adequately process the observations.

There are three fundamentally different approaches suggested to solve the present computing problem:

(a) Continue along the present path but achieve performance increases by upgrading existing systems,
(b) Replace minicomputers with mainframe machines in the so-called 'supercomputer' class - the massive computer option,
(c) Develop special-purpose hardware to perform critical tasks.

The recent emergence and ready availability of very large scale integration (VLSI) has had a major impact on the viability of special-purpose processing systems. This paper examines the potential for VLSI implementation of specialised digital hardware and reports on the development of a radio image forming computer which incorporates multiple VLSI back-projection processors. The architecture of this computer is an evolution of earlier designs for specialised processing hardware for the Fleurs Synthesis Telescope (FST) [Christiansen, 1973].

2 *THE FLEURS HARDWARE PROCESSING PROJECT*

The aim of the Fleurs hardware processing project was to develop digital hardware specifically oriented to the problems of processing data from rotational synthesis radiotelescopes, and in particular, the FST. Major problem areas in computer processing were identified in map formation, map display and map restoration. Special-purpose digital computers which overcame the computing limitations in each of these areas were constructed using medium-speed, standard integrated circuits. These computers achieve speeds up to two orders of magnitude faster than software solutions on a VAX 11/780 computer system. Completed hardware can be summarised as follows:

(a) Map Formation: A Fourier transform processor and a back-projection processor [Frater & Skellern, 1978; Skellern & Frater, 1980] together implement the task of map formation using a direct transform method.
(b) Map Restoration: A processor performs the iterative beam-subtraction technique CLEAN [Chen & Frater, 1982 and this Symposium].
(c) Map Display: A fast video display allows a 64 x 64 map window to be viewed on a 256 x 256 point display with contour, ruled surface or intensity format [Chen *et al.*, 1982].

Although these processors demonstrated the potential of specialised hardware, their high complexity argued against other groups undertaking a similar development. The ill-fated Culgoora synthesis project was a notable exception [McLean *et al.*, 1979]. The Fleurs group itself did not have the resources to see the processors past the developmental stage, and several production versions remain uncompleted.

Nonetheless, functionally similar computers can solve some of the current computing problems, which are associated with the same tasks addressed by the Fleurs machines. Newer, sophisticated mapping techniques, such as 'self-calibration', place a severe strain on present computer systems because they involve repeated application of these tasks. The difficulty does not lie in the identification of tasks to be performed in special-purpose processors but, rather, to offer these processors as viable alternatives to larger general-purpose computer systems. This is the promise of VLSI.

3 PROPERTIES OF GOOD ALGORITHMS FOR VLSI IMPLEMENTATION

Algorithms which are good for conventional computers are not necessarily good for VLSI, or indeed, for any specialised hardware implementation. The main issue for VLSI is one of computation versus communication. Designs based on achieving minimal devices are no longer valid. A more reasonable size/cost estimate comes from consideration of the area taken to route wires for data and control. The important optimisation is partitioning and placement of sub-systems based on minimising the amount of intercommunication. Good algorithms have regular and simple data and control flow to ensure subsystems can be connected by local and regular interconnections. The complexity of data flow in the FFT algorithm, for example, makes it a difficult candidate for VLSI.

Algorithms which can be constructed from only a few different functional cells are also desirable for VLSI. Hierarchical design techniques [Rowson, 1980] can be applied to expand a design in a way that takes advantage of commonalities and similarities of subfunctions. The concept here is similar to the use of subroutines in programming but with the added proviso that logically distinct functions must also be separated spatially (i.e. their physical layouts do not overlap). Using these techniques, an algorithm with few functional cells translates to one with only a few different layouts, and consequently, design and testing are reduced. In practice this may simply mean looking for algorithms for which bit-slices can be defined so that logic increases linearly with the number of bits being processed. An elementary example is the adoption of a carry-chain rather than a carry look-ahead scheme for addition. The speed of an individual sub-function often can be sacrificed for a reduction in design complexity, as it is in this example.

To achieve high speed in VLSI systems, algorithms which allow concurrent processing through the use of extensive pipelining and multiprocessing are chosen. For these algorithms, processor cycle-time can be made acceptably small, and performance increases are achieved by the operation of parallel, identical processors. In this way, high-speed systems are realised by design repetition, encompassing the use of multiple chips, the replication of a design within a single chip, and the emerging techniques of wafer-scale integration [Raffel, 1979]. The limitations to expansion frequently arise from wire length and pinout restrictions.

Consideration of these three properties - communication, functional complexity and concurrency - make the back-projection method a most favourable candidate for specialised VLSI hardware.

4 THE BACK-PROJECTION METHOD OF RADIO-IMAGE FORMATION

The essence of the back-projection method lies in the fact that where correlations arise from a group of parallel antenna spacings of any orientation, a one-dimensional Fourier transform for the group can be computed and used to determine the contribution made by the group to a sky radio-brightness map. The transform output constitutes a projection of the source distribution at some angle, and the contribution to the map is determined by back-projection. If no sensible grouping exists, the correlation measurements can be processed on an individual spacing basis. In this case, calculation of the projection from the correlation measurement reduces to a sequence of table look-ups, plus scaling. For wide-field mapping, it is necessary to take into account non-linear effects in the back-projection if spacings are not aligned East-West. The processing and grouping of spacings for this method are discussed by Frater [1978].

4.1 Processing requirements

Back-projection, rather than Fourier transformation, dominates the map formation time for a typical image. For an $n \times n$ image, back-projection requires $O\{n \times n\}$ operations, whereas the transform takes at most $O\{n\log(n)\}$ operations (the number of samples in the projection is not many times greater than n).

Using either special-purpose hardware or a commercial array-processor, projection generation times in the range 0.2 - 5.0 msec are readily achieved for a 1k-point projection. By comparison, a back-projection carried out in memory with an accumulation time of 350 nsec takes approximately 92 msec for a 512 x 512 image. This accumulation time is typical for high-density dynamic RAM, which offers the best (cycle time) x (cost-per-bit) product. Although lower times can be achieved with either parallel memory banks or faster memory, both techniques require less-dense memory chips which are unattractive because of area, power and cost penalties. It is unfortunate that the one-bit organisation of high density memory devices precludes their use in highly-parallel memory systems for typical useful image sizes.

The imbalance of the projection generation time against the back-projection time can be compensated by processing projections in parallel. VLSI makes this approach practical.

5 *A SYSTOLIC BACK-PROJECTION COMPUTER*

The back-projection is to be implemented by a special-purpose computer which consists of multiple back-projection processors connected in a linear systolic array [Kung, 1982]. A block diagram of a single processor is shown in Figure 1, and the configuration of the computer system is shown in Figure 2. Image points are read from the image memory and then pass through the processor array, collecting, at each processor, a contribution to the image from one spacing, or group of spacings. Projections and algorithm parameters are loaded into the processors via external interfaces to a supervising computer.

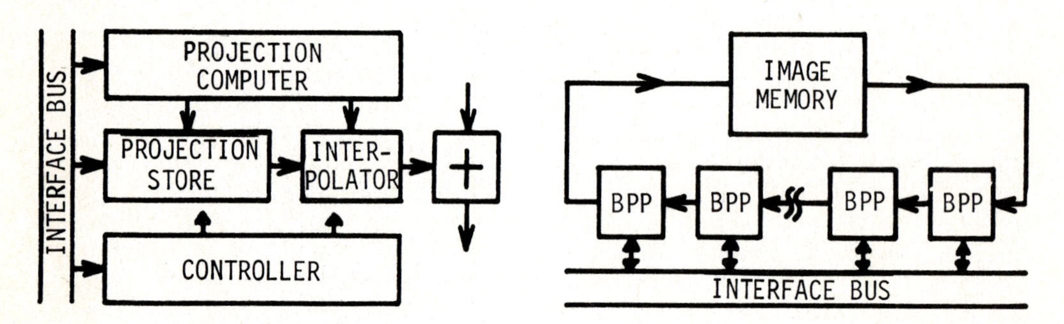

Figure 1: Block diagram of one back projection processor, (BPP);
Figure 2: A linear systolic array of back-projection processors. Each block labelled BPP is a processor shown in Figure 1.

A major advantage of this configuration is that input/output bandwidth limitations in the image memory are reduced because the design makes multiple use of the data from each access. Factors which influence the number of processors placed in the array include cost, array topology and the ability to load the processors from the supervising computer. For a regular, linear array, one processor per time sample allows map formation in a single pass through the image memory. The number of processors required to produce a map in one memory pass for an irregular array is (number of time samples) x (number of spacings), which may be prohibitively large. However, the approach allows a trade-off between the number of processors and map formation time. In addition, the array size can be modified easily to take advantage of future developments in both memory devices and transform chips for projection generation.

5.1 *A VLSI chip for back-projection*

The processor shown in Figure 1 will be implemented as a single-chip custom nMOS circuit together with two commercial memory chips for the projection store. Although it is desirable to incorporate the projection store in the processor, the area required for a 1k by 16-bit word store in the available 5 micron fabrication process is prohibitive.

It is also necessary to limit the functional operation of the processor to conserve space and to increase the likelihood of obtaining successful silicon. Thus, a parallel-ray back-projection algorithm is to be implemented in the first instance. Useful results can be obtained with this chip by limiting the map size in one reconstruction. Larger maps are then made from a mosaic of smaller ones but with a time penalty compared to forming a single larger map.

5.2 *Calculation of the back-projection*

Each nMOS processor computes the contribution of a projection at one angle to every image point before proceeding to the next projection. The outcome of a single back-projection is illustrated in Figure 3(a). In practice, the image and projection are sampled, as shown in Figure 3(b), and a method of assignment of projection rays to the image grid points must be determined.

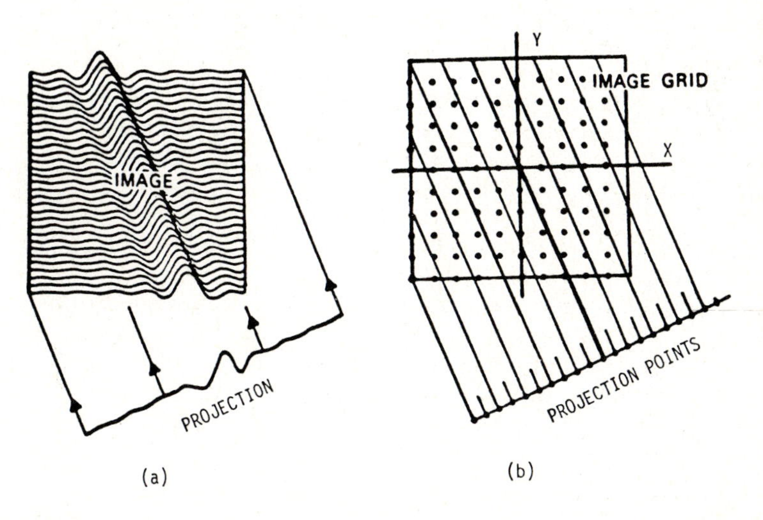

Figure 3: (a) Illustration of back-projection at one angle of view for a field-centre source. (b) The image and projection are sampled and a method of assigning the discrete projection rays to the image grid must be found.

If the image grid is a memory array with addresses defined in an (x,y) co-ordinate system, calculation of the back-projection for each point involves obtaining a measure, A(x,y), of the projection of the point onto the projection vector. By suitable scaling, A(x,y) is calculated as an address in the projection vector with a fractional part which specifies an interpolation between projection points. The appropriate projection value, or interpolated value, is then added to the image at the grid point with co-ordinates (x,y). In a parallel-ray back-projection, A(x,y) is simply given by

$$A(x,y) = ax + by \tag{1}$$

where a and b are constants which depend on the angle of view for the given projection and on address scaling.

The back-projection procedure is computed sequentially, commencing at a corner address and proceeding by rows. A schematic of a projection computer sub-system which includes non-linear terms has been given by Frater [1978] and Skellern & Frater [1980]. A reduced version including only linear terms has been designed. The operation of the projection computer is sequenced by 'start-of-image' and 'end-of-row' signals which are passed along the processor array with the image points. A word-size of 32-bits suffices for the projection calculation in all practical cases [Skellern & Frater, 1980].

5.3 *Implementation of the projection computer*

The projection computer consists of a microprogrammed data-path with an input latch, an addressable register array, an adder and an output latch. Data-path operation is synchronous using a standard two-phase, non-overlapping clock [Mead & Conway, 1980]. The data-path is designed as a strict separated hierarchy.

5.4 *The interpolator*

The interpolator accepts two adjacent projection values and is required to estimate an intermediate projection value by linear interpolation at one of sixteen, equally-spaced points. Prototype chips for the interpolator [Skellern & Frater, 1982] have been successfully designed, fabricated and tested.

5.5 *Fabrication and packaging*

It is planned to have the chip fabricated in Australia by the AWA Microelectronics Division. The chip size is approximately 9 mm x 6 mm in a 5 micron process using orthogonal line, non-optimised layouts. A standard 110-pin array ceramic package can be used satisfactorily to house the back-projection chip. Hybrid packaging of both the projection store and the back-projection chip is being investigated.

6 CONCLUSION

The back-projection method of radio image formation lends itself to the design of a systolic VLSI computer which can be configured to give very high-speed map construction. Development of a custom nMOS back-projection chip is well advanced. Early results of the project, which include the successful fabrication of prototypes for the interpolator and adder, indicate clearly that the development of VLSI processors to relieve general-purpose computers of demanding tasks is entirely feasible and practical.

7 *ACKNOWLEDGEMENTS*

This work is supported by the University of Sydney Research Grant and the Australian Radio Research Board.

8 *REFERENCES*

Chen, C.F., Frater, R.H., Skellern, D.J. & Jones, I.G. (1982). A Fast Image Display Processor. Proc. Radio Res. Board Sem. on Image Processing, Sydney, June 23-25, 3.1-3.6.

Chen, C.F. & Frater, R.H. (1982). High Speed Digital Image Restoration System with Special-Purpose Hardware for Two-Dimensional Convolution. Proc. Radio Res. Board Sem. on Image Processing, Sydney, June 23-25, 25.1-25.5.

Christiansen, W.N. (1973). The Fleurs Synthesis Telescope. Proc.Inst. Radio Electronics Engrs.Aust., 34, No. 8, 302-308.

Frater, R.H. & Skellern, D.J. (1978). Direct Transform Hardware Processing of Rotational Synthesis Data - I. Astron.Astrophys., 68, 391-396.

Frater, R.H. (1978). Direct Transform Hardware Processing of Rotational Synthesis Data - II. Astron.Astrophys., 68, 397-403.

Kung, H.T. (1982). Why Systolic Architectures?. Computer J., 15, No. 1, 37-46.

McLean, D.J., Beard, M. & Bos, A. (1979). A Proposed Correlator Backend for the Culgoora Radioheliograph. Proc.Astron.Soc.Aust., 3, No. 5 & 6, 371-375.

Mead, C.A. & Conway, L.A. (1980). Introduction to VLSI Systems. Reading, Mass.: Addison-Wesley.

Raffel, J.I. (1979). On the Use of Nonvolatile Programmable Links for Restructurable VLSI. Proc. Caltech Conf. on VLSI, January 22-24, 95-104.

Rowson, J.A. (1980). Understanding Hierarchical Design. Ph.D. Dissertation, California Institute of Technology.

Skellern, D.J. & Frater, R.H. (1980). A Microprogrammed Processor for Image Reconstruction from Profiles. Proc.Inst. Radio Electronics Engrs.Aust., 41, No. 1, 26-32.

Skellern, D.J. & Frater, R.H. (1982). An Integrated Linear Interpolator. Conference on Microelectronics, Adelaide, May 12-14, I.E. Aust., 112-115, (I.E. Aust.Nat.Conf.Pub. 82/4).

DISCUSSION

M.S. EWING

Self-calibration and related techniques require more than one spacing to be treated at once. How does your system accommodate this?

D.J. SKELLERN

The back-projection processor does nothing but construct a map - an operation which must be carried out in each iteration of self-cal. At present mapping involves a regridding procedure followed by inverse Fourier transformation. In self-cal, all correlations in one time-slot are considered each time the corresponding complex visibilities for the current source model are obtained. Obtaining these complex visibilities is another candidate for VLSI hardware. Indeed, the back-projection chip could be modified to carry out the projection operation. The desired complex visibilities are then estimated from the projection by filtering and Fourier transformation.

J.R. FORSTER

How flexible is the back-projection method compared to the FFT regarding field-of-view and angular resolution?

D.J. SKELLERN

The method is considerably more flexible than the FFT because we are completely free to choose the angular size and number of points in the map. In the FFT method, once the field of view is specified, the minimum number of points in the transform is determined if resolution is not to be degraded. Because it does not suffer from the same aliasing problems as the FFT, only the region of interest need be mapped. The centre of this region may be offset from the fringe-stop field centre.

10.5 A HIGH-SPEED HARDWARE "CLEAN" PROCESSOR AND ITS USE IN AN INTERACTIVE PROCESS

C.F. Chen
Ultrasonics Institute, Sydney, Australia

R.H. Frater
CSIRO, Division of Radiophysics, Sydney, Australia

Abstract. This paper describes a high-speed digital hardware CLEAN processor and its use to enhance the role of CLEAN as an interactive process. The processor consists of 80K words of RAM and achieves a speed of 12.8 ms per source component for a 128 x 128 map. A novel approach also enables the cleaning of a 256 x 256 map initially using the central section of the dirty antenna pattern and following this by a "tidying" process using successive sections stored on disk. A novel design of loop gain circuits is also described.

INTRODUCTION

Since its introduction by Högbom (1974), CLEAN has been widely used to restore images from the aperture synthesis techniques of radio astronomy. Mathematical aspects of CLEAN have been well covered by Schwarz (1978). But it is well known that the iterative process CLEAN is time consuming and expensive to run in a general purpose digital computer. The process becomes a major computing burden especially in cleaning extended sources, or in using a small loop gain or small grid spacing. Further, attempts at investigating the process under various conditions are also hampered. The other disadvantage of the process is the lack of an interactive environment because of slow processing time. Since image restoration is a singular problem and has no unique solution, problems are known to occur if CLEAN proceeds without intervention.

The actual CLEAN computing involves locating the maximum deflection in the dirty map, multiplication of the dirty antenna pattern by a scale factor, and subtraction or addition. These arithmetic and logic operations are, in fact, simple. But if we cannot accommodate an N X M dirty map to be cleaned and its 2N x 2M dirty antenna pattern in a computer main memory, data transfers to and from an external storage device will be involved in each iteration. If the data is stored on moving head disks, data I/O time can be several hundred times the actual computing time. To clean a 128 x 128 dirty map, the total data storage is 80K words and to clean a 256 x 256 dirty map, it is 320K words. This is beyond the main memory size of a general purpose mini computer. A 32 bit minicomputer like the VAX 780 or mainframe may be used to implement this simple but iterative operation by occupying a large amount of computer main memory. Because of the sequential search for the maximum deflection in the residual map, it is not a natural operation for an array processor. Therefore, the use of a special-purpose hardware processor appears to be an attractive and important solution.

This paper describes a high-speed hardware CLEAN processor and its use in conjunction with a fast hardware display processor to enhance the role of CLEAN as an interactive process.

"CLEAN-TIDY-REBUILD" PROCESS

Usually the implementation of CLEAN consists of two phases, i.e, the "clean" phase and the "rebuild" phase.In order to reduce the memory requirement for large maps, a new "clean-tidy-rebuild" algorithm has been developed.

To implement this process, the dirty map is first divided into segments selected on the basis of the radio source distribution. The segments need not be equal in size. The whole dirty map is then searched to locate a master segment in which the maximum deflection resides. Other segments are called slave segments. Correspondingly the dirty antenna pattern is divided into one master dirty antenna pattern and slave dirty antenna patterns. The master dirty antenna pattern is the central section of the dirty antenna pattern. The slave dirty antenna patterns are the outskirt sections of the dirty antenna pattern. The master dirty antenna pattern and the slave dirty antenna patterns are twice the size of their dirty map counterparts in both dimensions. A master cycle is implemented by cleaning the master segment for Dirac delta source components using the master dirty antenna pattern. The master cycle terminates when the maximum deflection of the residue in the master segment reaches 20% of its initial maximum. In the slave cycle, the dirty antenna pattern contributions due to these source components are subtracted from the slave segments using their corresponding slave dirty antenna patterns. This process is called "tidying". At the end of one full master-slave or clean-tidy cycle, the whole residual map is searched to locate a new master segment. If the residual map is above the noise level, the master-slave cycle iterates with this algorithm. At the end of this first processing phase, the clean-tidy phase, the usual rebuild phase follows.

THE STEP LOOP GAIN SCHEME

During the clean process only a fraction of the maximum deflection is subtracted as a source component. This is done by applying a loop gain which is less than 1 and serves as a damping factor in the iterative process. To implement it a normalised dirty antenna pattern multiplied by the fraction of the maximum deflection is centred at the maximum deflection and subtracted from the residual map. In order to avoid the need for a multiplier a novel step loop gain scheme is developed and implemented in the hardware processor.

The scheme scales the dirty antenna pattern in two's complement format by a binary right shift. Thus only a fixed number of the shifted dirty antenna patterns or of the values of source component is available. Any one of these values divided by the maximum deflection of the residual map becomes the loop gain. The scheme restricts the loop gain in a specified range. During the process the user no longer specifies a value for the loop gain, but its range. The scheme provides four ranges, $1 \geq K > 0.5$, $0.5 \geq K > 0.25$, $0.25 \geq K > 0.125$, and $0.125 \geq K > 0.0625$, where K is the loop gain. Within a specified range, as the process iterates, the loop gain varies as a function of the maximum deflection of the residual map or as a function of the content of the residual source distribution. This scheme has proved useful in detecting any significant digital quantization error during the iterative process (Chen 1981).

HARDWARE DESCRIPTION

A functional block diagram of the hardware CLEAN processor is depicted in Figure 1. The processor has one random access memory (RAM) of 256 x 256 x 16 bits (the large RAM) and another of 128 x 128 x 16 bits (the small RAM). The RAMS are of moderate speed with 400 ns access time. The data stored is in two's complement format. The processor provides two modes of operation, i.e. clean-rebuild mode and clean-tidy-rebuild mode. Clean-rebuild mode processes a dirty map up to 128 x 128 grid points and clean-tidy-rebuild mode up to 256 x 256 grid points. In the clean-rebuild mode the dirty map is stored in the small RAM and the dirty antenna pattern in the large RAM. In the clean-tidy-rebuild mode the whole dirty map is stored in the large RAM but logically divided into segments each of them $\leq$ 64 x 64 grid points. The small RAM stores a master dirty antenna pattern in the master cycle. The slave dirty antenna patterns are overlaid in the small RAM in the slave cycle.

The architecture of the processor aims at minimizing functional circuits and providing easy and flexible software supports. The interface circuit to the host computer is designed to make the processor less dependent on the host environment. Computing and comparison for the next maximum deflection are parallelly processed. The comparison circuit uses one comparator to search for the maximum deflection, either a positive value or an absolute value, and to compare the maximum deflection to the noise level or a user specified level. The addressing circuit centres the master dirty antenna pattern, or the clean antenna pattern at the maximum deflection. It also provides addressing information for the slave dirty antenna patterns. The loop gain circuit makes use of a position scaler to construct a 16 bit 15 shift binary right shifter for a positive or a negative value. The propagation delay for any number of shifts is one latch delay only. The computing circuit has an arithmetic and logic unit to implement subtraction or addition as required for cleaning or rebuilding a positive or a negative source component.

The circuits are pipeline structured and run by a micro-sequence control circuit consisting of a bit-slice micro-sequencer. A micro cycle is made to be 200 ns. Instructions are hand coded with 32 bits each. It takes only 4 micro-cycles to process one grid point per source component, i.e. 800 ns which is mainly the read-modify-write time of the RAM.

The cleaned source component is stored in a source component list in the host computer main memory by stealing time from the next iteration. In the clean-tidy-rebuild mode the list is appended during the successive master cycles. Every 1200 source components is transferred from the list to a contiguous disk file. During the rebuild phase a clean antenna pattern which is of limited extent is overlaid in the large or the small RAM storing the dirty antenna pattern. The source component is recalled from the list to the addressing circuit by stealing time in the previous iteration. The rebuild phase iterates until all the source components are reconvolved.

"CLEAN" AS AN INTERACTIVE PROCESS

There are good reasons to use the power and speed of the hardware CLEAN processor to enhance CLEAN as an interactive process. An interactive system, providing a high degree of human judgement in the processing, allows the user to see if the process is going well and to evaluate what remains after iterations. The organisation of the developed interactive system is depicted in Figure 2.

Figure 1. A Functional block diagram of the high speed hardware CLEAN processor.

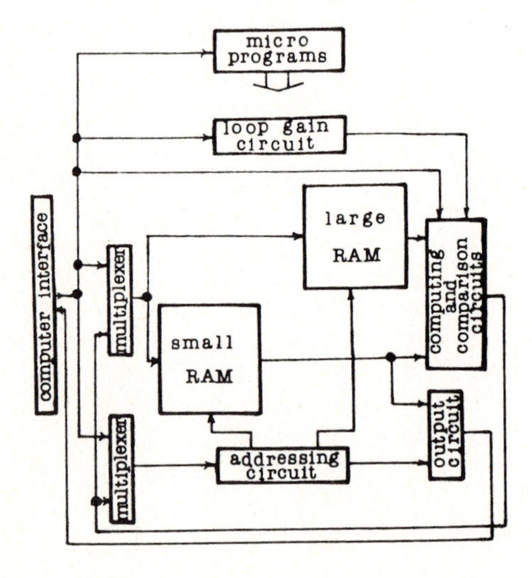

The system is hosted on a PDP 11/45 via the UNIBUS. It consists of two special-purpose hardware processors, i.e. the high-speed CLEAN processor and a fast display processor. The fast display processor (Chen et al. 1982 a) creates and modifies the presentation of radioastronomical maps at a television frame rate. The presentation format may be a contour display, a ruled-surface display, a log intensity display or a filled contour intensity display. The display processor allows instant change of the presentation format which can have an important effect on the effectiveness with which useful information in the map is identified.

An interactive program provides useful options and versatility in the processing. The program adopts conversational codes in selecting an option or assembling for a combination of options. The program has a processing module and an information module. Using the processing module the user can apply a limited search window inside the dirty map or the master segment or nominate a particular source component location. The maximum deflection searched can be specified as either a positive value or an absolute value. The processing module allows a discrete mode or a continuous mode cleaning down to a user specified residue level. In the continuous mode the user may display while cleaning or rebuilding, in reverse order as the source components were subtracted, with an option of presentation format. The rebuilt map can be either added to the final residue or to a clear background. The beamwidth of the clean antenna pattern can be specified as equal to or less than that of the dirty antenna pattern. The information module of the interactive program outputs message about the source component, the accumulated flux density, the step loop gain, the residue level and the mean of the residual map or their histograms. The information module also allows fast examination of the residual map in different presentation formats.

Figure 2. An interactive system of the CLEAN iterative process.

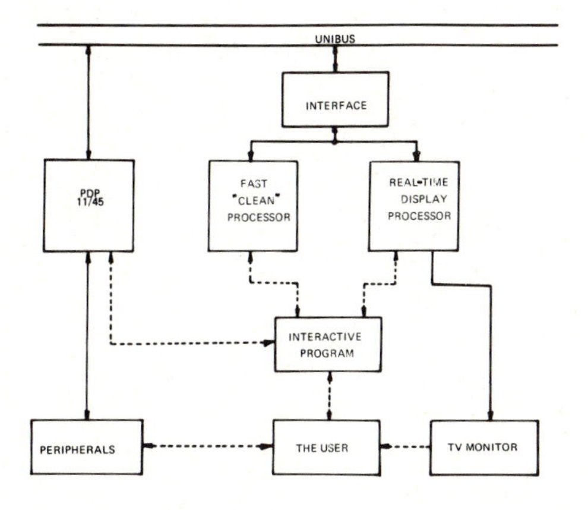

CONCLUSION

An original high-speed hardware CLEAN processor has been described. The use of step loop gain scheme avoids multiplication in the computational effort of CLEAN. A novel clean-tidy-rebuild algorithm was developed to reduce the size of the external memory. An interactive system was developed to enhance the role of CLEAN as an interactive process.

The hardware speed is 12.8 ms per source component for a 128 x 128 map. The speed gain over general mini-computer systems is in a factor of from 80 to 5000. Using the clean-tidy-rebuild mode for 256 x 256 maps its speed is 51.2 ms excluding the time of loading dirty antenna pattern to its RAM once every master-slave cycle during which many source components may be cleaned. Depending on the gain and the complexity of the map the speed gain over the conventional clean method is between one and two orders of magnitude.

ACKNOWLEDGEMENT

This work was supported by grants from the Australian Research Grant Committee and a University Research Grant of the University of Sydney. The system was completed in the Department of Electrical Engineering, University of Sydney in November, 1979.

REFERENCES

Chen, C.F. (1981). Quantization errors in the digital implementation of CLEAN. Proceedings, Astronomical Society of Australia, 4, no. 2, 256-259.

Chen, C.F., Frater, R.H., Skellern, D.J. & Jones, I.G. (1982 a). A fast image display processor. Proceedings, Radio Research Board Seminar on Image Processing, 3.1-3.6, Sydney, 23-25 June.

Chen, C.F. & Frater, R.H. (1982 b). High-speed digital image restoration system with special-purpose hardware for two-dimensional deconvolution. Proceedings, Radio Research Board Seminar on Image Processing, 25.1-25.5, Sydney, 23-25 June.

Högbom, J.A. (1974). Aperture synthesis with a non-regular distribution of interferometer baselines. Astron. & Astrophys., Suppl. 15, 417-426.

Schwarz, U.J. (1978). Mathematical-statistical description of the iterative beam removing technique (method CLEAN). Astron. & Astrophys., 65, 345-356.

10.6 WORKSHOP ON DIGITAL CORRELATORS

Chairman: R.H. Frater
Reporter: A.G. Little

Present. J.G. Ables, J.E. Baldwin, R.A. Batchelor, J. Bunton, J.W. Brooks, A. Bos, D.N. Cooper, P.E. Dewdney, Y. Chikada, C.H. Costain, M.S. Ewing, P.C. Egau, J.R. Forster, R.H. Frater, J.A. Galt, I. Jones, C.E. Jacka, A.G. Little, P.J. Napier, W.B. McAdam, J.D. O'Sullivan, B.J. Robinson, D.J. Skellern, D.G. Steer, A.J. Turtle, P.J. Warner, R.C. Walker, W.E. Wilson.

The first question to be discussed concerned the properties of existing or proposed correlator chips. The Nobeyama chips, 4 in number, were made by Fujitsu under the guidance of Y. Chikada. For a full description see the paper 'A digital FFT spectro-correlator for radio astronomy' in these proceedings. The chips operate at a 10 MHz clock rate and dissipate 400 mW/chip. They are basically gate arrays and have 2000 gates on one of the chips and 3900 on each of the other three. The cost is approximately $40 Australian per chip and they are commercially available. About 2 months was required for design and simulation with about 6 months per chip for production. About 80-90% of chips produced worked well.

The Caltech VLBI chip was next discussed and was described by M.S. Ewing. These chips are still in the development stage. As proposed, these chips should allow dual 8 lag complex correlations (at either 2 level or 3 level) with lobe rotation and 1 bit delay. The technology is expected to be NMOS but as yet there is no commitment to this. Estimated cost is $40/chip for ~4000 chips, with a 20% yield. Ewing felt that they might be forced to a gate array approach which would occupy more space. He also felt that their chip was not as far advanced as the Australia Telescope chip.

The Australia Telescope chip was described by J.G. Ables. The present design is for a 1 bit 8×8 lag correlator chip 6 mm $\times$ 6 mm in size. Some size reduction is possible (to quarter size) which could increase the yield. The technology is NMOS. The size might have to be increased for a 2 bit correlator in which case the yield might still be low. The 8×8 design is an extension of an earlier 2×2 version which worked. However it has not yet been possible to get a good sample of the 8×8 design. The clock rate is 10 MHz and Ables noted that the chip could be used to make a 1 bit spectrometer with a bandwidth of 1 GHz. The current expected cost is $40 per chip.

J.D. O'Sullivan reported on the Ferranti chip used at Westerbork. This is an older technology chip and is less ambitious than the others. By developing a custom chip for the prescaler it was possible to reduce the physical size of the correlator by about a factor of 2 over that of a full random-logic array design. However by the time chip, packaging, rack and assembly costs were included the cost was about the same. The cost of these special chips was around $15-$20 each.

Other correlators mentioned were a proposed 1 GHz correlator for the Anglo-Dutch mm telescope and a very large mm wave correlator for the US. There was no indication that special chips are being developed at present for these.

The main results of this discussion were first to raise the question of whether there is in principle a universal correlator chip and secondly to emphasize that good communications between groups developing chips is essential so that we get the best result for the astronomical community.

The next question discussed was how flexible a correlator should be. P.J. Napier commented that only about 2% of the total modes available at the VLA have been used because of the lack of software support. It is clear that the astronomer might have to provide his own software to activate some options. A. Bos pointed out that it was necessary to provide the flexibility otherwise it would be impossible to activate the options. He also made a plea for the use of software reconfiguration as a switch instead of using connectors, since these get worn out during testing and some don't work when you want them.

The configuration of correlators was then considered and the question was raised as to whether it is better to do the Fast Fourier Transform first or last. The answer seems to be that if done first the capacity of the correlator is effectively increased. This could mean of course a smaller correlator for the same ultimate capacity. There seem to be many ways of approaching this and the Nobeyama system (Chikada, these proceedings) is one. There is no clear cut answer as to how to do it and a study will have to be made of the degradation in signal/noise, if any, and of the number of bits on the input and output, and the levels of digitization required to minimize loss in sensitivity.

It was also suggested that the FFT could be done at the antenna. The Nobeyama system has a degradation of 5% over the normal digitization loss and uses 8 bits in the output of the FFT. Y. Chikada claimed that the full FFT approach as used by them is more efficient than a hybrid approach using filters and FFTs.

The question of phase switching in correlation systems was discussed briefly. P.J. Napier reported that although phase switches are used at the VLA the effects of cross talk are not completely eliminated. Bos mentioned that if the fringes are rotated some further suppression of cross talk is possible. The general opinion was that phase switching is a big help but is not the complete answer to removing spurious correlations.

Finally P.J. Napier put in a plea for testing of the whole system at frequent intervals. A testing facility has to be provided from the beginning.

11

Summary

11. SUMMARY

J.E. Baldwin
Mullard Radio Astronomy Observatory, Cavendish Laboratory,
Madingley Road, Cambridge CB3 OHE, England

This is the moment to look back not just on this symposium but on the last five years since the Groningen meeting to see what advances there have been. Are the problems that have engaged us here the same as those five years ago? Or has the whole subject changed?

The main problems of that time in making good images originated in three areas; the inadequacy of the instruments, the instability of the instruments and the messy nature of the reductions. The inadequacies of the instruments were many: for instance, large gaps in the uv plane or too few projections, poor signal-to-noise ratios, too few correlators and aberrations in the instruments. Instabilities afflicted the telescopes and receivers and, even more importantly, the phases were badly disturbed by the troposphere and ionosphere, whilst, of course, in X-ray studies the phases are completely lost and even the object of study is unstable under the X-ray beam. The reductions were inconvenient and laborious due to such features as gridding, non-planar telescopes and various non-linear techniques such as CLEAN and MEM which we apply to try to improve the pictures. Those were our main problems five years ago and their importance in practice gave weight to the discussion of them as matters of principle.

At this meeting most of these problems seemed less pressing. In many cases we have effectively solved them by very powerful but very simple techniques. For gaps in the uv plane the cure has been to build more telescopes to fill the gaps and for projection data the cure is to take more projections. For poor signal-to-noise ratios the cure is to make better detectors and use wider bandwidths; that too has been done. For too few correlators you build more, and we have heard in the last session in impressive detail just how large those correlators can now be. In summary, our instruments are far more adequate for their purpose now than then.

Next the problem of unstable instruments. It has always been traditional for a physicist to calibrate his instruments and so it is not a new idea that every interferometer should be calibrated, but the idea that you can calibrate by antennas is a new development which is the essence of self-cal, and which has proved to be extremely important. This, together with the addition of redundant base-lines in telescopes, has largely cured the effects of the unstable instruments.

The problem of messy reductions, which was a serious issue in our discussions at Groningen, has become less serious now just because computers are very much cheaper and because of the development of special-purpose hardware. Both changes have essentially made easy now things which were not possible then.

Overall we have a lot of solutions to the former problems and we should now ask whether there are any important outstanding questions. Here I address myself mainly to the radio astronomers because I feel slightly more competent in that field than in others. A few questions occur to me. We think we know how to make good array designs; we couldn't admit otherwise. But do we in fact make the best designs? If you ask what is meant by the best design, then you have to say what it is that you are trying to do. Before there is an answer to that question you cannot begin to say what is the best design. I think there is a case that in the future we will see the design of instruments moving in the direction of use for very specific purposes rather than for rather general purposes as at present. It is essentially impossible to optimize an instrument for all purposes and it may be sensible for economic as well as scientific reasons not even to try to do so.

In the calibration area there is a question which it is clear can be answered but certainly has not been tackled properly so far. It is how best to take account of the spectra of the scale sizes and time scales of the tropospheric and ionospheric irregularities in calibrating a particular instrument under particular circumstances.

As far as redundancy is concerned, we know that it is important but we don't really know how important it is relative to other factors. That is an area where more work should be done. The next question concerns CLEAN. Is a map clear or is it spotless? We have the same sorts of worries as television housewives have about their soap powders. My feeling from this meeting is that we are not dramatically further forward than we were at the Groningen meeting, in knowing what features in a cleaned map we can believe and what we can't believe in particular circumstances. Perhaps the general problem is too hard but it would seem worth while to have a large set of examples which would provide useful guidelines in practice.

My question about MEM is: Do we all have to become acolytes? I think I would be prepared to be one, provided that I can be a passive acolyte. We could all have our MEM gurus and let the gurus amongst themselves argue about matters of philosophy while we can get on with making maps. I suspect many of us would be happy to do that, but there was a warning in the section on MEM which I thought was rather ominous for all of us: that we might have to be active acolytes. The developments seem to show that MEM might enable one to answer very particular questions, but that the actual implementation of the software for those particular questions may well fall on those who ask the questions.

In some other fields of imaging which we've seen here I think we were enormously impressed by the quality of many of the images. In many cases the problems of imaging have been solved quite adequately for answering the question that those subjects raise. There is still some way to go in the optical field in restoring images with poor phase information. There have been quite significant advances since the Groningen meeting in computations from model data and also in restoring images from experiments in the physics laboratory. But we still have to get those same people into the dome to produce astronomical images which can be interpreted. There would be enormous interest in any successful experiments.

What I carry away from this meeting as being most impressive is the enormous range of new instrumentation which is either just finished or is being built or is at the stage of definite plans. That, for me, outweighted the new developments in image analysis itself. Finally I am left with a serious question which could be the starting point for future meetings where we can meet again to argue with our friends as we have done here so pleasantly. Are our instruments already too good? It seems to me that the major mis-match now is between the very high quality of our instruments and the very poor astronomical intuition we possess for interpreting the very beautiful pictures that we make. We have engineered things on the ground extremely well, and we now have to look at the interface to ourselves to see how to get the best out of the instruments that we have made. If we ask that question sharply enough, quite new types of instrument may assume importance.

Registrants

AUSTRALIA

Anglo-Australian Observatory
J.G. Robertson
K. Taylor

CSIRO Division of Applied Physics
P. Hariharan
Z.S. Hegedus
R. Oreb
W.H. Steele

CSIRO Division of Chemical Physics
S. Steenstrup
S.W. Wilkins

CSIRO Division of Atmospheric Research
J.E. Maguire

CSIRO Division of Mathematics and Statistics
R. Anderson
I. Koch
T. Speed

CSIRO Division of Radiophysics
J.G. Ables
J.D. Argyros
R.A. Batchelor
M.J. Batty
J.W. Brooks
G. Chmiel
D.J. Cooke
D.N. Cooper
J.R. Forster
R.H. Frater
R. Gough
G. Graves
C.E. Jacka
M.J. Kesteven
M.M. Komesaroff
R.N. Manchester
D.J. McConnell
R.X. McGee
D.J. McLean
L.M. Newton
D.I. Ostry
J.D. O'Sullivan
W.J. Payten
A.J. Pik
G.T. Poulton
P.T. Rayner
J.A. Roberts
B.J. Robinson
R.S. Roger
P.G. Rogers
K.V. Sheridan
A.E. Schinckel
C.S. Shukre
O.B. Slee
R.T. Stewart
R.H. Wand
K.J. Wellington
J.B. Whiteoak
W.E. Wilson

Defence Research Ctr, Adelaide
S.J. Anderson
G. Poropat

Mount Stromlo and Siding Spring Observatories
D.A. Carden
N.E. Killeen
W.L. Peters

Ultrasonics Institute
C.F. Chen
G. Kossoff
D. Robinson

University of Sydney
L.R. Allen
J. Bunton
M. Calabretta
T.W. Cole
D.F. Crawford
J. Davis
J.M. Durdin
P.C.B. Henderson
I.G. Jones
M.I. Large
A.G. Little
R.A. Niland
K.S.H. Ong
T. Percival
D. Rosenfeld
R.J. Sault
D.J. Skellern
W.J. Tango
A.J. Turtle
A. Tzioumis
A. Watkinson
N.L. Wu
K.C. Zhou

CANADA

Dominion Radio Astrophysical Observatory
C.H. Costain
P.E. Dewdney
J.A. Galt
D.G. Steer

WEST GERMANY

Max Planck Institut für Radioastronomie
E. Preuss
R. Wielebinski

Radioastronomisches Institut der Universitat Bonn
P. Steffen

INDIA

Raman Research Institute
R. Narayan

Radio Astronomy Ctr, Ootacamund
A.P. Rao

JAPAN

Nobeyama Radio Observatory
Y. Chikada

Waseda University
T. Daishido

KENYA

University of Nairobi
P.C. Egau

NETHERLANDS

Kapteyn Laboratory, Groningen
U.J. Schwarz

Netherlands Foundation for Radio Astronomy, Dwingeloo
A. Bos
W.M. Brouw
J.P. Hamaker
J. Noordam

NEW ZEALAND

University of Canterbury
R.H.T. Bates

SWEDEN

Stockholms Observatorium
J.A. Högbom

UNITED KINGDOM

Mullard Radio Astronomy Observatory, Cambridge
J.E. Baldwin
S.F. Gull
P.J. Warner

Nuffield Radio Astronomy Laboratory, Jodrell Bank
P.N. Wilkinson

USA

Environmental Research Institute Ann Arbor
J.R. Fienup

National Radio Astronomy Observatory
T.J. Cornwell
R.D. Ekers
W.M. Goss
P.J. Napier
F.R. Schwab
R.C. Walker

Owens Valley Radio Observatory
M.S. Ewing

Stanford University
R.N. Bracewell

University of Pennsylvania
N.H. Farhat